A series of student texts in

CONTEMPORARY BIOLOGY

General Editors:
Professor E. J. W. Barrington, F.R.S.
Professor Arthur J. Willis
Professor Michael A. Sleigh

An Introduction to Animal Behaviour

Third Edition

Aubrey Manning
B.Sc., D.Phil.

Professor of Natural History, University of Edinburgh

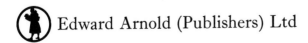 Edward Arnold (Publishers) Ltd

First published 1967
by Edward Arnold (Publishers) Limited
41 Bedford Square, London WC1 3DQ

Reprinted 1968, 1969, 1970, 1971
Second edition 1972
Reprinted 1973, 1974, 1975
Third edition 1979
Reprinted 1979
Reprinted with corrections 1981

Boards edition ISBN: 0 7131 2726 0
Paper edition ISBN: 0 7131 2727 9

Printed and bound in Great Britain by
Whitstable Litho Ltd., Whitstable, Kent

Preface to the First Edition

The study of animal behaviour is now becoming recognized as one of the major branches of biology. Not only does it serve to draw together disciplines as diverse as neurophysiology and ecology, which are otherwise studied in isolation, but it also offers problems of its own as intellectually challenging as any found in science.

There are, at present, few textbooks which approach behaviour from a biological point of view and link it to physiology on the one hand and psychology on the other. This book is an attempt in this direction. It is a personal selection of topics which I consider to be important, and makes no claim to cover the whole field—this is scarcely possible in a book of this length. I have tried to cut across any classification of behaviour by function—feeding, fighting, courtship, etc.—and have chosen topics which help in analysing the organization of behaviour within the individual animal. I have not attempted to make a thorough comparative survey through the animal kingdom. This is quite unprofitable at present, because our knowledge is very patchy and most topics are best exemplified from particular groups.

The book assumes some basic knowledge of physiology and genetics, but no more than would be provided in a school course. I hope the book will be useful for undergraduate courses and for school work. I hope, too, that it will live up to its title and serve as an *introduction* to the whole study of behaviour; there are fairly extensive references which should enable anyone to delve deeper where their interest is caught. I have tried to strike a balance between references to detailed, experimental papers and those to more general reviews of the literature.

It is a pleasure to acknowledge my indebtedness to Dr. Margaret Bastock, Dr. Philip Guiton, Mr. Peter Slater and Dr. David Wood-Gush, all of whom read parts of the typescript and helped greatly by their comments. I am also grateful to the various authors and publishers who gave permission to use their material for illustrations; they are acknowledged individually in the captions. Finally I must thank Dr. Peter Brunet, who was General Editor of the series during the preparation of this book, for all his help over the past 18 months.

Edinburgh   A.M.
1967

Preface to the Third Edition

The study of animal behaviour continues to expand, incorporating ideas and approaches from other branches of biology. As the subject develops, it not only involves new material but changes attitudes towards the old. This new edition contains a new section on communication and some discussion of various aspects of the new sociobiology, for these are areas in which striking developments have been made recently. I have also taken the opportunity to reorganize some of the material from the previous 'Hormones' chapter which is now included in the treatment of early development and motivation. The section on imprinting is now included within that on development. I decided to delete the chapter on the physiological basis of learning and memory. The importance of this field cannot be doubted, but it lacks a cohesive theme and it is no longer easy to see where the real advances are being made. Finally a certain amount of re-writing has been necessary in every remaining chapter. Despite the changes I have not let the book grow much for I am convinced that to be useful it must remain as a reasonably short introductory text.

The reference list is now quite extensive and rightly so for a book which aims to help readers explore a subject more fully. It contains a diverse mixture of old and new literature—books, reviews and specialist papers. I cannot claim to have been either systematic or consistent in my choice. The field is expanding so rapidly that any selection is, on one type of judgement, soon out of date. I hope that a good proportion of my references will still be worth reading in ten years time. I doubt if any behavioural text book could expect more.

It is a pleasure to acknowledge the help of my colleagues John Deag and Arthur Ewing in preparing some of the new sections. In the overall planning of this edition I was much encouraged to find that my own ideas for revision coincided closely with those of Pat Bateson. His comments and advice were a great help.

In conclusion, my gratitude to the team from Edward Arnold for help and patience over a long, long period.

Edinburgh, 1978   Aubrey Manning

Contents

Latin Names of Species where not given in text

INVERTEBRATES

Mollusca
limpet	*Patella* sp.
sea slug	*Tritonia* sp.
octopus	*Octopus vulgaris*
cuttle-fish	*Sepia officinalis*

Arthropoda
water flea	*Daphnia* sp.
fiddler crab	*Uca* spp.
jumping spiders	*Salticidae*
cockroach, American	*Periplaneta americana*
cricket	*Gryllus* sp.
praying mantis	*Mantis religiosa*
termites	*Isoptera*
aphid	*Aphis* sp.
ladybird	*Coccinella* sp.
silver-washed fritillary	*Dryas paphia*
cinnabar moth	*Hypocrita jacobaeae*
blow fly	*Phormia regina*
mosquito	*Aëdes* sp.
ant	*Formica* sp.
digger wasp	*Ammophila* sp.
wasp	*Vespa* sp.
bumble-bee	*Bombus* sp.
honey-bee	*Apis mellifera*

VERTEBRATES

Fish
minnow	*Phoxinus phoxinus*
three-spined stickleback	*Gasterosteus aculeatus*
Siamese fighting-fish	*Betta splendens*

Fish (cont.)

gourami	*Colisa lalia*
goldfish	*Carassius auratus*

Amphibia

frogs	*Rana* spp.
newts	*Triturus* spp.

Reptiles

turtle	*Chrysemys picta*
garter snake	*Thamnophis sirtalis*
water snake	*Natrix sipedon*
rattlesnake	*Crotalus* sp.

Birds

great crested grebe	*Podiceps cristatus*
cormorant	*Phalacrocorax carbo*
emperor penguin	*Aptenodytes forsteri*
great blue heron	*Ardea herodias*
mallard	*Anas platyrhynchos*
garganey	*Anas querquedula*
shelduck	*Tadorna tadorna*
mandarin duck	*Aix galericulata*
wood-duck	*Aix sponsa*
Canada goose	*Branta canadensis*
goose, grey lag or domestic	*Anser anser*
hawk	*Falco* sp.
turkey, wild or domestic	*Meleagris gallopavo*
black grouse	*Lyrurus tetrix*
red grouse	*Lagopus lagopus*
pheasant	*Phasianus* sp.
quail (Japanese)	*Coturnix coturnix*
chicken, domestic and jungle fowl	*Gallus gallus*
moor- or water-hen	*Gallinula chloropus*
coot	*Fulica atra*
oyster catcher	*Haematopus ostralegus*
ruff	*Philomachus pugnax*
gulls	*Laridae*
black-headed gull	*Larus ridibundus*
lesser black-backed gull	*Larus fuscus*
glaucous-winged gull	*Larus glaucescens*
common gull	*Larus canus*
Galapagos swallow-tailed gull	*Larus furcatus*
herring gull	*Larus argentatus*
laughing gull	*Larus atricilla*
kittiwake	*Rissa tridactyla*
terns	*Sterna* spp.
sandwich tern	*Sterna sandvicensis*
wideawake tern	*Sterna fuscata*
royal tern	*Sterna maxima*

Birds (cont.)

black noddy	*Anous tenuirostris*
guillemot	*Uria aalge*
razorbill	*Alca torda*
pigeon	*Columba livia*
ring dove	*Streptopelia risoria*
cuckoo (European)	*Cuculus canorus*
owl	*Strix* sp.
love-birds	*Agapornis* spp.
yellow-shafted flicker	*Colaptes auratus*
lark	*Alauda* sp.
swallow	*Hirundo rustica*
red-winged blackbird	*Agelaius phoeniceus*
carib grackle	*Quiscalus lugubris*
tanagers	*Thraupidae*
warblers (American)	*Parulidae*
Steller's jay	*Cyanocitta stelleri*
Florida scrub jay	*Aphelocoma coerulescens*
crow	*Corvus* sp.
blue titmouse	*Parus caeruleus*
great titmouse	*Parus major*
thrushes	*Turdinae*
robin, European	*Erithacus rubecula*
pipit	*Anthus* sp.
tree-pipit	*Anthus trivialis*
bou-bou shrike	*Laniarius aethiopicus*
Arabian babbler	*Turdoides squamiceps*
starling	*Sturnus vulgaris*
finches	*Fringillinae*
buntings	*Emberizinae*
chaffinch	*Fringilla coelebs*
canary	*Serinus canaria*
songsparrow	*Melospiza melodia*
white crowned sparrow	*Zonotrichia leucophrys*
Oregon junco	*Junco oreganus*
indigo bunting	*Passerina cyanea*
zebra finch	*Taeniopygia guttata*
Bengalese finch (striated finch)	*Lonchura striata*
spice finch	*Lonchura punctulata*
cutthroat finch	*Amadina fasciata*

Mammals

INSECTIVORA

water shrew	*Neomys fodiens*
tree shrew	*Tupaia belangeri*

RODENTIA

Pygmy mouse	*Baiomys taylori*
mouse, domestic	*Mus musculus*

Mammals (cont.)

brown rat, also domestic rat, laboratory rat	*Rattus norvegicus*
flying squirrel	*Glaucomys volans*
guinea pig	*Cavia porcellus*
squirrel	*Sciurus* sp.
LAGOMORPHA	
rabbit	*Oryctolagus cuniculus*
CETACEA	
Humpback whale	*Megaptera novaeangliae*
PROBOSCIDEA	
African elephant	*Loxodonta africana*
PERISSODACTYLA	
white rhinoceros	*Ceratotherium simum*
horse	*Equus caballus*
ARTIODACTYLA	
hippopotamus	*Hippopotamus amphibius*
pygmy hippopotamus	*Choeropsis liberiensis*
red deer	*Cervus elaphus*
moose	*Alces alces*
duiker	*Cephalophus* sp.
wildebeeste	*Connochaetes taurinus*
Uganda kob	*Adenota kob*
musk ox	*Ovibos moschatus*
sheep, domestic	*Ovis aries*
goat	*Capra* sp.
CARNIVORA	
polecat	*Mustela putorius*
stoat	*Mustela erminea*
skunk	*Mephitis mephitis*
spotted hyena	*Crocuta crocuta*
cape hunting dog	*Lycaon pictus*
fox	*Vulpes* sp.
coyote	*Canis latrans*
dog, domestic	*Canis familiaris*
wolf	*Canis lupus*
leopard	*Panthera pardus*
tiger	*Panthera tigris*
lion	*Panthera leo*
cat, domestic	*Felis domesticus*
cheetah	*Acinonyx jubatus*
PRIMATES	
lemurs	*Lemuridae*
marmoset	*Callithrix* sp.
squirrel monkey	*Saimiri sciurea*
howler monkey	*Alouatta* sp.
spider monkey	*Ateles* sp.
rhesus monkey	*Macaca mulatta*

xiv   LATIN NAMES OF SPECIES WHERE NOT GIVEN IN TEXT

Mammals (cont.)

Japanese monkey	*Macaca fuscata*
Barbary macaque (Barbary ape)	*Macaca sylvanus*
patas monkey	*Erythrocebus patas*
chacma baboon	*Papio ursinus*
yellow baboon	*Papio cynocephalus*
hamadryas baboon	*Papio hamadryas*
gelada baboon	*Theropithecus gelada*
black and white colobus	*Colobus polykomos*
red colobus	*Colobus badius*
mangabey	*Cercocebus albigena*
langur	*Presbytis entellus*
gibbons	*Hylobatidae*
orangutan	*Pongo pygmaeus*
chimpanzee	*Pan troglodytes*
gorilla	*Gorilla gorilla*

I

Reflexes and Complex Behaviour

There are two main approaches to the study of behaviour, the physiological and the psychological. Physiologists are primarily interested in mechanisms and aim at giving an explanation of behaviour in terms of the functioning of the nervous system. Psychologists are more concerned with the behaviour itself, studying those factors in an animal's environment and history which affect the development and performance of overt behaviour. A good deal of the work described in this book is that of ethologists, who are sometimes contrasted with psychologists because of their different fields of interest (see p. 27). However both study the behaviour of whole organisms and in this they share an approach distinct from that of physiologists.

The two approaches are both essential and complementary. Physiologists sometimes like to emphasize that their methods are the more fundamental, and it is true that ultimately we shall hope to be able to explain behaviour in terms of the functioning of the basic units of the nervous system—the neurons. However, the main function of the nervous system is to produce behaviour and we must investigate the end product in its own right. Many of the most important aspects of neural organization can be expressed only in behavioural terms. Even if we knew how every neuron operated in the performance of some pattern of behaviour, this would not remove the need for us to study it at a behavioural level also. Behaviour has its own organization and its own units which we must use for its study. Trying to describe the nest building behaviour of a bird in terms of the action of individual neurons, would be equivalent to trying to read a page of a book with a high powered microscope.

Yet physiology and behaviour studies are often closely linked and can help one another. This book does not itself contain much formal physiology but we must always keep it in mind. Indeed one of the most flourishing areas of psychological research at present is that called 'physiological psychology', in which attention is focused simultaneously on an animal's behaviour and its associated physiology. One of the obstacles to closer links between physiologists and other behaviour workers is the small overlap between the types of behaviour each group studies. Complete physiological analysis is not yet possible save in the simplest of responses involving only a handful of neurons. This opening chapter will examine some possible ways of bridging the gap between such simple responses and the more complex ones which will be our main concern.

'Behaviour' includes all those processes by which an animal senses the external world and the internal state of its body, and responds to changes which it perceives. Many of such processes will take place 'inside' the nervous system and may not be directly observable. What the animal does may involve violent activity or complete inactivity, but all are equally behaviour. This immediately faces us with the problems of observation and measurement. In the physical sciences, and even elsewhere in biology, we have recognized units—amps, molecules, pH, genes, action potentials, etc.—for measuring and classifying our observations. When we watch an animal moving about its environment we have no such framework. Behaviour is continuous for as long as the animal lives and, strictly speaking, every movement counts. Put this way the task becomes impossible and in fact we cannot begin to study behaviour unless we abstract and simplify. We have to make some decisions as to what it is important to record, given the problem we are interested in, and what can be safely ignored.

If we are studying the courtship behaviour of sticklebacks or pheasants we will probably decide that there is no need to record the number of times the animals breathe. Breathing is a part of that continuous activity which constitutes behaviour but it will often be judged irrelevant for behavioural studies. In fact when animals are highly aroused their breathing rate goes up (see Chapter 5), and thus it may be a useful indication of internal changes. If such changes are the focus of study then breathing could be important. So it may in other circumstances. Male newts court females on the bottom of ponds or streams and their courtship is punctuated by trips to the surface to breathe. Halliday and Sweatman[198] have found that males can postpone breathing up to a point if courtship is proceeding smoothly, but if there are delays or the female moves off, for example, they catch up on their breathing. Thus records of breathing are an important part of the study of newt courtship patterns, unlike those of pheasants. All this is simply to emphasize that the behavioural measures we use must be chosen to suit the animal and the problem under investigation. We must be flexible and prepared to change our units of measurement as we learn more about the situation. Hinde[217] discusses these questions in more detail in the opening chapter of his advanced textbook.

Reflexes are often considered as among the simplest units of behaviour. The reflex which causes us to close our eyes when something flashes towards them, or withdraw our foot when we step on a sharp object, is functionally very important. Yet it seems so very different in scale from the kind of behaviour with which most of this book will be concerned—building a nest, displaying to a mate, running through a maze to get food, and so on. Such patterns are certainly more complex in form than reflexes, but it is the aim of this chapter to show that they are both behaviour and that we learn some of the basic features of behavioural mechanisms from studying properties which both share.

It is not possible to draw a firm line between reflexes and complex behaviour. Neither term can even be defined satisfactorily, but common sense will determine their use in what follows. Clearly complex behaviour can incorporate many reflexes; the swallowing reflex is the culmination of complex food-seeking behaviour, and the reflexes controlling balance and walking are involved in almost all complex behaviour. The two represent the ends of a continuous scale and somewhere intermediate along this scale are the 'tropisms' by which animals orientate themselves with respect to light, gravity and other environmental factors (see Fraenkel and Gunn[152]). Tropisms have some characteristics of reflexes, but they involve movements of the whole body and not just a single group of muscles. Whilst we can often describe the neural pathways and properties of a reflex in a rather exact manner, we can scarcely ever do this for complex behaviour. In the latter, the neurons involved must be numbered at least in hundreds and the impinging variables become legion. Nevertheless we should not be put off by quantitative differences no matter how great. There is much to be learnt from a comparison between the properties of reflexes and more complex behaviour, and it is a useful way to give some physiological foundations to a study of behaviour.

In 1906 C. S. Sherrington's book *The Integrative Action of the Nervous System*[435] was published. Sherrington, more than any other single person, can be regarded as the founder of modern neurophysiology. In his book he considered the way in which reflexes operate and how the central nervous system integrates them into adaptive behaviour, combining information gathered from different sources, arranging sequences of action and allocating priorities. *Integrative Action* is a scientific classic which can still be read for pleasure and profit. Sherrington had to work with apparatus we now regard as crude. There were no electronic stimulators or oscilloscopes, simply induction coils and levers, attached to the limb of the animal and writing on the smoked drum of a kymograph.

In the first few chapters of his book Sherrington discusses some of the properties of reflexes and contrasts them with those of the same movements when elicited by direct stimulation of nerves to the muscles concerned. We may, in turn, use part of his classification to compare the properties of reflexes and more complex behaviour.

Latency

Reflexes and complex behaviour both show latency in response—there is a delay between giving a stimulus and seeing its effect. Sherrington calculates that, allowing time for conduction along axons, the dog flexion reflex in which it withdraws its leg in response to painful stimuli on the skin, should have a latency of about 27 msec. In fact the latency usually lies between 60 and 200 msec.

Latencies are harder to measure for complex behaviour because there are often difficulties in fixing precisely the time of onset of the stimulus, but they are none the less vivid. Wells[492] describes how, when a tiny shrimp (*Mysis*) is presented to a newly-hatched cuttle-fish (*Sepia*), there is no detectable response for perhaps as long as 2 minutes. Then the nearest eye of the cuttle-fish turns to fixate on the shrimp. There is a further delay, but usually only a few seconds, before the cuttle-fish turns towards the shrimp so that both of its eyes are brought to bear. Another brief delay follows, and then it launches an attack and seizes the shrimp with its tentacles. A male fruit-fly (*Drosophila*) is often not aroused by its first encounter with a female. He may touch her with his fore-legs, but stands still if she moves away. Only at a subsequent encounter some seconds later does he begin his courtship display.

The latency of reflexes is known to be due to delay in the transmission of impulses across the synapses (a term we owe to Sherrington) between one neuron and the next. It is hardly surprising to find delays between stimulus and response in complex behaviour for in the chain between receptors and effectors there must often be dozens of synapses to cross.

With reflexes it is found that the stronger the stimulus, the shorter the latency (see Fig. 1.1), but measurements of the latency of complex behaviour are inextricably intermingled with factors affecting thresholds. Reflexes have a relatively stable and rather low threshold. They are one of the body's protective mechanisms and need to be constantly on tap if required. With complex behaviour, in contrast, thresholds vary enormously. A food stimulus which produces an immediate and intense response from a hungry dog may be ignored an hour later after it has fed. Consequently any attempt to investigate how latencies change with stimulus strength is hedged about with difficulties. There are two ways to get an estimate of latency in such cases. One can try to get all one's subjects equally motivated, i.e. equally hungry if investigating latency to a food stimulus. Secondly, one can use a large number of animals with varying thresholds and then

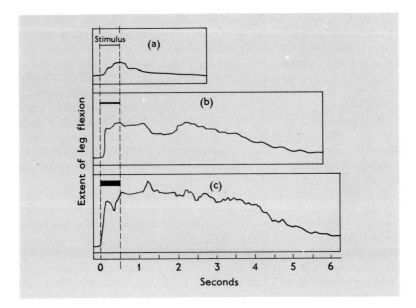

Fig. 1.1   The extent and persistence of the dog's flexion reflex with three strengths of stimulus, each of 0·5 sec duration. The area enclosed between the line and the X axis gives a measure of the 'amount' of the response. Even with a weak stimulus (a) the after-discharge represents 75% of the total 'amount'; with the strongest stimulus (c) it represents over 90%. Note that the latency of the response decreases as the stimulus increases in strength. (Modified from Sherrington,[435] 1906, *The Integrative Action of the Nervous System.* Charles Scribner's Sons, New York.)

compare statistically the range of latencies given to one stimulus with that given to another. This method was used by Hinde[214] who measured the latency between presenting various frightening stimuli to chaffinches and their first alarm call. Just as with reflexes, the stimulus known on other grounds to be strongest, produced the shortest latency.

After-discharge

When a motor nerve is stimulated directly the muscle it supplies contracts and then relaxes within a few seconds of the stimulus ceasing. The same is not true for muscle contractions elicited through a reflex arc; they often persist at full intensity for many seconds after the stimulus ends and then relax gradually. Sherrington found that the length and strength of this after-discharge in the dog's flexion reflex was related to the strength of the stimulus (see Fig. 1.1).

After-discharge is a familiar phenomenon in complex behaviour. Hinde[210] studied the 'mobbing' response which chaffinches make towards owls and other predators. The bird comes quite close to the owl and appears restless, making frequent short flights sometimes towards, sometimes away from it. Repeatedly the chaffinch gives the 'chink' alarm call. In the wild this behaviour attracts other birds who also begin mobbing until the owl is surrounded by a flock of harrying small birds and it often beats a retreat. Hinde found that the rate of calling 'chink' was a good measure of the strength of this mobbing response in an individual chaffinch alone in a large cage and found that calling persisted long after the owl was removed. The stronger the chaffinch was calling at the time, the longer was this after-discharge.

A male *Drosophila* who is courting an unreceptive female sometimes loses contact with her if she jumps or flies away. In such instances he often continues courtship movements for several seconds, orientating himself to the spot where the female was last perceived. With complex behaviour such as this it is difficult to state categorically that all stimuli are removed. When Sherrington switched off his induction coil he could then be certain that no stimulus relevant to the reflex was reaching the dog. But perhaps after she has gone, some of the female *Drosophila*'s smell remains behind and the male's continued courtship is not strictly an after-discharge. Again, Hinde could show that whilst the owl was in the cage the chaffinch associated it with other objects close by. Some of the apparent after-discharge may have been a response to these objects which remained after the owl was gone. In spite of these qualifications there can be little doubt that complex behaviour does show what Lorenz[305] aptly called 'reaction momentum'; it takes time to slow down once aroused. Prolonged excitability at synapses must be involved just as it is in reflex after-discharge. In Chapter 3 we shall discuss how a stimulus can 'arouse' an animal and thus, not only do its effects persist, but the animal may become more responsive to subsequent stimuli.

Summation

One of the most obvious integrating properties of the central nervous system is its ability to summate stimuli coming at different times (temporal summation) and from different places (spatial summation). Sherrington gives several beautifully clear examples. The scratch reflex of the dog is elicited by irritating stimuli anywhere on a saddle-shaped area of its back. The hind leg on the same side is brought forward and rhythmically scratches at the spot. With weak stimuli, a series of 5 or 10 given in rapid succession may not evoke any response, but after 20 or 30 scratching appears—the stimuli have been summed in time. Fig. 1.2 shows the spatial summation of stimuli from two areas of skin 8 cm apart; neither is strong enough alone to provoke scratching, but they are effective when given together.

Sherrington found that points up to 20 cm apart can summate in this way, but the effect diminishes with distance up to this maximum.

Dethier[130] studied the stimuli which cause blow-flies to extend their proboscis preparatory to drinking. The flies can detect sugars and other food substances with sensory hairs on their fore-tarsi. They search for food by running over a surface and extending their proboscis when the front legs encounter anything suitable. As measured by this proboscis extension, the

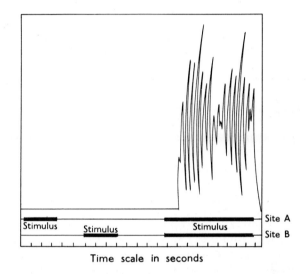

Fig. 1.2 Spatial summation leading to the appearance of the scratch reflex in the dog. The tracing represents the movements of the dog's leg when scratching. A and B are two points on the shoulder skin and weak stimuli given singly, first at A, then at B, do not evoke the reflex. When both points are stimulated simultaneously the reflex appears with a latency of about 1 sec. (Modified from Sherrington,[435] 1906, *The Integrative Action of the Nervous System*. Charles Scribner's Sons, New York.)

flies can detect sugars at very low concentrations. Dethier found that when only one leg is dipped into a solution the lowest concentration of sucrose to which 50% of flies respond is 0·0037M. However, if both legs are stimulated together there is summation and now 50% of flies respond to 0·0018M. He further found that if one leg was stimulated by unpleasant substances such as salt or acid which tend to make a fly retract its proboscis, whilst simultaneously the other leg is dipped into sugar, a kind of algebraic summation occurs. The stronger the acid on one side, the stronger the sugar has to be on the other to overcome the negative effects and cause a fly to extend its proboscis.

With more complex behaviour summation frequently occurs between stimuli of quite different types perceived by different sense organs. We all know how the sight and smell of food summate when we are hungry. Beach[41] showed that male rats respond sexually to a combination of olfactory, visual and tactile stimuli from a receptive female. Young males do not respond unless two such sources are available—it does not matter which two. Mature males, with previous sexual experience, will respond to one type of stimulus alone.

'Warm-up'

Sherrington found that some reflexes do not first appear at full strength but, with no change to the stimulus, their intensity increases over a few seconds. Hinde's chaffinches show a similar type of 'warm-up' effect. Fig. 1.3 is a record of the rate of calling by a chaffinch in successive 10-second periods after presenting an owl and the maximum rate is not reached for about $2\frac{1}{2}$ minutes. Sevenster-Bol[433] counted the courtship movements made by male sticklebacks to a female confined in a glass tube (see p. 82 for more details of this experiment). She found that over a 5-minute period the number of movements rose steadily in each successive minute of the period. Sherrington was able to show that 'warm-up' in some reflexes is due to the summation of stimuli which come to evoke a response from more and more motor nerve fibres, producing a stronger contraction. He called this phenomenon 'motor recruitment'.

Some analogous process probably occurs with complex behaviour, but we commonly see that it is not only the intensity of a response which changes with persistent stimulation; its nature changes as well. Again, in some later work, Sherrington[436] provides an excellent example from what

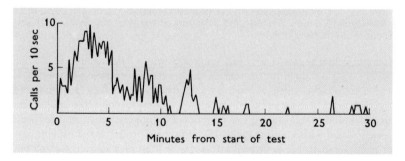

Fig. 1.3 The 'warm-up' and subsequent 'fatigue' of alarm calling when a chaffinch is presented with a stuffed owl in its cage. The maximum rate of calling occurs after about $2\frac{1}{2}$ min. (From Hinde,[210] 1954, *Proc. R. Soc., B.*, **142**, 306.)

he calls the cat's 'pinna reflex'. Repeated tactile stimuli to a cat's ear first
cause it to be laid back. If stimulation persists the ear is fluttered, thirdly
the cat shakes its head and when all else fails to remove the irritation it
brings up its hind leg and scratches. Clearly there is more involved here
than the recruitment of extra motor nerve fibres. Mechanisms which
control patterns of movement such as ear-fluttering and head-shaking must
be recruited. Perhaps all these mechanisms are activated in some way by
the stimuli to the ear but their thresholds are different. That for laying back
the ear will have the lowest threshold, with successively higher ones for the
other three patterns. As the persistent stimuli summate, and the threshold
for activating each pattern is passed, it replaces the pattern which preceded
it. Workers studying complex behaviour frequently rank the various
patterns they observe on a similar 'intensity scale' of increasing thresholds.
A system similar to that suggested for the cat has been used by Bastock and
Manning[33] to explain the fact that a male *Drosophila* switches from one
courtship pattern to another when courting a female whose behaviour
remains constant.

Fatigue

When a muscle is made to contract by regularly repeated stimulation of
its motor nerve it continues to respond for a very long time—several hours
or more. This is not the case when the same muscle is stimulated through a
reflex arc. Sherrington found that the dog scratch reflex begins to wane
after about 20 seconds of continuous mechanical or electrical stimulation at
a single point on the skin. The movements of the leg become weaker and
lose their rhythm. This fatigue is certainly not due to the muscles because
if the flexion reflex which uses the same muscles is now evoked, it appears
at full strength. Further, if the site of stimulation on the skin is switched a
few centimetres the scratch reflex recovers and, after a short rest, the first
site once more becomes effective when stimulated (see Fig. 1.4).

This means that the origin of the fatigue must lie somewhere between
the sense organs in the skin and the origin of the motor nerve. Sherrington
believed that it was caused by increased resistance to transmission across
the synapses between the internuncial neurons, which convey impulses
down the spinal cord, and the motor neurons.

Franzisket[154] has confirmed that the internuncials are the site of fatigue
in the leg-wiping reflex of the frog. When its back skin is touched lightly
the frog brings up its hind leg and makes a single wiping movement. This
fatigues after some dozens of stimuli but, as with the dog, shifting the site
of stimulation will keep the leg working. Franzisket observed that tiny
twitches of the frog's skin around the point where it was touched continue
after the leg has ceased to respond. This shows that the sensory nerves from
the skin are still working and impulses are being transmitted to the motor

nerves which supply the local muscles of the skin. The leg-wipe has disappeared because the internuncial neurons linking the sensory nerves to those of the leg muscles are not transmitting.

Fatigue is always a feature of complex behaviour. Compare Fig. 1.3, which shows how the calling rate of the chaffinch declines in the mobbing situation, with the left-hand portion of Fig. 1.4, showing the decline of the

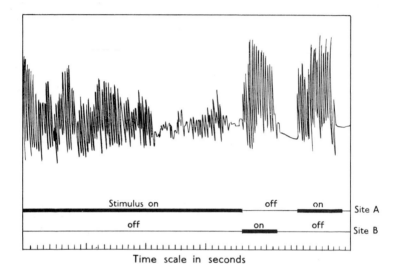

Fig. 1.4   The 'fatigue' of the dog's scratch reflex. Prolonged stimulation at site A eventually ceases to be effective, but stimulation at site B immediately after brings the response back and after only 9 sec further stimulation at A evokes the response again. (Modified from Sherrington,[435] 1906, *The Integrative Action of the Nervous System*. Charles Scribner's Sons, New York.)

dog's scratch reflex. In the chaffinch example, however, 'fatigue' is merely a name for what we observe—that the response declines even though the stimulus is maintained. Taken alone, such observations tell us nothing about the mechanisms involved and certainly these are not the same in every case.

Sometimes, as with reflexes, a change in stimulus brings about a reawakening of a response which has fatigued. Young passerine birds beg for food when the parent bird arrives at the nest by stretching up their necks and gaping with wide-open beaks. Prechtl[383] found that any dark object appearing over the rim of the nest evokes begging and so does a jolt to the nest, which normally signals the parent alighting. If begging is repeatedly evoked by visual stimuli it becomes weaker and eventually there is no further response. If the nest is now jolted begging appears at full strength again. In such a case we may call the original fatigue 'stimulus-specific', i.e. other types of stimulus will overcome it.

Hinde's[211] work with the chaffinch mobbing response shows that there are other types of fatigue which we can describe in behavioural terms. Fig. 1.5 shows the results of one of his experiments. An owl is presented to a bird for 30 minutes; by the end of this time it has usually stopped mobbing and as a measure of the strength of its response Hinde counts the number of 'chink' calls in the first 6 minutes. The owl is removed for a period

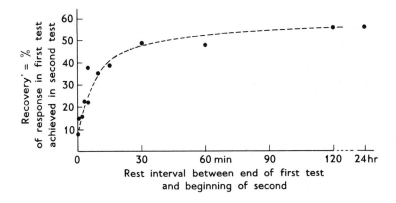

Fig. 1.5 The recovery of the owl-mobbing response after prolonged exposure to a stuffed owl; further explanation in text. (From Hinde,[211] *Proc. R. Soc., B.* **142,** 306.)

varying in length from half a minute to 24 hours. Then it is put back for a second mobbing test and again the calls made in the first 6 minutes are counted. Obviously the shorter the gap between the two tests the less we expect the bird to respond on the second occasion. Fig. 1.5 shows that the strength of calling is up to 50% of its initial level if 30 minutes elapse between the tests, but a rest of 24 hours scarcely improves on this and Hinde finds that even longer rests never bring it back to its original level.

Now since an owl was the stimulus for both tests some of this lack of recovery could be due to stimulus-specific fatigue as with the begging of the young birds. However, Hinde found that using a stuffed stoat for the second mobbing test—a stoat is just as strong a stimulus as an owl—while it produced a small increase in calling, still left essentially the same picture as that shown in Fig. 1.5. Recovery from fatigue is rapid over the first 30 minutes, but thereafter it is extremely slow and is 'response-specific', i.e. it is some property of the mobbing response itself. Fatigue of this type cannot result from the same mechanisms as does stimulus-specific fatigue. Hinde suggests that in fact there are two distinct processes here, one of which recovers rapidly whilst the other shows very little recovery. To these we must add the small degree of stimulus-specific fatigue demonstrated by exchanging owl for stoat. This means that perhaps three distinct fatigue and recovery processes are going on simultaneously when a chaffinch responds.

This is complexity enough, and although reflexes and complex behaviour both exhibit fatigue, the situation revealed by the chaffinch experiments must make us beware of jumping to conclusions about mechanisms. Each case must be studied separately to unravel the processes involved and only careful behaviour tests can do this.

Inhibition

This category is not so much a property of reflexes as a whole aspect of the functioning of the nervous system. One of Sherrington's greatest contributions was the demonstration that sometimes impulses, passing along axons, inhibit transmission by the nerve cell with which they synapse, even though this cell may be simultaneously receiving excitatory impulses from other axons. Inhibition operates at every level within the central nervous system. Since every neuron is potentially in contact with every other, it is quite reasonable to consider why an animal is not thrown into a convulsion by every stimulus it gets. Part of the answer, as we have already seen, lies with the delay and fatigue that occur in synaptic transmission, but active inhibition is also involved.

Muscles are commonly arranged in antagonistic pairs, such that one flexes a portion of a limb and the other extends it. Sherrington showed that excitation of one member of a pair is accompanied by inhibition of its antagonist. Such inhibition is not absolute, and an inhibited muscle does not simply go limp. Once it is stretched by its antagonist contracting then its own 'stretch reflex' (see p. 18 for a fuller description) will tend to make it contract. Although the antagonist may override it, it will take up the slack, so to speak, in an active fashion. Much finer control of movement is possible if muscles can be made to work against one another in this way. Mutual inhibition allows them to take the lead in turn during limb movements. Sherrington found that it is not only antagonists on the same limb which inhibit each other but also muscles located on opposite limbs and which have antagonistic effects during locomotion. When the flexors of one limb are contracting, the flexors of the opposite limb are inhibited. *Reciprocal inhibition* of this type is one of the basic integrating mechanisms for walking.

With reflexes the active inhibition of a movement is easily distinguished from fatigue. The fatigue of the scratch reflex (Fig. 1.4) is marked by a gradual weakening. When the same reflex is inhibited by stimulating the antagonistic flexion reflex, it stops abruptly with no previous weakening and begins again the instant that the flexion stimulus is removed (Fig. 1.6). Inhibition serves to bring about a smooth and rapid transition from one reflex action to another.

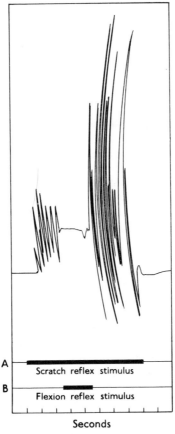

Fig. 1.6 Inhibition of the scratch reflex by the flexion reflex. The stimulus denoted on line A evokes the scratch reflex, but this response is inhibited when the stimulus on line B evokes the flexion reflex. The moment B is removed the scratch reflex returns, and much more vigorously than before—an instance of 'reflex-rebound'. (Modified from Sherrington,[435] 1906, *The Integrative Action of the Nervous System*. Charles Scribner's Sons, New York.)

The role of inhibition in complex behaviour is superficially less obvious than that of excitation. We stimulate an animal and the most conspicuous result is that it makes a response. But in doing so it has made a swift transition which requires the inhibition of its behaviour prior to the stimulus until it adjusts its responses to the new situation. Sherrington considers the way in which reflexes appear to 'compete' for control of the final common pathway, i.e. that which controls the muscles whose action is common to several different reflexes. In an analogous way we may regard the different systems controlling patterns of complex behaviour like feeding, fighting and sleeping, as competing for control of the animal. Such patterns are obviously incompatible and only one can occur at a time. Which one does appear will depend on various factors both inside and outside the animal. When, say, the feeding system gains control, all others must be inhibited for the time being.

Sometimes we can identify parts of an animal's nervous system which have an inhibitory function, serving to suppress activity which would otherwise become incessant. Recent work on the brain of insects has revealed mechanisms of this kind. The brain selects appropriate responses by 'removing' inhibition from the relevant parts of the ventral nerve cord and its ganglia which then excite the required muscles. One of the most clear-cut examples of this comes from Roeder's[396] work with the praying mantis (Fig. 1.7). When a male mantis is decapitated, cutting off the brain

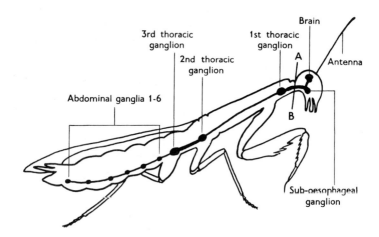

Fig. 1.7 Diagram of the central nervous system of the praying mantis. In Roeder's experiments the nerve cord was cut along the line AB. (From Roeder,[396] 1963, *Nerve Cells and Insect Behavior*. By permission from Harvard Univ. Press, Cambridge, Mass. © by the President and Fellows of Harvard College.)

from the ventral nerve cord, the body begins incessant stepping movements which carry it in a circle. In addition, the abdomen and genital appendages begin incessant copulatory movements and a headless male can often mate successfully. In nature males are sometimes decapitated as they approach the large, predatory females and these arrangements are highly adaptive!

In vertebrates, too, we often find that the result of destroying some area of the brain is to make one type of behaviour much easier to evoke. As we shall see in later chapters, some caution is needed in interpreting the results of brain damage experiments, but nevertheless there can be no doubt that parts of the vertebrate brain regulate the activity of others by inhibition.

Sherrington found that when inhibition was removed from a reflex it returned at a higher intensity than it had previously. Fig. 1.6 shows this phenomenon, which Sherrington called 'reflex rebound', for the scratch reflex. We commonly observe that when a particular type of complex behaviour—for example courtship—has not been elicited for some time, it has a lowered threshold and is performed with high intensity when it is, at last, evoked. It is possible that the system controlling courtship has been inhibited by those of other activities and shows something akin to reflex rebound when this inhibition has been removed. Kennedy[265-6] has interpreted some aspects of the behaviour of aphids along these lines. The behaviour of winged aphids alternates between periods of flight and periods of settling and feeding upon leaves. If an aphid settles on an 'unattractive' surface—an old leaf, for example—it does not stay long and soon takes off but flies relatively weakly and soon settles again. Conversely, if it has settled on an attractive young shoot it stays for a long period but, when it takes off, flies vigorously and for a long time.

By an elegant series of experiments, Kennedy[266] has been able to exclude any simple explanation for this relationship based on physical exhaustion during flight and recovery after resting and feeding on a young leaf. He suggests that there is mutual inhibition between the systems controlling flight behaviour and those controlling settling. As with the reflexes, activation of the settling system may temporarily inhibit the expression of the flight system but, at the same time, gradually lower the threshold for flight. Kennedy calls this relationship 'antagonistic induction' and it clearly resembles reflex rebound in some ways.

In Chapters 4 and 5 we shall discuss other evidence that the systems controlling complex behaviour do inhibit one another and that particular patterns 'break through' when their inhibition is weakened. However, reflex rebound cannot provide the whole answer to the problem of how one type of complex behaviour comes to 'take over' from another. Other factors can cause thresholds to be lowered. For example, it is quite inappropriate to speak, in physiological terms, of the 'inhibition of drinking' if a dog is physically prevented from reaching its water bowl, but the dog nevertheless becomes increasingly responsive to water.

Feed-back control

We have just discussed how inhibitory mechanisms can lead to the switching of activity from one set of controls to another. But very commonly reflex or complex behaviour consists, not of such sudden switches, but rather of a steady output of some activity which has to be held at a given level. When we 'stand at ease' our body is evenly balanced over the pelvic girdle and easily corrects for any slight jostling we may receive. To do so the muscles of the legs and back must be held at a constant level of tension and if shifted from this level, they must correct to bring the body upright again. Analogously, animals under normal circumstances maintain a very constant body weight, and they do this by seeking out food and water in the correct proportions. They eat and drink sufficient for their needs at regular intervals. If a surplus is available they do not overeat. In times of scarcity they spend a higher proportion of their time in feeding, and consume more when the chance arises so as to replace any deficit.

Both these examples show us behaviour acting as a homeostatic system, designed to preserve the *status quo*. In the first case this was achieved through relatively simple reflex systems controlling the leg and trunk muscles; and in the second, by a series of more complex systems regulating the search for food, feeding and satiation. In both cases the operation requires that the end result (posture and balance whilst standing, state of nutrition) is monitored in some way. When it deviates from a set value a signal is sent to the control mechanisms to correct the imbalance and bring the end result back to the set value again.

Engineers who design the control systems for machinery are familiar with mechanisms which have to function in a comparable way, and refer to them as operating by feed-back control. Some diagrams will help to explain the basic principles of feed-back control and show how we can apply them to reflexes and complex behaviour. We can distinguish between two types of control mechanism—open-loop and closed-loop. Fig. 1.8a and b represent the two systems.

In both systems some mechanism (represented here simply by a rectangle) produces an output in response to inputs of some type, and this output is affected by disturbances of various kinds. The open circle divided into quadrants represents some process of interaction whereby the disturbance acts.

Now in the system represented by the upper diagram disturbances to the output are not corrected. A boat is headed along a particular compass course with the tiller tied down; as cross currents move it off course there is no means of changing its direction to compensate—this is an open loop

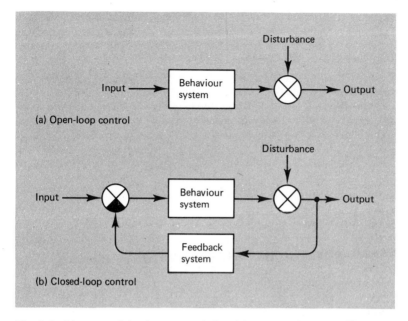

Fig. 1.8 Diagrams of simple open- and closed-loop control systems. The output from the behavioural system is affected by disturbance factors (the crossed circle represents interaction). With open-loop control (a), if the output is affected by disturbance no correction takes place. In the closed-loop arrangement (b) the results of the disturbance feed back to affect the input to the behavioural system. The dark segment of the crossed circle represents an interaction between the feedback mechanisms and the input which tends to bring back the output to its original value. The feedback system produces its own output which is proportional to the disturbance and changes the input to the behavioural system both to the right amount and in the right direction.

situation. In a closed loop situation, represented by the lower diagram, the output is measured and deviations are fed back to affect the input. The input is normally changed to the correct extent and in the correct direction to restore the output to its original level. Our boat now has an automatic pilot device which responds to the cross currents by turning the rudder so as to keep the boat on course. This closed loop situation includes feed-back control and is obviously the one most applicable to reflex or complex behaviours which are homeostatic in the effect.

This doesn't mean that all behaviour involves feedback control. Open-loop control has been shown to operate in some cases, particularly when a movement must be made very rapidly and, using neurons and synapses as a living system must, there simply is not time to modify the movement during its progress. The 'strike' of the mantis, some of whose predatory tendencies we have just been discussing, is such a case. The mantis moves towards a fly and orientates its body, slowly and precisely (operations which certainly involve feed-back control) but once aimed the strike is an all-or-nothing movement. If the fly moves after the strike is initiated this makes no difference to the form of the movement.

But to return to our original examples, the control of posture and the regulation of food and water intake clearly both involve feed-back which

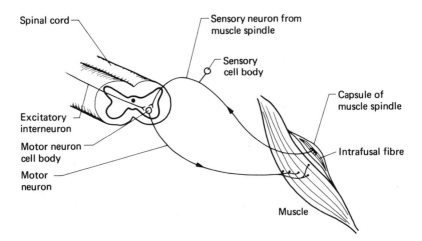

Fig. 1.9 A simplified diagram of some of the neural pathways involved in the stretch reflex. Explanation in text.

enables a steady state to be maintained. The mechanisms are obviously very different in the two cases, but the principles of operation are similar.

The control of posture is just one example of fine muscular control and physiologists have made a very complete analysis of the mechanisms involved. Indeed in some cases we know the paths of the neurons involved and can actually identify the structures which function as parts of the closed-loop control system. One such is illustrated in Fig. 1.9 which represents a typical muscle on the limb of a mammal, such as would be involved in maintaining posture. Motor neurons which have their cell bodies in the ventral horn of the spinal cord run to the muscle; it is their activity which determines the tension developed by the muscle. In parallel with every skeletal muscle, and embedded within its fibres so that they contract and relax with it, are muscle spindles. These are specialized sense organs for recording the degree of tension in the muscle. Their sensory nerves run back to the spinal cord and, entering through the dorsal root, synapse with the motor neurons to the muscle. Thus a loop is closed which forms the basis of the stretch reflex, already referred to on p. 12. When a muscle is stretched by the contraction of its antagonists, the muscle spindles are stretched also and their sensory fibres increase their rate of firing, stimulating the motor neurons so that the muscle contracts. It is easy to equate the units of the closed loop control system on Fig. 1.8b with those of this real muscle mechanism. The output (state of tension in the muscle) is affected by a disturbance (being stretched by other muscles) and a feed-back mechanism (muscle spindle) records the change and feeds back to change the input (motor nerve) and restore the original output.

This is a simplified picture of the real situation, which in fact includes other regulatory mechanisms allowing for very fine graded control over the muscle contractions involved both in the maintenance of posture and in movements, but it serves to illustrate the reality of feed-back control at a reflex level (for further details there is a clear account in Hinde,[217] Chapter 3).

It is much more difficult to identify the neurons or receptors involved in the control of food or water intake, nor is it very important to do so for a behavioural analysis. The feed-back principle is clear in the feeding behaviour of most animals which do not feed continuously (as some filter-feeding invertebrates do). In fact as we shall discuss more fully in Chapter 4 there are quite well defined parts of the brain in mammals which can be equated with the feeding control system. Again referring back to Fig. 1.8b, we may take the output to be represented by an adequate level of nutrients in the blood stream. Hunger tends to deplete this level and this disturbance is detected by cells in one part of the brain—the hypothalamus (see p. 142)—which initiates activity in the neural mechanisms controlling feeding. This activity ceases when the blood sugar level rises or even earlier when other signals—from a full stomach for example—record that feeding is completed. Fig. 4.5 and the discussion on pp. 123–6 describe a model for motivation which incoporates a feed-back control system of this type.

Certainly the feed-back principle can help us to understand how many different types of behaviour are organized. As we have just seen it can be applied to reflex and complex patterns within individuals. It can also be applied to groups of animals to some extent. For instance we may regard the manner in which animals respond to so-called 'density-dependent factors' as a form of feed-back control. As the population rises beyond the normal level fighting between animals may increase, food become shorter, or disease spread more rapidly—some combination of these and other factors will all act more strongly as density increases. The animals' responses to these changes include reduced breeding, emigration etc. and the net result is to reduce the population back to a lower level.

Because it has such wide application, we shall have to refer again to feed-back mechanisms at various points in this book—see, for example, the discussion of the development of bird song on p. 56. Most advanced texts of physiology or behaviour give more details and McFarland[315] provides a complete review.

This chapter has tried to show that reflexes and complex behaviour share many properties. In fact as we have mentioned previously, it is often possible to break down complex behaviour patterns into smaller units, some of which are immediately equatable with reflexes. Behaviour is frequently organized in a hierarchical way such that we can regard simple patterns like standing, walking, biting or swallowing as being subordinated to higher control systems, which we might label attacking, fleeing, feeding etc. Sometimes subordinate reflexes may be unique to one type of higher control—the various reflexes by which an insect cleans itself, rubbing its legs together, wiping the eyes, antennae etc., normally occur only during a cleaning session. But commonly reflexes must be 'available' for several higher systems. The reflexes which are involved in flight or walking, for example, must be employed when the insect migrates, searches for food, a mate and so on. Thus the higher control systems must compete for control of reflexes rather in the manner that reflexes compete for control of muscles, as Sherrington described.

Of course we cannot always explain behavioural observations using reflex terminology and such an attempt is often pointless. There is a difference in complexity and behaviour has to be analysed in its own terms not simply as a branch of neurophysiology. Nevertheless any study of behaviour which is not mindful of the underlying physiology will miss much that can be valuable because there are some continuities between the different levels of analysis. Even if much of the physiology remains in the background, in what follows we must always bear in mind the constant activity of the nervous system. It does not lie dormant requiring a stimulus to provoke it into a response. The sense organs and the proprioceptors in the muscles provide a continuous background flow of impulses whose level is affected both by external stimuli and by efferent impulses from the brain. Information from the external world and the muscles is combined by the central nervous system with information on various aspects of the body's metabolic state—temperature, carbon dioxide concentration and water balance, among others. Any shift from the optimum is rapidly corrected, and most of the overt behaviour we observe as a result of all this neural activity is immediately adaptive to the animal's survival. As we have seen behaviour can function as a homeostatic mechanism, one that is both sensitive and powerful.

2

The Development of Behaviour

Nearly all the behaviour we observe in animals is adaptive. They respond to appropriate stimuli in an effective manner and thereby feed themselves, find shelter, mate and rear families. Animals are certainly not infallible, but when they do make mistakes it is often because they have been transported into an unnatural environment. We are not surprised if birds make futile attempts to escape when first put into a cage.

INSTINCT AND LEARNING

How does an animal's behaviour become so well fitted to its normal environment? There are two basic ways. Firstly, it may be born with the right responses 'built in' to the nervous system as part of its inherited structure. Honey-bees inherit the ability to form wings and wing muscles for flight; they also inherit the tendency to fly towards flowers and seek nectar and pollen. Such responses are popularly called 'instinctive'—a term which has often been abused but remains useful. Instinctive behaviour evolves gradually, as do structural features, and natural selection modifies it to fit the environment in the best way. It forms a kind of 'species memory' passed on from each generation to its offspring.

Alternatively an animal may have no inherited responsiveness with regard to a situation but instead have the ability to modify its behaviour in the light of experience. It learns which responses give the best results and changes its behaviour accordingly.

Instinct and learning both ensure adaptive behaviour, the former by selection operating during the history of a species, the latter during the history of an individual. Stated in this way there is a clear dichotomy, which is unrealistic when actual examples are examined. But before we do this it is worth considering the importance of instinct and learning in a general way through the animal kingdom.

Instinct and learning in their biological setting

Instinct can equip an animal with a series of adaptive responses which appear ready-made at their first performance. This is clearly advantageous for animals with short lifespans and little or no parental care. The arthropods, for example, show a remarkable development of instinct for no other course is open to them. A female digger-wasp emerges from her underground pupa in spring. Her parents died the previous summer. She has to mate with a male wasp and then perform a whole series of complex patterns connected with digging out a nest hole, constructing cells within it, hunting and killing prey such as caterpillars, provisioning the cells with the prey, laying eggs and finally sealing up the cells. All this must be completed within a few weeks, after which the wasp dies. It is quite inconceivable that she could achieve this tight schedule if she had to learn everything from scratch and by trial and error.

Contrast the digger-wasp's situation with that of a lion cub. Born quite helpless, it is sheltered and fed by its mother until it can move around. It is gradually introduced to solid food and gains agility in playing with its litter mates. It has constant opportunities to watch and copy its parents and other members of the group as they stalk and capture prey. It may catch its first small live prey when 6 months old, but it is 2 years or more before it has grown sufficiently to feed itself. Its behaviour, and particularly the methods and stratagems it uses in hunting, may change according to circumstances throughout its life.

The digger-wasp which must rely on pre-set instinctive behaviour and the lion who can learn in relative leisure represent two extremes on the behavioural scale. In fact the preceding descriptions have greatly oversimplified their actual behavioural development. The digger-wasp can and must learn many things during its brief life—the exact locality of each of its nests, for example, so that it can return to them after its hunting trips. The young lion possesses some predatory tendencies which are certainly instinctive, even though it has to learn how to direct them.

All animals above the annelid worm level show both kinds of behaviour and each has its own special advantages. This is clearly illustrated from studies on bird calls. Birds often show a strong development of both instinct and learning ability. As we shall discuss later, the song of a male bird often requires the experience both of singing and listening to other males before it takes on its final form. By contrast, in every bird species studied both the production of and response to the alarm calls of the species appear perfect at the first showing. Natural selection has favoured an inherited response where the delay of learning may prove fatal, and such contrasts suggest that the pattern of development itself may evolve by selection, see p. 59.

One advantage of learning over instinct is its greater potential for changing behaviour to meet changing circumstances. Such a consideration is obviously more important to a long-lived animal than to an insect which lives only a few weeks. A further relevant factor may be body size, because highly-developed learning ability requires a relatively large amount of brain tissue, insupportable in a very small animal. Usually body size and life span are positively correlated to some extent and large animals live longer than small ones. Apart from these physical considerations it is clear that natural selection can produce different degrees of learning ability to match a species' life history. The two most advanced orders of the insects, the Hymenoptera (ants, bees and wasps) and the Diptera (two-winged flies), are comparable in size and life span. The Hymenoptera, in addition to a rich instinctive behaviour repertoire, show an extraordinary facility for learning, albeit of a specialized type, and this plays an important role in their lives. During her brief 3 weeks of foraging a worker honey-bee will learn the precise location of her hive and the locations of the series of flower crops on which she feeds. She may move from one to another of these during the course of a day's foraging because she also learns at what time of day each is secreting the most nectar. After three visits to a food dish marked by a particular colour, a honey bee worker retains her memory of this colour for the rest of her foraging life. Even after one visit it is five or six days before her responses to the colour are back to the neutral level (see Menzel).[342]

The Diptera *can* learn. Hover flies can learn something of the position of the flowers they visit, and house flies tend to return to the same places to settle in a room. It has proved possible using some ingenuity to get *Drosophila* to learn some simple discriminations. But for the most part the Dipteran memory is brief and their learning powers very limited. Unlike the Hymenoptera they do not reproduce in fixed nests to which they must return regularly. Their short lives are largely governed, with complete success, by inherited responses to stimuli which signal the presence of food, shelter and a mate.

The characteristics of instinct and learning

We must now examine more closely the characteristics of instinct and learning. There are two conspicuous features of instinctive behaviour which may seem, at first sight, to be unique. First, that it consists of rigid, stereotyped patterns of movement which are very similar in all individuals of a species; all digger-wasps of the same species build their nests in the same

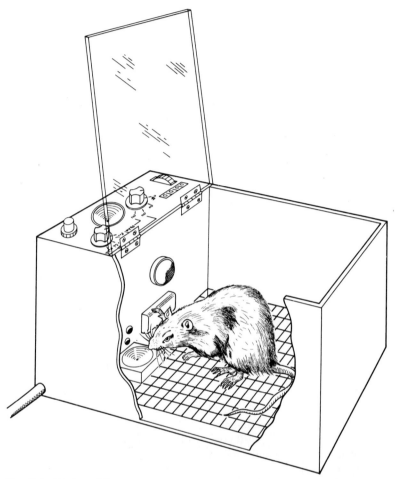

Fig. 2.1   Rat in a Skinner box pressing the lever which delivers a small pellet of food into the cup.

way, domestic cockerels all use the same series of movements when courting hens, and so on. Secondly, instinctive patterns can often be evoked most readily by very simple stimuli. When presented with a complex situation the animal responds to one part of it and virtually ignores the rest. A robin displays more aggression to a tuft of red feathers from the breast of a rival male than to a complete bird which lacks only these feathers.

However striking, such characteristics are quite inadequate to distinguish instinctive from learnt behaviour. The latter is often described rather vaguely as being 'more flexible', but in fact the patterns of movement involved may be just as stereotyped as those of instinct. Rats placed in a box where they must learn to press on a projecting lever in order to obtain a pellet of food (Fig. 2.1) will develop a particular manner of doing so. Some use always their left paw, others press with their chins, and individual rats tend to be conservative in the method they use. In his delightful book, *King Solomon's Ring*, Lorenz[306] describes how water shrews learn the geography of their environment in amazing detail. If at one point on a trail they have to jump over a small log, this movement is learnt with such fixity that they continue to make the jump in precisely the same fashion long after the obstacle is removed. There are plenty of other examples where animals have comparable difficulty in 'unlearning' something. The movements appear to become almost 'automatic' and may persist even if they are no longer effective. Learnt patterns nearly always involve responding to particular cues in the environment and, just as with instinctive behaviour, other features may be ignored. It is relatively easy to train animals to discriminate one key stimulus within a complex changing situation.

If we are to make any critical distinction between instinctive and learnt behaviour, it must be based not upon their overt characteristics but upon their development within the individual. Rats learn to press a food lever and each does so in a stereotyped way, but this way is unique to the rat and varies between rats. All sexually receptive female rats show the same stereotyped acceptance posture when mounted by a male (Fig. 2.2). Furthermore if we put a hungry rat into a box with a food lever, it may be hours before,

Fig. 2.2 Copulation in rats. The female's posture with raised pelvis and deflected tail is very stereotyped in form. The pressure of the male's fore-limbs on her flanks is one of the stimuli necessary for the female to respond. (From Barnett,[28] 1963, *A Study in Behaviour*. Methuen, London.)

by chance, it pushes against the lever. But if a virgin female rat is brought into receptive condition by a hormone injection, it assumes the acceptance posture after very little experience with a male.

Here we have the basis for developmental criteria of instinctive behaviour and the commonest way of testing them has been the so-called 'isolation experiment'. Animals are kept individually out of contact with others from as early an age as possible. When mature, their responses to a variety of stimuli are tested and compared with those of animals reared normally. Rather few animals have been tested under really rigorous conditions but some fish and birds have been shown to perform various feeding, sexual and alarm patterns of behaviour quite normally after being reared in isolation. The life histories of many insects are natural isolation experiments which demonstrate vividly that no practice or learning are involved in the behavioural development of the adults. However easily isolation experiments enable us to eliminate the possibility that animals learn how to do something, they offer only a very restricted view of the factors that may be operating during development—learning and practice are only two from a wide range of possibilities.

The point is that isolation can tell us only what factors are *not* important for the development of behaviour, it tells us nothing of what *is* involved and this may not be at all obvious. For example, young mallard ducklings hatched from an incubator respond preferentially to the call notes of mallard ducks over other related calls the first time that they hear them. They clearly show an instinctive preference for their own species call. However Gottlieb[171-2] has shown that one factor which contributes to the development of this preference is the embryo duckling's ability to hear the calls— quite unlike those of the mother—that itself and other ducklings make whilst still inside the eggshell.

Nobody would doubt that the development of many behaviour patterns must be under genetic control and result from an inherited potentiality of the central nervous system. But this is the point from which studies of development should take off. We should certainly not be content to label such patterns as 'instinctive' and leave it at that. Genes may control behavioural development, but to do so they must interact with the developing animals' environment. Gottlieb's ducklings need some auditory stimulation if they are to be primed to respond to the maternal calls. It will be our task to discover such factors in the behavioural environment of animals which affect the development of their instinctive patterns. Clearly we must avoid any simple dichotomy which ascribes instinct to the genes and learning to the environment. Both must be involved in the development of all behaviour. This point might not seem to need such emphasis, but in fact misunderstandings here have been, until quite recently, the basis for a considerable dispute among animal behaviour workers.

Ethologists and psychologists

After World War II there were two main schools of animal behaviour in the West. The first consisted largely of American experimental psychologists who could claim descent from J. B. Watson,[488] whose book *Behaviorism*, published in 1924, was a landmark in the modern experimental approach to behaviour. The second school—largely European—was founded in the late 1930's by Konrad Lorenz, an Austrian working in Germany. He was joined by Niko Tinbergen from the Netherlands, who collaborated with him in some early work. When the War ended both men rapidly built up groups of research workers and others, usually zoologists, became associated with their approach. This group call themselves 'ethologists', and ethology has been defined simply as 'the scientific study of behaviour'.

Ethologists and American psychologists approached their subject from quite opposite ends. Ethologists studied a wide range of animals under near natural conditions, often in the field with the use of hides. They were impressed by, and concentrated on, the rich variety of adaptive behaviour patterns shown by insects, fish and birds. Reproductive behaviour which is often conspicuous and easy to observe formed a large part of their studies, and ethologists often have great success in keeping animals and getting them to breed in captivity.

The original ethological approach is set out by Tinbergen[460] in his book, *The Study of Instinct*, and he later developed his ideas in an important review[466] of the aims of ethology. They are, in essence, concerned with the way behaviour fits animals for survival. Ethologists have studied the stimuli which evoke and the motivation which controls instinctive behaviour, and latterly have often studied the development of such behaviour in some detail. Yet function and evolution have always been a central issue for them. They study behavioural evolution by comparing the behaviour of closely related species and deducing the evolution of particular behaviour patterns, just as a comparative anatomist does for structural features. Ethologists were not much interested in problems related to learning and, though they recognized that their animals could learn, they generally ignored or eliminated this fact from their theories and experiments.

Experimental psychologists, on the other hand, were almost exclusively interested in learning. They largely ignored an animal's behaviour in its natural environment and deliberately confined their subjects to experimental conditions. A maze; a 'shuttle-box' in which it learns to move from one side to another to avoid electric shock; a 'Skinner box' in which it learns to press a lever or peck at a key to receive a reward of food or drink; these were the situations that interested the psychologists. Their aim was to construct 'laws of behaviour' which would describe how an animal's behaviour changes after given levels of practice, reward or punishment and to make predictions, for example, about the efficacy of one or another experimental situation for producing learning.

It is not surprising that the experimental psychologists used a narrower range of animals than the ethologists. Mammals are the best animals for their type of learning study, and were used almost exclusively, with domesticated albino rats providing the great majority of the subjects.

At first the two schools took little note of each other's findings, but after some years cross-fire began and each tended to adopt rather extreme positions. See, for example, Lehrman[286] for an extreme statement of the case against the ethological concepts then current.* The psychologists accused the ethologists, among other things, of grossly underestimating the role of the environment in the development of behaviour, of regarding the label 'instinctive' as an adequate explanation in itself and of applying this label far too readily and without sufficient evidence. Conversely, ethologists complained that most experimental psychologists were ignorant of the behaviour of any animal other than the white rat. They suggested that the reality of instinct would become obvious to anybody who could drag himself away from a Skinner box to watch a colony of honey-bees or a nesting male stickleback.

Both sides scored good points and both have subsequently benefited from a dispute which for once did generate more light than heat. Ethologists are now much more cautious in their use of terms and increasingly turn their attention to the development of behaviour. Experimental psychologists are employing a wider range of animals and studying them in a wider range of situations. At one time to label someone 'an ethologist' implied a good deal about the theoretical framework within which he worked and also his background—usually zoological—but this is now much less true. The ethological approach which aims to relate an animal's behaviour to its natural environment has been adopted by workers from a wide range of backgrounds, zoological, physiological and psychological.

THE DEVELOPMENT OF BEHAVIOUR WITHIN THE INDIVIDUAL

The majority of workers would now agree that any rigid classification of behaviour into 'instinctive' or 'learnt' is inadequate, but that all behaviour presents us with problems of development. What processes are involved in the emergence of fully-formed adult behaviour? This is a very wide topic

---

* Quoting him here might seem to place Lehrman in the psychologist's camp, but this would be quite wrong. He had a strong biological background and, though critical of much of the early ethological theory, his sympathy for the ethological approach was matched by a highly critical view of much contemporary experimental psychology. For an account of Lehrman's very influential work in this context, see Beer.[44]

and here we can only take a few examples to illustrate some of the complex range of influences that must be considered. They will serve to emphasize how meaningless it is to try to press behaviour into categories.

The importance of early experiences

A great deal of work on a variety of animals, birds and mammals especially, has shown how sensitive they are to events when young. Sometimes changes in their environment which we might consider trivial turn out to have effects which persist and can be detected right through their lives. We must remember that development starts when the egg is fertilized although the majority of behavioural experiments begin much later with animals which though young, are relatively independent. Consequently we may overlook the earliest stages of development such as the young bird goes through in the egg or the young mammal inside its mother.

It is certainly not immune from factors that affect its behaviour even then. We have just mentioned Gottlieb's work on the effects produced by young ducklings hearing their own and others' calls whilst still in the eggs. Such responsiveness may have surprising results. Vince[477] has shown that embryo quails respond to the clicking calls made by the other members of the clutch. Remarkably, this communication between embryos serves to synchronize their hatching. Slow developers are accelerated by hearing the calls characteristic of advanced embryos and, to a lesser extent, advanced embryos are slowed down by the calls from less advanced neighbours.

In mammals the embryo is more insulated from the external world but, of course, more directly dependent on its mother's physiological state. If female rats are kept under stress during pregnancy the behaviour of their offspring is affected.[258,454] The type of change produced concerns the degree of fearfulness that they show when first put into a strange environment. Similar effects can be produced by a whole variety of treatments to young animals after birth. Mild electric shock, cooling or even the simple act of lifting them from the nest and then replacing them can affect their subsequent behaviour as adults. Rats which have been handled during infancy mature more quickly—their eyes open sooner and they are heavier at a given age than unhandled controls. In addition they appear less 'emotional' in frightening situations. By this we mean, for example, that when placed in a large, brightly lit arena they move around and explore more than unhandled rats, which spend more time crouching and cling to the edges of the arena if they do move.

Neither the nature of such effects nor the way in which early experience produces them is at all clear and this is currently an active field of research. It is not a very easy subject to follow because, as with many complex questions, the results obtained are not always consistent. The nature of emotionality and its relationship to fearfulness and other stress-related responses is certainly not as simple as some of the literature would imply. Archer[16] provides a critical review of the various measures of emotionality used with rats and mice. It has been common to consider the effects of extra stimulation as beneficial, and to regard the laboratory environment as monotonous and unstimulating to the young rodents. This presumes that the natural environment is more stimulating but as Daly[117] points out most of the evidence suggests the opposite. In the wild, young rodents are often reared in total darkness in burrows with very constant temperature and humidity and considerable insulation from noise. Beside this, the average laboratory animal house must offer a riot of extra stimulation. However the genetical differences between wild and laboratory rats are so great that it is not easy to make valid comparisons between them on such difficult measures as emotionality. The 'extra' stimulation of young animals may act on them directly so as to change their subsequent development, but more probably it also affects them through their mother. She reacts to their changed condition after handling or shock by giving them more attention than untreated young. Barnett and Burn[30] have some direct evidence of this with baby rats that had been marked by having their ears punched. Just after they were returned to the nest such babies received three times as much nuzzling and licking from their mothers as did untouched babies. There is now evidence that the effects of handling rats in infancy can be transmitted at least as far as their grandchildren.[128] Presumably handled rats have altered maternal physiology and behaviour which affects the early experience of their offspring whose maternal behaviour is, in turn, affected with results still detectable in their own young.

If stimulation affects young rodents via changes in the behaviour of their mother towards them, it is worth remembering how variable is the amount of close maternal contact different mammals receive. Quadagno and Banks[386] describe how pygmy mice remain attached to their mother's nipples throughout the day and are separated from her only at night when she forages out of the nest. This probably represents a higher level of contact than the part-time attention received by rats and mice. At the other end of the scale are rabbits, where Zarrow et al.[519] have shown that the mother visits her litter only once per day and then for only a few minutes. Milk is pumped into the babies, they are briefly groomed and then the nest is covered and left. Presumably this behaviour helps to minimize the chances that predators will find the nest by picking up the mother's scent. A similar function is suggested for the still more remarkable maternal behaviour of tree shrews (insectivores probably quite similar to the early ancestors of the primates, see Fig. 8.8, p. 287). Martin[336] (see also D'Souza and Martin[139]) observed pairs breeding in captivity and found that they build two separate nests in one of which the adults sleep, whilst the female produces her litter, usually of two, in the other. The young are visited briefly on every second day, they suckle rapidly and become distended with milk. The mother then leaves them with the very minimum of grooming and typically does not return for forty-eight hours. Martin discovered that the young are capable of grooming and licking themselves from the day of birth. They stay alone in the nest until they emerge at 33 days of age. This is a very remarkable type of development for a mammal and forces us to recognize that natural selection can produce a range of specializations in the pattern of development which are matched to other aspects of a species' life history. The influence of the mother on the earliest stages of behavioural development must be very different in the pigmy mouse, rat, rabbit and tree shrew. Even close relatives can vary enormously in their stage of development and independence at birth: we can contrast the helpless blind naked young of rats and rabbits with the active young, fully furred and with eyes open, born to guinea pigs and hares.

As we shall discuss more fully in Chapter 8, all the primates have very long periods of development—years in the apes and man—in which they stay in close association with their mothers. Separation, particularly in the early months, has very severe effects on the infants. In the experiments of Harlow and his group[203] young rhesus monkeys were isolated at birth and reared artificially. Although they grew well enough they were behaviourally crippled and when subsequently put with other monkeys, showed almost none of the normal social responses. In particular they showed totally inadequate sexual and parental behaviour. This is another illustration of the subtle factors that must be at work during behavioural development. If we watch a young monkey with its mother, it is almost impossible to identify in its behaviour those factors which are contributing to the infant's social development, yet we know they are there.

The work of Hinde's group, well described in his book,[219] has followed in great detail the normal development of rhesus monkeys reared by their mothers living in small groups. They could trace the gradual growth of independence due in part to the infant, in part to the mother. At first the infant is scarcely ever out of contact. The mother rarely allows her infant to move beyond arm's reach and even when it leaves its mother to explore it returns to her frequently—using her as a secure base from which to investigate new objects. Gradually, in parallel with the infant's increasing independence, the mother becomes less solicitous and even begins to reject some of the infant's approaches.

Knowing the normal course of development, Hinde and his co-workers then went on to study how it was disturbed by forced separation of mother and infant. They studied the effects of levels of deprivation far less drastic than those imposed by Harlow. In a typical set of observations a baby rhesus and its mother, living in a group, are watched regularly for some weeks. Then when it is 6 months old—well able to feed itself—its mother is removed from the group for a few days. Far from being isolated, the baby is usually 'adopted' by other females of the group and given a great deal of attention. Nevertheless its behaviour shows a great change, distress calls increase, it moves around less and spends long periods in a very characteristic hunched posture. When the mother is returned, there is usually an instant reunion and the infant spends much more time clinging to her than just before separation. The pattern of their relationship is different from the normal with a 6-month infant and it takes several weeks to recover.

Several fascinating conclusions, with obvious human parallels, have emerged from systematic studies of this type. For instance, the infants worst affected by the brief separation are those whose relationship with their mother before separation appeared to be less good. From our point of view, in the discussion of early experience and development, we may note how profoundly the normal smooth adjustment of mother/infant relations was affected by even a brief interruption. The effects on the young monkey were both general and persistent, for Hinde could detect even years later that monkeys which had been separated were more fearful in strange environments.

Imprinting

So far the effects of early experience which we have been considering have been fairly general in their nature, affecting a wide range of behaviour, some quite drastically. Sometimes particular early experiences can be shown to have rather clearly defined results. Of course whether we call the results of an experience 'general' or 'specific' depends to a great extent on how closely we look for effects. It is very easy to label an effect as specific if only a few aspects of behaviour are studied. Nevertheless some examples are quite striking: playing a tape recording of song to a young bird will affect the song it sings when it becomes sexually mature months later (see p. 55); so far as we know nothing else in its behaviour is changed by this experience. Some of the results of 'imprinting' appear to be equally specific.

Imprinting cannot be precisely defined. As we shall use the term here it refers to various behavioural changes whereby a young animal becomes attached to a 'mother figure'. It was Lorenz[303] who introduced the topic to most behaviour workers by experiments with geese in which he got broods of goslings to follow him and treat him as their mother figure. A good deal of the work on imprinting has been carried out using birds like geese, ducks, pheasants and chickens, which have precocial young (i.e. those that can walk at hatching and do not stay in the nest). Imprinting may be measured by the amount of attention paid to the mother, time spent close, latency to approach, time spent following if she moves, and so on. This type of response to the mother figure is usually called 'filial imprinting' to contrast it with 'sexual imprinting' which we shall discuss later. Imprinting usually takes place soon after hatching or birth and often results in a very fixed attachment, difficult to change. Lorenz described it as a unique form of learning which, unlike other forms, was irreversible and restricted to a brief 'sensitive period' just after hatching. He also claimed that a young bird's choice of a mother also affected its choice of a sexual partner when it matured. Although some would still agree that imprinting should be studied as a unique form of learning (notably Hess[209]), most workers in the field now consider that it is best regarded as one aspect of the whole developmental process whereby very young animals come to adjust to the world they meet.

As a consequence imprinting has aroused great interest amongst ethologists and psychologists. We can concentrate our discussion on birds but the importance of imprinting-like developmental processes in insects and mammals is also recognized. Sluckin[438] and Bateson[35-6] provide excellent surveys of the whole subject.

Although, as we have just noted, it is probably best not to press imprinting into the mould of learning, it can be investigated in much the same way, i.e. young birds are given a period of training with a mother figure and then tested for their retention of the experience. For example, in one of his tests Guiton[187] used a circular runway around which an object could be moved suspended off the ground (see Fig. 2.3). Chicks were put

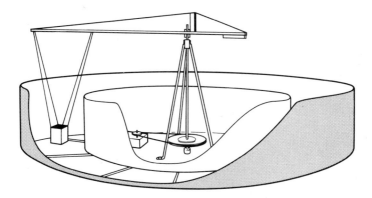

Fig. 2.3 Apparatus to study the following response of young birds. Different models can be attached to the long arm which rotates, moving them slowly along the circular runway. (After Hinde,[219] *Biological Bases of Human Social Behaviour.* © R. A. Hinde. 1974. Used with permission of McGraw-Hill Book Company.)

into the runway singly when 1 to 2 days old and exposed for 30 minutes to one of two moving objects, a red cardboard box or a green box, which differed in shape and size. The chicks usually followed their box for a good proportion of the time. Three days later, they were put back in the runway at a point equidistant between both boxes, which were then set in motion. Guiton recorded how much time was spent following each box or standing close to it, and found that after a single exposure the chicks discriminated almost completely, and ran at once to the model with which they had previously been trained. In this, as in a number of other studies of a similar type, the fact that the chicks follow the model is used as a convenient measure for the strength of the response. But following is not an essential part of imprinting—the chicks revealed their imprinting by their choice of the model to which they had been previously exposed.

We can organize our brief discussion of imprinting by considering three questions:

   1. What are the stimuli which direct the young birds' approach?
   2. Is there a sensitive period and, if so, what ends it?
   3. What are the results in adult life of imprinting during infancy?

   1. WHAT ARE THE STIMULI WHICH DIRECT APPROACH? Lorenz's early observations were most dramatic because he got broods of greylag goslings to imprint on himself. There seems to be no limit to the range of visual stimuli for imprinting. Birds have been imprinted upon large canvas 'hides' inside which a man can move, down through cardboard cubes and toy balloons to matchboxes. Colour and shape seem to be equally immaterial. Nor is movement necessary; a stationary object will attract young birds provided it contrasts with its background, so will flashing lights. Bateson and Reese[39,40] have shown that within a few minutes day-old chicks and ducklings will learn to stand on a pedal in order to switch on a flashing light, which they then approach. The rapidity with which they acquire this response suggests that the light has reinforcing properties even before the young birds are imprinted to it. They actively seek such stimulation and such attraction forms the basis for the imprinting attachment.

Clearly the young chick hatches already prepared to respond to visual objects and this sensitivity can be further affected by its environment. Bateson[37] discusses a number of experiments showing that exposing chicks to light just after hatching, markedly increases their responsiveness to conspicuous objects over chicks kept in the dark. Just 18 minutes of constant light before returning them to the dark had effects which could be detected up to 12 hours later. This effect is not just due to the light arousing the chicks in a general way, because playing them sounds, or stroking the chicks in the dark did not make them more responsive to visual objects, if anything the reverse.

This specific arousal of the visual system reminds us of Gottlieb's similar experiments described on p. 26 where he heightened the auditory responsiveness of ducklings by playing them sounds. Auditory stimuli are themselves attractive to young birds and will often enhance imprinting to objects close to their source. Sound is almost essential to induce following in mallard ducklings, for example. In some cases the response to sound goes beyond this and Klopfer[269] has demonstrated that young wood-ducks show a form of auditory imprinting, moving towards a sound source. They will respond to a wide range of sounds on the first exposure but subsequently discriminate against unfamiliar sounds. Wood-ducks nest in holes in trees, and normally the young hear their mother calling from the water outside the nest hole before they have ever seen her.

2. IS THERE A SENSITIVE PERIOD AND, IF SO, WHAT ENDS IT? By using the term 'sensitive period' we imply that only within this period will exposure to an object result in the bird becoming attached to it; 'critical period' is often used synonymously. The sharpness of the sensitive period depends to some extent on how exacting are the criteria adopted. If the criterion 'follows a moving object at first exposure' is used, then many precocial birds will satisfy this for 10 days or more after hatching. However, if one takes as a criterion 'the formation of a lasting attachment following a single exposure to an object', then the sensitive period may appear much more limited. Figure 2.4 illustrates some results using both types of criteria. Those from Ramsay and Hess[390] measured how well mallard ducklings discriminate and stay close to a model, 5 to 70 hours after a single 30-minute exposure to it. Boyd and Fabricius[63] simply scored birds which followed a model on their initial exposure.

Guiton[187] found that chicks kept in groups ceased to follow moving objects 3 days after hatching, but chicks reared in isolation remained responsive much longer. He could show that the socially-reared chicks became imprinted upon one another. This may be important in the natural situation. Boyd and Fabricius[63] point out that mallard ducklings do not normally leave the nest until the second day after hatching, well past the peak of the sensitive period. By this time some of them may be, at least partially, imprinted upon each other. Even if only a few of the brood actively approach and follow the mother bird as she leads them to water, the brood will act as a group and stay together because the rest will follow the maternally imprinted ducklings.

Since social rearing restricts the sensitive period it may be that the imprinting process itself plays a part in ending this period. Bateson[34] reared chicks in small pens whose walls had conspicuous patterns of horizontal or vertical stripes in black and white or red and yellow. After 3 days of isolation in a pen the chicks were tested in a plain runway with moving objects of similar striped patterns. They preferred to follow the model whose pattern matched that of their rearing pen. In a rather similar experiment Taylor *et al.*[449] found that exposure of young chicks to coloured walls in their pen shifts their subsequent choice of colours towards that

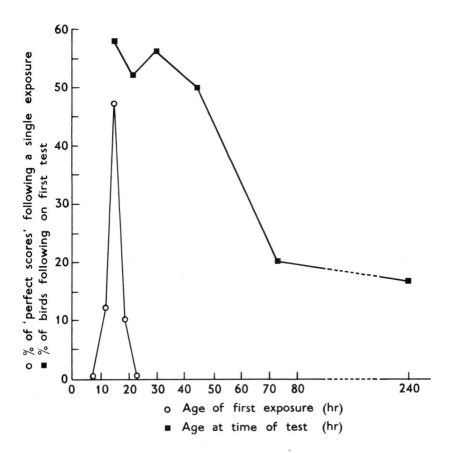

o % of 'perfect scores' following a single exposure
■ % of birds following on first test

o   Age of first exposure (hr)
■   Age at time of test   (hr)

Fig. 2.4 Two ways of expressing the sensitive period for imprinting. The open circles show the very sharp peak obtained when mallard ducklings were tested for discrimination and following some time after a single training exposure at the ages shown. The black squares show the much broader peak obtained by scoring simply the percentage of birds which follow a moving object on the first exposure. (Data from Ramsay and Hess,[390] and Boyd and Fabricius.[63])

which is familiar, even if the initial exposure was only 15 minutes. Such experiments strongly suggest that imprinting involves the young animal learning to discriminate the familiar from the unfamiliar and approaching the former. In the absence of any conspicuous moving object the chick, in effect, becomes imprinted on its surroundings and its reponsiveness to new objects declines.

Other workers have laid more emphasis on the growth of fear itself as the cause of waning responsiveness. There is no doubt that escape responses do increase from the first day; newly-hatched chicks or ducklings will approach new objects, but by the third day they will flee and crouch. The point at issue is whether the escape tendency grows only when the mother figure becomes sufficiently familiar through imprinting to make other environmental stimuli recognizably unfamiliar, or whether it grows anyway, independently of such experience, as a result of maturation.

Moltz and Stettner[351] reduced the visual experience of ducklings by rearing them with translucent hoods which meant that their eyes received light but no pattern. Such birds when tested at 2 or 3 days of age without their hoods, followed a moving box more than control birds, and showed less fear. This certainly suggests that previous visual experience of the surroundings does affect the growth of fear.

Loud noises or mild electric shocks, which would be frightening to older birds, make chicks follow their mother figures all the harder. This is probably because heightened arousal leads to a more intense response to both familiar and unfamiliar objects.

3. WHAT ARE THE SUBSEQUENT RESULTS OF IMPRINTING? Lorenz and others observed that young birds reared by a foster mother of another species ceased to follow her as they became independent, but later courted and attempted to mate with birds of the foster mother's species when they became sexually mature.

The conclusion that such a highly specific behavioural change can follow from an early experience has been amply confirmed by later experiments, reviewed by Bateson.[38] For example, both male and female turkeys have been sexually imprinted on humans[417] and cockerels imprinted onto cardboard boxes which they courted and attempted to mount.[188] Systematic cross-fostering experiments designed to study sexual imprinting have been carried out both with various species of ducks and of Estrildine finches.

With the finches, Immelmann[250] placed a single egg of one species e.g. zebra finch, in a clutch of a second, e.g. Bengalese finch, and allowed the Bengalese parents to rear the whole brood. Subsequently, cross-fostered zebra finch males were isolated until they were sexually mature. Immelmann then gave them a choice between a zebra female and a Bengalese. The results were quite unequivocal; the zebra male directed all his courtship towards the Bengalese female. This preference was all the more striking because when a zebra male was put in with the two females the zebra female usually responded at once with all the usual con-specific greeting calls. The Bengalese female was, at best, neutral and usually showed avoidance as he approached her.

Immelmann found that if cross-fostered males were obligatorily paired with females of their own species most would eventually mate with them and raise broods of young. Astonishingly, such experience did *nothing* to alter the preference described above. When once again given a choice, the males totally ignored con-specific females and courted the foster species. Sonnemann and Sjölander[443] have studied female zebra finches similarly reared by Bengalese. Although the results are not so striking as with the males, the females do show a definite preference for males of the foster species, and choose to perch close to Bengalese males rather than males of their own species.

In his duck experiments Schutz[425] found that only males imprinted on the foster-species and he suggested that this is because whilst each duck species has a distinctive male plumage pattern, the females are all much alike and cryptically coloured. It may be 'easier' to evolve an inherited discrimination for a particular male pattern than for a female (see p. 71 for an analogous example from fish). Supporting this conclusion, Schutz found that in the Chilean teal where both sexes are alike and have cryptic plumage, the female also imprints. However, this does not explain why geese, which resemble the Chilean teal in this respect of plumage, do not show sexual imprinting in either sex. Nor does it explain why male shel-duck, whose females are brightly coloured like themselves, do imprint upon females of other species.

In general the effects of cross-fostering were not so strong in the ducks as in the finches, at least in the sense that not all cross-fostered males were imprinted. Thus of 34 mallard males reared with other species (either foster siblings or a foster parent) 22 subsequently tried to pair with the foster species. They paired with females if they had been reared with a foster mother, but formed homosexual pairs if the foster parent had been a male. In nature, imprinting upon the mother as a figure to be followed and sexual imprinting normally go together, but we do not yet know whether they are identical processes. The sensitive period for sexual imprinting extends far beyond that for the following response. Schutz reared some ducklings with their own species for up to 3 weeks and then cross-fostered them. One-third of such birds still became sexually imprinted on their foster species, although this is to be compared with the two-thirds which responded when cross-fostered from hatching. Immelmann's evidence suggests that sexual imprinting is complete in the finches by 33 days of age, i.e. when fledglings are just able to look after themselves. Keeping males with the foster species for 94 days, i.e. until just before sexual maturity, had no extra effect.

The strength of imprinting in the finches was not affected by their experience after leaving the nest—as we have seen no amount of subsequent contact with their own species modified their preference. It is unlikely that sexual imprinting is so strong in all birds, but we lack much comparative evidence. Imprinting of this type certainly occurs in pigeons but note that it *cannot* occur in the European cuckoo and other parasitic birds whose young are always reared by foster species.

The experiments just described reveal the critical nature of early experience in determining the young animals choice of a sexual partner. Both Immelmann and Schutz point out that, dramatic though the results of cross-fostering may be, it is still easier to imprint a male on his own species than a foster species. Probably there are certain inherited tendencies which normally strengthen attachment to the con-specific female in the natural situation, but as yet we know very little about the interaction between inherited and learnt responsiveness towards sexual partners.

Imprinting in mammals

Imprinting-like phenomena are clearly involved in the social development of mammals. It is common knowledge that the younger we take and rear a litter of wild mammals the easier it is to tame them. Orphan lambs reared by humans follow them about and often show little attraction towards other sheep. This is not just a filial attachment, a form of sexual imprinting must also occur and zoo authorities know to their cost that hand reared animals are often useless for breeding when they are mature.

The behaviour of mammals is very much dominated by their sense of smell, and it is perhaps not surprising to find that their early olfactory experience often affects their choice of a mate. In mice, rats and guinea-pigs, mature animals are more attracted by others whose scent matches that which was present in the nest during the time they were being reared (see Carter and Marr[86]).

The development of social responses is also highly dependent on early experience. Scott and Fuller[428] summarize the extensive work of their group on dogs (see also Scott[427]). They have found that there is a sensitive period from about 3 to 10 weeks of age during which a puppy is forming normal social contacts. If isolated beyond the age of 14 weeks they no longer respond and their behaviour is very abnormal. A very short exposure at the height of the sensitive period is sufficient for them to form a normal relation with human beings. Dogs, like some sexually imprinted birds, seem perfectly capable of accepting both man and their own species as social partners.

Imprinting has always attracted the attention of psychiatrists because there is no doubt that human infants are extremely sensitive to early experience of many kinds. In particular, Bowlby[61-2] has developed a theory of the attachment of a baby to its mother which draws extensively from the animal experiments. In earlier work he suggested that the period from 18 months to 3 years was especially sensitive and that separation from, or lack of an adequate mother figure at this time led to a greatly increased risk of psychological disturbance in adolescence and later life. The idea of a restricted sensitive period now receives less support but there is a considerable amount of evidence to confirm the basic conclusion that maternal separation in childhood has long-term effects for human children.

Inevitably the situation is not as clear cut as with simpler animals. Separating a human baby from its mother will involve a whole series of changes to its life, almost always deleterious, and it is difficult to link up cause and effect very closely. The whole topic is well discussed by Hinde[219] in his Chapter 13.

There is some dispute over the recognition of sensitive periods in the social development of mammals. (Papers relating to this dispute are introduced and collected together in McGill[317] see also Denenberg.[127]) As we have seen when discussing sensitive periods for imprinting in young birds, it is often very difficult to distinguish between a change in responsiveness that occurs as a result of previous experience and a change involving *maturation* (see the next section) which occurs irrespective of what has gone before. We can certainly point to clear sensitive periods at particular turning points in the behaviour of *adult* mammals. The time around birth requires a crucial switch in the responses of the mother who must accept her newborn and start lactation. Klopfer *et al.*[270] have shown that shortly after she gives birth, a mother goat has a brief period when she is sensitive to the smell of her kid. Unless she has access to the kid within an hour of birth it is subsequently rejected, but 5 minutes' contact during the sensitive period will prevent rejection, even after a separation of 3 hours. Similarly, unless a ewe has close contact with her lamb during the first day of birth she will not begin secreting milk, and it now seems increasingly likely that the same is true of human mothers. Skin contact with the newborn baby over the first day is probably the most important factor.

Birth is a distinct event and there is no difficulty in marking off a critical period in the mother mammal. It is much less easy to determine critical events in the much slower course of behavioural development. The study of sensitive periods nearly always involves either isolating the young animal from supposed influences or alternatively attempting to add extra experience at different periods (as in the 'early handling' experiments we discussed earlier on p. 29). In either case the young mammal is developing in a changed environment and this often means there are loopholes in any argument about sensitive periods.

To some extent this argument is merely academic, but it relates to one's view of behavioural development. Is it a totally continuous process with interactions between the growing animal and its environment possible at every stage? Or is it more like embryonic development where we know that certain events must take place within a critical period if they are to take place at all? Once the critical period is past the embryonic cells are no longer competent to respond.

The development of behaviour probably exhibits both sets of characteristics and much will depend on a particular animal's life history. With an infancy period as brief as that of a zebra finch, which is sexually mature at 3 months, behavioural development is highly compressed and imprinting takes on an all-or-nothing, irreversible appearance. When we come to discuss the development of bird song on p. 55 we shall certainly find evidence of sensitive periods beyond which experience has no effect. But the infancy of some mammals is very prolonged and development is far more a continuous process—we have only to think of the work with rhesus monkeys which was discussed earlier on p. 31. The behaviour of the young monkey gradually develops over months or even years. Experience has effects at all stages, some transient, some permanent, but rarely appearing as irreversible since there is so much more time for later experiences to change the direction of development.

Development involving growth or maturation

The development of an animal's behaviour must obviously be linked to its normal growth processes. For example, the development of sexual behaviour in most vertebrates is linked with the growth of the gonads. Sometimes improvement in the performance of a behaviour pattern can be associated with the development of the animal's nervous system and such improvement is usually called 'maturation'.

Young birds can often be seen making vigorous flapping movements with their wings whilst still in the nest and it is commonly supposed that they are practising flying. Human parents often support young babies on their legs and encourage them to 'practise' walking. In fact there is no evidence that the early development of bird flight, or human walking are affected in any way by such activities. Over a century ago Spalding[444] showed that young swallows reared in cages so small that they could not stretch their wings flew just as well as normally reared birds when released. By the age of 18 months, when the majority of children are walking, their skills are very similar whether they walked first at 10 months or 15. In both cases it is the maturation of the central nervous system and its co-ordination with muscular development which count. Practice, of course, eventually adds all the finer points of skill—young fledglings are notoriously clumsy fliers—but the basic pattern matures without it.

The behaviour of embryos is determined by their stage in development and their increasing complexity of structure as they develop is paralleled by an increasing repertoire of behaviour, both spontaneous and in response to external stimuli. (For a particularly well studied example, see Oppenheim's[375] review of work on the development of behaviour in chick embryos.) Some of the clearest studies concern behaviour patterns in embryos which can be readily identified with patterns in the adult. Loco-

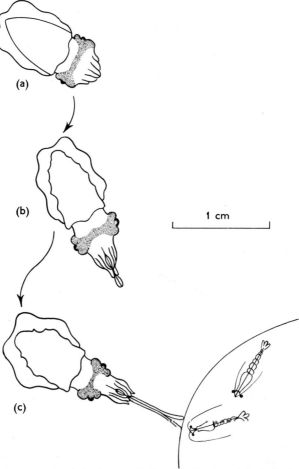

(a)

(b)

1 cm

(c)

Fig. 2.5 The attack of *Sepia* on *Mysis*: (a) Newly-hatched *Sepia* at rest; (b) swimming towards the prey, the eyes turned forward and fixating it; (c) stabbing with the long pair of tentacles at the *Mysis* behind glass. (From Wells,[492] 1958, *Behaviour*, **13**, 96.)

motory movements of various types are obvious examples and their development can often be related to maturational changes. One of the classic examples is Carmichael's[83-4] study on developing frog tadpoles. Whilst still in the jelly they begin flexing their tails back and forth as in swimming and these movements get more vigorous and complete with time. Is this improvement due to practice or simply to growth? Carmichael kept some developing tadpoles under continuous light anaesthesis which did not slow their growth but prevented all movement. When brought round and released at the age of emerging from the jelly they swam just as proficiently as unanaesthetized tadpoles, which implies that maturation and not practice is the cause of the changes seen in the behaviour of normal tadpoles.

The particular nature of development in crickets enabled Bentley and Hoy[49] to study one type of maturation in great detail. Crickets go through a series of 9–11 larval or nymphal stages (instars) which become increasingly similar to the adult with each successive moult. The final moult usually involves the biggest changes, with the full development of wings and adult genitalia. Since they have no wings cricket nymphs do not fly, but some elements of the flight pattern can be detected as early as the 7th instar. Nymphs at this stage will adopt the typical flying posture (holding their body rigidly with antennae and hind legs stretched out parallel to it) when suspended in a wind tunnel. All the muscles necessary for flight are present from an early stage, albeit reduced in size. Bentley and Hoy inserted fine wire electrodes into some of these muscles of the thorax and recorded action potentials from them. Since the relationship between the input from motor neurons and the output from the muscles is relatively simple in insects, the muscle activity gives a very accurate picture of the nerve activity responsible for it. Bentley and Hoy could detect signs of the rhythmic nerve impulses characteristic of flight from 7th instar nymphs, but it was incomplete and not sustained. It becomes more complete with each successive instar—the elements always appearing in the same sequence in different individuals—until by the last instar before the final moult, the entire pattern was complete. At this stage then, the maturation of the nymphs nervous system is finished and it awaits only the development of wings. Substantially the same story is true for the cricket's song, another activity which requires the wings of an adult (they are rubbed together rapidly to produce sound) for proper expression. For both flight and singing Bentley and Hoy found that they could not elicit neural activity from the thoracic centres of nymphs unless they made lesions in the brain. This presumably removed inhibition, as in a mantis whose head is removed (see p. 14), which normally prevents the nymphal muscles from being stimulated into useless activity until the rest of the body's development has caught up.

The cricket's behaviour develops along very precise lines in the absence of any practice, but there are cases in which practice is required from the outset if behaviour is to develop properly. In Chapter 1 Wells'[492–3] experiments with young cuttle-fish were mentioned. It will be recalled that their response to a tiny shrimp presented in a glass tube can be divided into four stages (see Fig. 2.5):

1. A latency before there is any observable response (Fig. 2.5a).
2. The nearest eye of the cuttle-fish fixates the shrimp.
3. The cuttle-fish turns towards the shrimp and both eyes fixate (Fig. 2.5b).
4. It attacks and seizes the shrimp (Fig. 2.5c).

Stages 2, 3 and 4 usually occupy about 10 seconds and this time varies very little with age or experience. The duration of stage 1 shows a rapid decline with successive tests as shown in Fig. 2.6. After five trials at a rate of one per day, the latency is reduced from about 120 seconds to 10 seconds or less. This change is the same whether the attacks are successful or unsuccessful, whether made with a 1-day-old cuttle-fish or one starved for 5 days before its first test. The only common factor appears to be practice in attacking shrimps. Learning is ruled out because in some other tests Wells

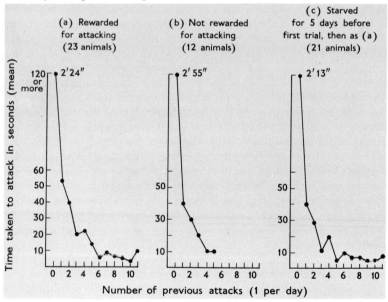

Fig. 2.6 The rapid decline of latency to attack *Mysis* in newly-hatched *Sepia*. The latency depends on the number of attacks that have already been made and is not affected by reward (a & b) food deprivation or age (c). (From Wells,[492] 1958, *Behaviour*, **13**, 96.)

finds it impossible to show any sign of learning ability until the cuttle-fish are a month old. Baby cuttle-fish go on battering away at a glass plate behind which they can see a shrimp, until they are physically exhausted. Nor are they deterred from such attacks by electric shocks. With similar treatment an adult will stop attacking after a couple of trials. The vertical lobe of the cuttle-fish brain, which is known to be concerned with learning in adults does not develop until relatively late after hatching.

The development of pecking in newly-hatched chicks provides another example of the interaction between maturation and practice. Young chicks have an inherited tendency to peck at objects which contrast with the background, but their aim is at first rather poor, and various workers have studied how it improves. One of the most complete studies was made by Cruze.[111] He hand-fed chicks in the dark on powdered food for periods of up to 5 days before testing the accuracy of their pecking. Whilst in the dark they are inactive and have no chance to practise the movement. Cruze measured accuracy by putting the chicks individually into a small arena with a black floor, on to which he scattered two or three grains of millet. Each chick was allowed 25 pecks scored for miss or hit and grains were replaced if the chick swallowed them. After accuracy tests the chicks were allowed to feed naturally in the light and the effects of practice on their accuracy measured again after 12 hours.

Table 2.1 The pecking accuracy of chicks at different ages before and after 12 hours of practice. Each figure represents an average from 25 chicks. (Modified from Cruze.[111])

Age (hr.)	Practice (hr.)	Av. misses (25 pecks)
24	0	**6·04**
48	12	1·96
48	0	**4·32**
72	12	1·76
72	0	**3·00**
96	12	0·76
96	0	**1·88**
120	12	0·16
120	0	**1·00**

Table 2.1 shows the results of one experiment. There is a steady improvement with age (heavy type), but at any age 12 hours of practice greatly improves accuracy (light type). Much of this improvement must result from maturation although we must bear in mind that mechanisms other than those specifically controlling pecking will influence accuracy. The chicks' legs grow stronger and perhaps their stability improves which would help their aim.

Hormones and the behavioural development of mammals

It is appropriate to discuss this rather specialized aspect of development here because it seems certain to involve specific maturational changes in the nervous system. We must briefly summarize a fairly complex story, Ward[483] provides a clear review with more detail. The genetic sex of an animal is fixed at fertilization and cannot later be changed: the sex chromosome complement determines whether the gonads develop into an ovary or a testis. However, the development of external genitalia and those neural mechanisms responsible for initiating sexual behaviour in mammals, about which most is known, are not determined until much later in development and in some rodents behaviour is not fixed until after birth. Both males and females inherit a brain which can mediate both male and female behaviour; which pattern becomes dominant depends on the secretion of hormone at a critical period in development. This comes from the testis of male embryos or newborns which shows a brief period of secretory activity and produces minute quantities of the hormone *testosterone*. Testosterone is picked up by particular regions of the *hypothalamus*, a part of the brain much involved in regulatory activities of all kinds and which controls the functioning of the pituitary gland lying immediately below it. (The hypothalamus is described in more detail in Chapter 4, p. 142).

If the hypothalamus picks up testosterone its development is switched along masculine lines, if not then it becomes feminine. The ovary of embryo females does not secrete, but if females are given small quantities of hormone at the right stage their behaviour is almost completely masculinized. When adult they will mount and thrust on oestrus females just like males and will even show the complete pattern of ejaculation. Often they are also de-feminized and cease to behave or function as females and are usually sterile. It is interesting to note that either the male or female sex hormones, testosterone and oestrogen respectively, will serve to masculinize male or female embryos—they are both steroid hormones and closely related in chemical terms. Denying male embryos their normal pulse of hormone by removing the testes before they secrete, or injecting a chemical which inhibits the action of testosterone, results in de-masculinization. We may note that this is a clear case of a critical period in development (see p. 40) for hormone treatment earlier or later has no effect. The die is cast once the period is over and, for example, a male de-masculinized by early castration will not have his masculinity restored by large doses of testosterone given later in life.

The timing and extent of the critical period varies greatly between different mammals. In sheep it extends for some weeks during mid-pregnancy and there are different critical periods for masculinization of the genitalia and for sexual behaviour itself. In rats and mice the critical period is a day or two just after birth. (See Austin and Short's[18,19] admirable short books for more comparative information.) The masculinization of females, which results from early hormone treatment, does not just lead to their having changed sexual behaviour. Masculinized female rats show increased aggressiveness, another male characteristic.[71] Masculinized female monkeys indulge in more than the normal amount of 'rough and tumble' play which is typical of male infants. In humans too there are clearly detectable effects when female embryos are exposed to male hormones during pregnancy. This happens if the foetus has a rare genetic abnormality which leads to the so-called 'adreno-genital syndrome' (AGS). The adrenal glands of the foetus function abnormally and release a masculinizing hormone into the bloodstream. When AGS girls are born their genitalia are often masculinized but this can be corrected with surgery and with the right treatment they can grow up to be fully functioning females. However it is fascinating that the effects of masculinization can be detected in their behaviour as children. Ehrhardt and Baker[141] have shown that AGS girls are more 'tomboyish'. For example, they indulge in more rough play, prefer boys to girls as playmates and tend to reject dolls as playthings. Of course we know that such preferences can be greatly influenced by parents and the particular culture in which children grow up (see Money and Ehrhardt[352]). However there can be little question in this case that the AGS girls were treated as girls by their parents and that the changes to their behaviour were spontaneous.

Fixed action patterns

The changes produced by hormones in behavioural development are very striking and may appear to resemble a switch mechanism, directing the animal's development into one of two distinct channels, masculine or feminine. In fact, as far as behaviour is concerned it is probably better to regard the changes as a shift in bias. Both sexes inherit the potential to perform the motor patterns of male and female sexual behaviour and probably retain this whatever the course of development. For example normal female mice particularly when in oestrus, quite commonly mount other females but androgenized females do so much more readily. Almost certainly one aspect of the change relates to the increased responsiveness of androgenized females to stimuli from other females.

This is a very common feature of developmental changes, it proves easier to modify the stimuli to which animals will respond than to alter the form of the responses themselves. Cockerels raised under hens or in incubators, isolated in battery cages or kept in flocks may differ markedly in the range of objects they will court and the frequency with which they do so, but they all show the same stereotyped movements immediately recog-

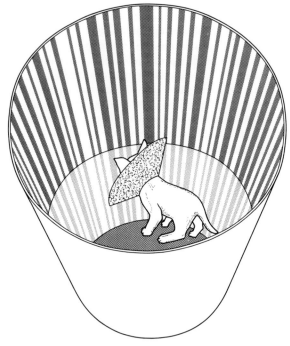

Fig. 2.7 Apparatus used by Blakemore and Cooper to restrict the visual environment of kittens. The kittens wore a black collar so that they couldn't see their
bodies and stood on a glass plate inside a high cylinder whose walls carried vertical
or horizontal stripes of varying thickness. (After Blakemore and Cooper,[54] 1970,
Nature, Lond., **228**, 477–478.)

nizable as 'waltzing', 'tidbitting' and 'mounting' by those who study the
behaviour of the fowl.[512]

In the next chapter we shall discuss evidence that animals may inherit
the tendency to respond to particular types of stimuli. Even where there is
such an inherent bias it is often easy to alter it by learning—gull chicks
without experience prefer to peck at red and avoid blue, but if responses to
blue are rewarded and those to red are not, they soon shift their preference.
We now know some cases where it is also possible to shift the original bias
by altering early experience even if learning is not involved.

Blakemore and Cooper[54] reared a number of kittens in complete darkness from birth to 5 months of age. From 2 weeks when their eyes opened
they spent an average of 5 hours each day in a very special environment
(see Fig. 2.7) so that almost all they see is a pattern of horizontal or vertical
stripes. When first brought out into the normal world at 5 months the
kittens were, understandably enough, disorientated. However they quite

quickly managed to cope but their vision remained abnormal. This was revealed dramatically when a 'horizontal' and a 'vertical' kitten were put together. If the experimenters presented them with a rod held vertically and shaken the 'vertical' kitten moved forward to play with it, but the 'horizontal' kitten ignored it. When the rod was turned to the horizontal, the 'horizontal' kitten immediately took interest and approached whilst the 'vertical' kitten no longer responded—it behaved as if the rod had suddenly disappeared.

Subsequently Blakemore and Cooper used neurophysiological techniques to look at the functioning of neurons in the kitten's visual cortex. (This is the part of the brain which receives, in coded form, the visual information from the cat's retina.) Each of these neurons has a 'preferred orientation' which means that they fire when lines or edges at that orientation are presented to the eye. In a normal cat the orientations are distributed all around the clock, but vertically or horizontally reared kittens showed very strong bias towards the corresponding orientation. Thus horizontal kittens had *no* units responding to vertical or to lines or edges within 20° either side of vertical, they literally cannot see a rod when it is held vertically.

These experiments show clearly that not only learning but also the early visual environment acts to shape the development of some aspects of the cat's visual system. The same is true of other sensory systems.

As we mentioned above, this lability does not appear to be matched on the motor side. Many ethologists, particularly those working with insects, fish or birds are familiar with the extreme constancy of movement and postures used in feeding, fighting or courtship displays. The displays of the cockerel referred to above are good examples, so are the courtship displays of ducks, see Fig. 6.4 p. 207. Such movements are characteristic of each species, indeed so much so that they can often be used as taxonomic characters in the way morphological structures can. Following Lorenz these units of behaviour have often been called 'fixed action patterns'.

We know little of how such patterns develop and it is commonly supposed that they appear in their typical form at the very first performance, perhaps not until an animal is fully mature and reproducing. However it is not nearly so easy to study the development of motor patterns as it is to modify early sensory experience, and accordingly we may have an over simplified view. There are very few detailed studies which either compare the exact form of a fixed action pattern in different members of the same species, or examine the way it is performed at different stages through the life history of an individual. This is a laborious task involving the analysis of slow motion film or videotape. Where we have such records for gulls and ducks and *Drosophila* they do reveal some variation between individuals and some changes in the exact form of a movement as an animal grows up— often one supposes because its muscles and skeleton are growing too.

For this, and other reasons, Barlow[25] in a most interesting review of all aspects of such behaviour suggests replacing the adjective 'fixed' by 'modal'. This would reflect more accurately that 'modal action patterns' probably show a range of variability similar to most other behavioural characters. Barlow's point is sound, though he recognizes that the role the patterns play in the life of animals often dictates that they keep variation within strict limits. For example they are often involved in intra-specific signalling as in courtship displays or alarm calls. We shall discuss communication in the next chapter, but the need for reasonable constancy is obvious if signals are to be clearly understood, a point well argued by Morris[356] and Bekoff[45]. Again some patterns have a very precise function which must be met at the first performance. Provine[385] has studied the patterns cockroach (*Periplaneta*) larvae perform when hatching from the ootheca, the communal egg case. Peristaltic movements run from the tail to the head in order to propel the larva out of the ootheca and its earliest cuticle. This pattern is repeated rhythmically until the larva is clear; it is not performed at any other occasion during the cockroach's life. Provine could show that it appears at the correct time in development, whether the larvae are inside the ootheca or have already been removed and cultured artificially.

Such an example certainly suggests that the neural network necessary to control such movements matures spontaneously and occasionally we can observe young animals perform recognizable fragments of fixed action patterns long before they become integrated into the full adult pattern. Nestling cormorants, holding nothing in their beaks, make the head and neck movements which form part of the 'tremble-shove' pattern used by the adult during nest-building. They may also perform isolated fragments of preening movements, even before their eyes are open or they have any feathers.

Because of the spontaneity and lack of obvious practice in such cases, ethologists have tended to regard fixed action patterns as being the closest thing to pure inherited behaviour. Certainly there must be an important inherited component in the development of such patterns and we shall discuss some genetic evidence for this in Chapter 6. However it is important to remember that it is virtually impossible to say anything firm about the inheritance of behaviour which varies so little between individuals. Genetics depends on the analysis of how *differences* are inherited.

To some extent the capacity to perform fixed action patterns is determined along with the developing structure of the animal. The precise and very constant form of the courtship wing movements of *Drosophila* males is certainly determined in part by the structure of their thorax and its musculature.[151] But in addition to structural considerations, the developing nervous system must lay down circuits which predispose the animal to perform this or that particular sequence of muscle contractions that go to make up a fixed action pattern. We can see this most clearly in the nervous systems of some invertebrates which have relatively small numbers of cells. In favourable animals some of the neurons can be mapped together with their interconnections. Modern techniques allow a dye to be injected into a neuron through the micro-electrode which has just stimulated it or recorded its activity. Thus it is possible to explore the relationship between the structure of the nervous system and behaviour in a most exact way. Hoyle[235] provides an excellent account of this approach which is beginning to yield some exciting results.

It is now clear that there is great regularity in the central nervous structure of molluscs and insects. Reasonable numbers of homologous neurons can be mapped with absolute confidence from one individual of a species to the next, and this constancy of structure underlies a constancy of function. Stimulating homologous motor neurons, for example, always affects the same muscles. Willows[502] (see also Willows, Dorsett and Hoyle[503]) has been able to map many of the very large cell bodies in the brain of the sea

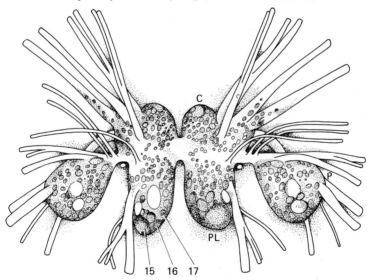

Fig. 2.8 Drawing of dorsal view of the 'brain' of the sea slug, *Tritonia*. It consists of three fused ganglia on each side, cerebral (C), pedal (P) and pleural (PL), from which nerves run to all parts of the slug's body. Large nerve cell bodies are clearly visible at the surface and those numbered are some of those which can be reliably identified in all specimens. Some of the cells are almost 500 μ (0·5 mm) across ; the whole complex is 6–7 mm across. Cells 15, 16 and 17 are in the area concerned with escape swimming. (Redrawn from Willows *et al.*,[503] 1973, *J. Neurobiol*, **4**, 207–237.)

slug *Tritonia* (see Fig. 2.8). Stimulating cells in the region of the group 15–17 leads to the production of the characteristic pattern of escape swimming which *Tritonia* normally performs when it detects a starfish— its natural predator. Such a pattern is quite complex and increasing evidence of this type[239] shows that the potential to perform fixed action patterns must develop along with the nervous system. We may recall the work of Bentley and Hoy[49] on the development of flight in crickets (see p. 43) where they produced evidence of just this type. We discuss the evolution of such patterns in relation to their underlying neural mechanisms in Chapter 6, p. 217.

Although we know little about the factors that are involved in the development of fixed action patterns it is clear that conventional learning or practice are unlikely to be among them. Lorenz[308] considered that learning can affect development only in a limited way and that in particular it cannot modify the fixed action patterns themselves. Certainly an animal's ability to learn novel patterns of movement—new motor skills as we might call them—is circumscribed by its inherited behaviour. Many ground-nesting birds retrieve an egg which has slipped out of the nest in the same way. They extend the neck and hook the lower mandible over the egg which is then rolled back by drawing in the head (see Fig. 2.9). This fixed action pattern can be elicited on the first occasion that an incubating bird is presented

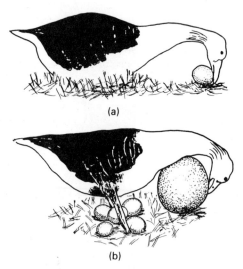

(a)

(b)

Fig. 2.9 Greylag goose retrieving an egg which is outside the nest. This movement is very stereotyped in form and used by many ground-nesting birds. The goose attempts to retrieve a giant egg in precisely the same fashion. (From Lorenz and Tinbergen,[310] 1938, *Z. Tierpsychol.*, **2**, 1)

with an egg outside the rim of the nest. For a goose to use the bill in this way is reasonably efficient because it has a broad one, but birds with narrower bills such as gulls and waders often have considerable trouble because the egg tends to slip out to one side. It would seem much easier for these birds to use their wing or foot as a scoop, but they show no variation from the fixed pattern.

Perhaps the most vivid example of how resistant inherited fixed action patterns are to change, comes from the work of Dilger[136] with love-birds. These are members of the parrot family which breed readily in captivity. Within the genus *Agapornis* two types of nest-building behaviour are represented. All species tear strips of material from leaves to build with (in the laboratory newspaper forms an excellent substitute) but whilst some tuck the strips into their rump feathers and fly back to the nest carrying several pieces at once, other species carry the strips singly in their bill. Dilger crossed two species which differed in this respect and watched the nest-building behaviour of the hybrids. For some time such birds were incapable of building a nest at all because they attempted to perform some kind of compromise between the two collecting methods. They might start to tuck a strip between the rump feathers, but either failed to let go of it or failed to tuck it properly. The end result was usually that the strip fell to the ground and the whole process began again. The only success the hybrids had was when they managed to retain a strip in their bill after the attempted tucking procedure. Dilger found that even after months of practice the birds managed successful carrying in only 41% of their trials. Two years later they were successful in nearly all trials, but before carrying a strip in their bill they still made the turning movement of the head which is the preliminary to tucking. Parrots are intelligent birds which are known to learn quickly in other situations, but here the inherited predisposition to perform the tucking sequence outweighed learning for a long time. More recently Buckley[76] has repeated and extended Dilger's observations, and he describes a number of other ways in which the behaviour of the hybrid birds is disrupted.

Such examples of the relative impotence of learning are very convincing, but we must remember that stereotyped patterns of behaviour certainly can be acquired and modified by learning. Mammals and birds can learn new motor skills which persist—circus animals offer many examples— although it is usually easier to train them to perform patterns which correspond in some way with their natural behavioural repertoire. When they are confined in cramped quarters animals often develop stereotyped patterns of movement, head-swaying, rearing etc. which they repeat endlessly and with great constancy of form. These stereotypes must involve motor control very similar to that which controls the naturally evolved patterns.

Fixed action patterns play an important role in the lives of many animal species, providing them with adaptive pre-set responses to the normal environment which they encounter. Because they have been shaped by natural selection they form convenient 'phylogenetic units' for various types of behavioural study and we shall refer to them again at other places in this book. We can now turn to consider some examples of behavioural development in which learning and inherited tendencies can be shown to interact.

Development involving the interaction of inherited tendencies with learning

We have previously mentioned the manner in which the basic pattern of bird flight matures in the absence of any practice whereas the finer skills of flying develop later. By close observation of young animals and controlling the time at which they are first exposed to otherwise natural situations, it is sometimes possible to identify analogous stages in their behavioural development.

Eibl-Eibesfeldt[143] describes a number of examples from his studies of young mammals. He reared young polecats in isolation and never gave them any opportunity to catch live prey. Such animals showed varying degrees of interest when first presented with a live rat, but no hint of attack unless the rat ran away from them. If it did so, they instantly pursued and seized it in their mouth, usually shaking it rapidly in a characteristic way. Their bites were at first badly orientated, but after a few trials they were seizing prey by the nape of the neck and killing with a single bite. Clearly there are inherited components to the killing pattern which are completed by learning. Eibl-Eibesfeldt found that normally polecats picked up the necessary practice during play sessions with their litter mates. In a similar series of observations with hand-reared squirrels, he has shown that although they respond to nuts and try to open them on the first exposure, their efforts are uncoordinated. They have to learn to direct their gnawing to the thinnest portion of the shell and to confine it to this one area.

One of the most beautiful examples of the dovetailing of inherited and learnt components during development is that of bird song. Until recently song was difficult to study because there was no way of expressing it graphically. The development of the sound spectrograph has turned this situation inside out and made detailed analysis easier for song than for most behaviour. Bird song is just like any other behaviour pattern—a controlled sequence of muscle contractions which in this case we perceive as sound. It can be recorded on tape and played into the spectrograph which produces a chart showing how much energy was emitted at the various sound frequencies at any time. Some examples of sound spectrographs of bird song are shown in Fig. 2.10.

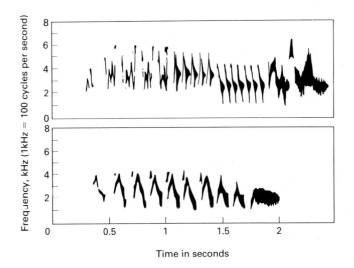

Time in seconds

Fig. 2.10 Sound spectrographs of chaffinch song. The upper diagram shows the song of a normal male, the lower is the song of a male reared in isolation from the nest. The isolate's song is at the right pitch and fairly normal in length, but it is far simpler in form and in particular lacks the chaffinch's characteristic 'terminal flourish'. Isolated white-crowned sparrows show very similar types of song defect (see text and Fig. 2.11 for further details). (From Thorpe, W. H.,[456] 1961, *Bird Song*. Cambridge.)

Marler and Tamura[335] have worked on the song development of the American white-crowned sparrow. This small finch has a wide range on the Pacific coast, and birds from different regions have recognizably different song 'dialects'. If young males are taken immediately after hatching and reared alone in sound-proof chambers then, no matter which region they come from, they all eventually sing very similar and simplified versions of the normal song. Obviously they must pick up the local dialect by listening to adult birds and modifying their own simple song pattern accordingly. Marler and Tamura found that this learning process usually takes place during the first 3 months of life and thus before the bird has ever sung itself. Males captured in their first autumn and reared alone begin to sing for the first time with a recognizable version of the local dialect. Up to 3 months of age, isolated males can be 'trained' to sing their own or other dialects by playing to them tape-recorded songs, though the results of such training do not show until the birds begin to sing themselves some months later. Beyond 4 months of age the birds are unreceptive to any further training and their songs when they begin singing are not affected. Here then we have a simple, inherited song pattern which is sensitive to modification by learning but only during early life. The young birds 'carry' the memory of the songs they hear and reproduce them when they first sing.

The analysis has been taken further by the experiments of Konishi.[274] He completely deafened birds at various ages by removing the cochlea of the inner ear. If this is done to a young fledgling just out of the nest, it will subsequently sing, but it produces only a series of disconnected notes. These are quite unlike the song of isolated birds which, though simple in form, would still be recognizable as 'white-crowned sparrow' to an ornithologist. The bird has to be able to hear itself in order to produce this inherited song pattern. In other words, it would be more accurate to say that it is not the capacity to produce the simple song which is inherited, but rather some kind of neural template representing this song, against which the bird matches the notes which it produces and adjusts them to fit. It requires auditory feed-back if it is to realize this inherited potential.

In some birds, such as the chaffinch, the way the song develops perhaps reveals this feed-back control in actual operation. They begin the breeding season singing a 'sub-song', a soft rambling pattern of notes varying in pitch and length. As this is repeated the notes become louder and less variable, always approaching closer to the final pattern which presumably corresponds to that held in the template.

With the white-crowned sparrow Konishi found that if he deafened young birds *after* they had been 'trained' with normal song but *before* they had themselves sung, their subsequent song resembled that of birds deafened as fledglings. They need to hear themselves in order to match up the song they produce with that which they have stored in their memories. Presumably the songs they have heard as fledglings have modified the inherited template so that it conforms to the more complex characteristics of normal adult song. Once the birds have matched their own output with this and sung the adult song, they can go on singing normally even when deafened. At this stage song development comes to an end in the white-crowned sparrow; after its first spring the bird is no longer susceptible to further experience and keeps much the same song pattern for the rest of its life.

It is not possible to generalize too much from this example, for one striking feature of bird song development is how much it varies between species. The song of chaffinches (see Fig. 2.10) develops in much the same way as the white-crowned sparrow, but Oregon juncos[334] and indigo buntings[394] continue to modify their songs according to experience for at least another year, and blackbirds may change the details of their songs throughout their life[196] (see also reviews in Hinde[216]).

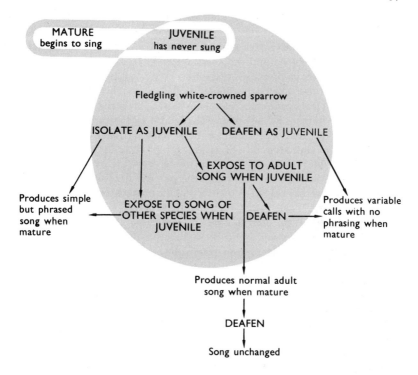

Fig. 2.11 A summary of the results obtained by Marler, Tamura and Konishi on song development in the white-crowned sparrow.

The results of the experiments with the white-crowned sparrow are summarized in Fig. 2.11. We have gone over them in some detail because they are an object lesson in the study of behavioural development. They show how subtle and complex are the interactions between inherited tendencies and the environment which finally produce behaviour as we observe it.

Fig. 2.11 also indicates the result of one more experiment with the sparrows which introduces the last aspect of behavioural development to be considered.

The inheritance of a predisposition to learn particular things

White-crowned sparrows only learn to sing the white-crowned sparrow song. If the songs of other species are played to males during their 'sensitive period' for learning they have no effect and the birds' subsequent songs sound like those of isolated males. The chaffinch behaves very similarly, and Thorpe[456] points out that this can scarcely be a result of limitations in its sound-producing organ or syrinx. The related bullfinch and greenfinch have syrinxes of almost identical structure. They have only a poorly developed natural song but are good mimics and will learn to reproduce the songs of many other birds.

Some clue to the chaffinch's limitation is provided by the song of the one species it will learn to imitate reasonably well. This is the tree-pipit, whose song to the human ear also has a chaffinch-like tone, though it is very different in pattern. The chaffinch must have an inherited tendency to single out this tone from all others and to reproduce it.

A remarkable variation on this theme is provided by the zebra finch where Immelmann[251] has shown that the young males learn the song of the male bird which has helped to rear them. One expresses it this way because although normally this means they acquire zebra finch song, if a clutch of zebra finch eggs is fostered to a pair of Bengalese finches the males copy Bengalese song. They will do this even if there is plenty of zebra finch song to be heard in the aviary room where they are being reared. Thus the young birds do not show the inherited selectiveness towards particular types of song model as do chaffinches or white-crowned sparrows. Their selectiveness relates to the bond which the developing bird has with one adult amongst many potential models.

There are a number of other cases on record where the singular facility which animals show for particular types of learning suggests an inherited predisposition. Of course without the possibility of proper genetical analysis it can remain only a suggestion, but it is none the less a powerful one. Tinbergen's[462] work with the herring-gull has shown that whereas these birds do not learn to recognize their own eggs, after a few days they do recognize their own chicks and react aggressively towards strange ones. From an adaptive viewpoint this makes sense; eggs cannot wander from the nest-site, chicks can. Nevertheless it can rarely be possible for a pair of herring-gulls personally to test the wisdom of this, they probably have an inherited tendency to learn details of the plumage of their own chicks. It is certainly not adequate to suggest that learning the chicks' plumage is an inevitable outcome of rearing them. In kittiwakes, cliff nesting gulls whose young are confined to tiny ledges, transfer of chicks is possible up until the time of fledging. The parents do not distinguish between their own and foreign chicks and it can hardly be a coincidence that here is one species where neither eggs nor chicks can wander from the nest. But egg recognition can sometimes be vital and Tschanz[471] has shown that guillemots, which build no nest but lay their single egg on open cliff ledges, do learn their own egg. The eggs can roll around and it is clearly adaptive to know which one to retrieve. This may be the reason why guillemot eggs are so variable in ground colour and pattern. The royal terns studied by the Buckleys[77] also have extremely variable egg patterns and also learn their individual egg. They nest in tightly packed colonies—more than 7 nests in each square metre—on featureless terrain and attachment to the nest site alone is clearly not enough to ensure that the parents incubate the right egg.

Turning to other types of situation which reveal a predisposition to learn, Hinde and Tinbergen[225] describe how young titmice learn to use their feet to hold down large pieces of food in order to break off pieces with their bills. Young chaffinches do not learn to do this even when they are reared by titmice foster-parents and this difference is probably inherited. The facility with which polecats learn to seize their prey by the back of the neck is a similar example. So is the learning capacity of bees, for they learn the features of their nest and food sources (together with landmarks around them) with extraordinary rapidity and once learnt they do not forget. Lindauer[298] describes differences between the way two geographical races of honey bee learn the details of a food source—one concentrates on detailed close-range features, the other on more distant land-marks. These differences seem certain to be genetic, representing particular specializations of inherited learning ability. As we discussed earlier (p. 23) the whole remarkable learning ability of the Hymenoptera has certainly evolved in response to their way of life.

This is really the important conclusion to draw from these diverse examples of inherited predispositions to learn. Learning capacities are one further way in which animals are equipped to cope with their environment. As we shall discuss more fully in Chapter 7, natural selection has led to specializations and it is no longer profitable to consider learning in isolation from the natural situations which face the species we are studying. Nor can we isolate learning from the animal's instinctive behaviour. For the most part the examples of behavioural development which we have been considering in this chapter give little encouragement for attempts to classify behaviour into that which is inherited and that which is learnt. It may be possible to do this in some cases but it does not seem a very helpful exercise unless it yields information on how the behaviour develops. In the constant struggle to be well adapted natural selection usually operates on the end result.

Sometimes it matters little how this end is achieved but sometimes the path of development itself may be the subject of selection. We have already considered some of the factors that may predispose certain groups to rely mainly on inherited behaviour, others on learning. However, even among close relatives there may be wide variation; as mentioned above, the songs of different birds vary widely in their mode of development and this is true even of quite close relatives. Nottebohm[372] and Marler[332] both discuss possible reasons for this diversity. Some quite elaborate songs, such as those of doves and pigeons, develop without any auditory feed-back—

deafened isolates sing identically with normal birds. Such a pattern of development yields a clear and unambiguous signal. On the other hand, the variation which results from the copying of adults allows local 'dialects' to develop and be perpetuated. This may allow a local population to interbreed and discriminate against outsiders, which may sometimes be advantageous. Since the copying process is never perfect, this type of song development also allows individual males to be recognized. This may be valuable for females and for other males who live in adjacent areas. If everybody's song is slightly different, males can learn the songs of neighbours with whom territorial boundaries are now settled and it is easier to detect new males who may be trying to invade the settled area. Which type of development occurs in a species may depend on other aspects of its life history, which will determine whether song stereotypy or song variability gives the best advantage.

Similarly as already discussed on p. 39 European cuckoos cannot develop their sexual preferences by imprinting, yet pigeons, ducks and game birds do so. It may be, as argued on p. 38, that inherited responsiveness to the subtly cryptic patterns of female ducks or game birds is difficult to evolve. Equally it is possible that development by sexual imprinting allows males to adjust immediately to any changes which may evolve in the plumage of females. Again, life history determines which course of development is to be followed (see Immelmann[252] for a further discussion of the evolutionary advantages of imprinting).

The study of behavioural development is one of fundamental value. The most important property of the nervous system is its ability to store information. Some of this is fed in by genes during the development of the nervous system, other information is added later from the environment and by learning. There seems no good reason to regard these two processes as distinct; they may have almost everything in common. Galambos[158] concludes a paper on mechanisms of learning: 'It could be argued, in brief, that no important gap separates the explanations for how the nervous system comes to be organized during embryological development in the first place; for how it operates to produce the innate responses characteristic of each species in the second place; and for how it becomes reorganized, finally as a result of experiences during life. If this idea should be correct, the solution of any one of these problems would mean that the answer for the others would drop like a ripe plum, so to speak, into our outstretched hands.'

3

Stimuli and Communication

Stimuli can affect behaviour in a number of ways. It is possible to recognize three categories of effect, although they are not sharply separated. Stimuli may *arouse*, they may *elicit* a response and they may serve to *orientate* the animal as it responds. We may briefly examine the orientating role of stimuli before concentrating on the others.

Orientation is, of course, a constant background feature of any animal's life. The term can refer—as in Fraenkel and Gunn's[152] famous book *The Orientation of Animals*—to the mechanisms whereby animals respond to the basic physical qualities of their environment. Light, gravity, air and water currents etc. are perceived and the animal positions itself so as to be in a correct relationship to them. Blowfly maggots crawl directly away from light when they move out of their food source to pupate. Fish rest with their head facing into the water current. From an early age young rats, even in total darkness, show a 'righting reflex' which keeps them upright with respect to gravity, and so on.

Orientation of this type is certainly not just a simple reflex process for it often includes quite complex mechanisms of perception and control. Consider the way honey-bees guide themselves using the sun as a compass or the remarkable abilities of some migrating birds to orientate by the pattern of stars in the night sky.

Orientation can also refer to the way in which stimuli control responses to more transitory features of the environment. Drakes direct their courtship displays according to the position of the ducks on the water. Some displays are given when lateral to the female, others when the male is directly in front of her. The experiment shown in Fig. 3.1 illustrates the use of landmarks in the orientation of the digger wasp, *Philanthus*, to its nest burrow. The wasp learns some characteristics of the ring of pine cones when she leaves the nest and, after they have been moved, the cones orientate—or rather misorientate—her response on a return trip.

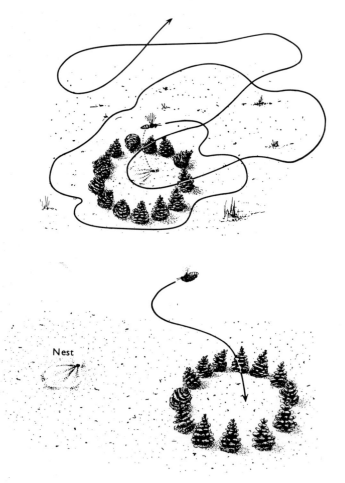

Fig. 3.1 The digger wasp, *Philanthus triangulum*, builds a nest burrow in sand. Whilst she was in the burrow Tinbergen placed a circle of pine cones around the entrance. When she emerged, the wasp reacted to the new situation by a wavering orientation flight (upper picture) before flying off. Returning with prey (lower picture), she orientated to the circle although it had been moved during her absence. (From Tinbergen,[460] 1951. *The Study of Instinct.* Oxford University Press, London.)

Recent physiological work with mammals reveals something of the interaction between the other effects of stimuli—the elicitation of responses and arousal. Every stimulus evokes two types of response within the brain. The first via what may be called 'specific sensory pathways', is one directly related to the stimulus. Visual stimuli evoke activity in the visual centres of the brain, sounds in the auditory centres, and so on. The second type of response is less specific because each incoming sensory pathway also gives off side branches or collaterals which go to a diffuse series of fibre tracts called the 'reticular formation'. This connects via so-called 'non-specific pathways' to all the brain's higher centres and 'arouses' them into action. This means that any stimulus may not only evoke a response pertaining to itself, but also change the animal's state of arousal and responsiveness to other stimuli, both related and unrelated to the first. Parts analogous to the reticular formation of mammals are now being revealed in the brains of lower vertebrates and also in insects.

At the behavioural level we can observe the arousing and motivating effects of stimuli in the phenomenon of 'warm-up' discussed on p. 8, where a response becomes stronger as it progresses. Of course, it may not be just the external stimulus which is arousing the animal, but also the actual performance of the response itself. However, Noirot[368] has shown that the maternal response of mice to a weak stimulus (a dead 1-day-old baby mouse) is stronger if they have previously been given a powerful stimulus (a live baby). She found in some later experiments[369] that this holds true even when the live baby is presented inside a perforated box so that, although the adult can perceive its presence, it cannot make any appropriate response. The arousal effects of a single 5-minute presentation of the live baby could be detected even when it was several days before the dead baby was presented. This shows that the stimulus alone is sufficient to arouse maternal behaviour and make the mouse persistently more responsive to subsequent stimuli.

Perhaps, following Hinde,[217] we should say that the maternal responses were 'primed' by the initial stimulus, and retain the term 'arousal' for effects which are more short-lived. Sometimes we know that long-term changes as the result of a stimulus are caused by the induction of hormone secretion. The sight of a male dove courting leads to hormonal changes in the female which make her more ready to take part in nest-building and here the stimulus is effective only if it is repeated many times. It requires a few hours of courtship usually spread out over several days to bring the female dove into reproductive condition. She is able to summate the stimuli over a long time (far beyond the second or two which apply in most cases of temporal summation which we discussed in Chapter 1, p. 6). Stimuli which act in this way are not uncommon in reproduction and Adler[1] describes a number of interesting examples. He refers to stimuli which act over a long period as 'pumps', and contrasts them with stimuli acting as 'triggers'. Pumps act cumulatively to accelerate (or less usually slow down) some relatively slow process, as in the growth of the female dove's reproductive system. In a similar way the smell of males accelerates the sexual development of female mice. Triggers act rapidly as their name suggests, and often in a once-for-all fashion. Female cats and rabbits stay in a sexually receptive state until they mate with a male. Stimuli from the act of copulation trigger off a pulse of luteinizing hormone from the pituitary gland which causes the female to ovulate. Similarly the stimulus of suckling triggers off the final let-down of milk from a female rat, although the prior growth of the mammary glands is accelerated by the cumulative pump action of the pregnant female grooming and licking her nipples. Both trigger and pump effects are also mediated by hormones; milk let-down involves the release of oxytocin from the posterior part of the pituitary, mammary gland growth involves prolactin secretion from the anterior pituitary.

Whichever way we choose to classify the modes of action of stimuli we must recognize that the context in which a stimulus occurs will affect an animal's response to it. A male bird shows heightened aggressiveness when at home on its territory. Here the male is receiving a variety of stimuli from its surroundings and these have been associated with victory over rivals during previous encounters. As a result the bird is much more responsive to aggressive stimuli than when it strays outside the boundaries of its territory. Here context affects motivation and, as we shall discuss in both this and the succeeding chapter, stimuli and motivation almost always interact to determine the nature and strength of an animal's response to changes in both its internal and external environment.

DIVERSE SENSORY CAPACITIES

Every animal inhabits a world of its own whose character is largely determined by the information it receives from its sense organs. As observers and interpreters of animal behaviour we would be greatly handicapped if we had to rely solely on the evidence of our own senses. Modern instruments enable us to explore the world of animals whose sensory capacities may extend beyond our own. It is here that the study of sensory physiology and of behaviour come to overlap.

For example, the main visual receptors of insects are their compound eyes, whose construction and properties are very different from those of vertebrates. They provide poor image formation, but are excellent for detecting movement and often have a very wide field of view. All insects so far tested prove to have colour vision, but their sensitivity is shifted towards the short-wave end of the spectrum compared with ours. Thus, with a few exceptions, insects cannot see red as a colour, they confuse it with black or dark grey, but on the other hand they can see into the ultra-violet.

We know that foraging bees are initially attracted by the colours of flowers which contrast with the background, but they are aware of contrasts which are invisible to us and vice versa. Bees fly to the flowers of white bryony (*Bryonia dioica*) because the petals reflect large amounts of ultra-violet, although they appear to us as pale green and provide little contrast to the leaves. Further, not all flowers which appear white to us do so to bees. One type of white reflects ultra-violet along with all the other wavelengths and the bees see this as some equivalent of 'white'. The other—equally white to us—reflects all wavelengths except ultra-violet and appears to bees as the colour complementary to ultra-violet which is blue-green. The nectar-guide patterns of flowers provide patterns of colour contrast which help bees to locate the flowers' nectaries. It used to be something of a puzzle why some flowers lacked nectar-guides. Once a system which can show up ultra-violet was used to look at such flowers the patterns, adapted to the eyes of bees not our own, were revealed (Fig. 3.2).

Another visual faculty outside our own which bees possess is their sensitivity to the plane of polarization of light. Light which reaches us from areas of blue sky is vibrating predominantly in one plane. The angle of this plane changes in a regular fashion with respect to the position of the sun.

Fig. 3.2 Flowers seen in two different lights. The left hand picture records what the human eye can see and the flower has no nectar guide pattern. On the right is the view which approximates to what a bee sees, photographed through an ultra-violet sensitive system, a striking colour pattern is revealed. (From a photograph by Thomas Eisner.)

Von Frisch[156] has shown how bees can use this to locate the sun's position even when it is obscured directly by clouds or a screen.

Again, the flicker-fusion frequency of the human eye is about 50 per second. This means that a standard filament lamp worked off the A.C. mains supply of 50 cycles per second appears as a steady light source to us. In fact it shows fluctuations of up to 5% of its intensity as the filament heats up and cools down 100 times per second. Insect eyes which may have a flicker-fusion frequency of up to 250 per second can follow these fluctuations, and recordings of nerve impulses coming in from their eyes show corresponding cyclical changes under A.C. lamps. We have no evidence that this fluctuating sensory input affects the behaviour of insects, but it must be borne in mind when performing behavioural experiments under artificial light.

It is not only insects which inhabit a sensory world different from our own; so do much closer relatives. Our world is dominated by sight, but a dog or a cat probably gets more useful information from its sense of smell. The experiments of Griffin[179] and others have shown how bats locate objects and hunt insects on the wing using an auditory echo detection system of extraordinary accuracy. The tropical fish investigated by Lissmann[300] live in water of extreme turbidity where vision is useless. They orientate by setting up an electric field around themselves, using specially modified muscle tissue to generate pulses of electrical energy, and measuring how this field is distorted by other objects in the water.

Information of the type provided by these examples is a necessary preliminary to any thorough behavioural study. However, even if we know an animal's sensory capacities in considerable detail, this alone does not necessarily tell us much about the stimuli to which it will react.

SIGN-STIMULI

In the last chapter we mentioned that it is common to find animals responding only to a part of the stimuli presented to them. They can be taught to discriminate one aspect of a complex situation in this way, but responsiveness to such sign-stimuli is a regular feature of instinctive behaviour in situations where learning can be ruled out. We considered the case of the stimuli which evoke aggressive responses from a male robin on its territory. Here the red breast feathers of the rival are a far stronger stimulus than all the rest of the bird. Pied fly-catcher males have a correspondingly high responsiveness to white feathers—the breast colour in this species. Then there is the classic work of Tinbergen[460] on the sign-stimuli to which a male stickleback responds during its reproductive cycle. Fig. 3.3 shows some of the models which were effective in evoking aggressive

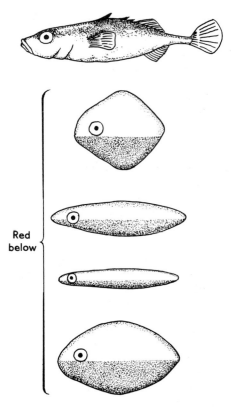

Red
below

Fig. 3.3 A series of models used in tests of aggression with male sticklebacks. The four crude imitations coloured red below released more attacks than the accurate model which lacked red. (From Tinbergen,[460] 1951. *The Study of Instinct*. Oxford University Press, London.)

responses. They are astonishingly crude. For aggression the red throat of the rival is most important; for courtship responses on the other hand the swollen belly of the female, ripe with eggs, is the most important stimulus.

There are many examples of auditory and chemical sign-stimuli too. Turkey hens which are breeding for the first time will accept as chicks any object which makes the typical cheeping call. On the other hand they ignore visual stimuli in this situation, and deaf turkey hens kill most of their chicks because they never receive the auditory sign-stimulus for parental behaviour.[422]

Minnows have an extraordinary sensitivity to chemicals from their own species. If a minnow is scratched or wounded in any way so that some of its blood gets into the water, other minnows show panic flight. They show far less fear when the blood of other types of fish is shed. A similar specificity for particular chemicals is shown by many species of moth, whose males are attracted to the females by their scent, which they can detect in minute concentrations.[423]

To label something a 'sign-stimulus' does not imply that a response will never occur in its absence. Some responses, such as that of the male moths, are extremely specific, but a really aggressive male stickleback will attack almost anything, including a ripe female. Even so, we consistently find that the more red there is on a model, the more it is attacked. We are quite justified in calling the red throat a sign-stimulus too.

Similarly, there is often more than one sign-stimulus which will evoke a response. In such cases the lack of one stimulus can be compensated for by an increase to another. Honey-bees guarding their hive entrance detect potential robber bees partly by colour and partly by their characteristic hovering flight. Little balls of wool evoke an attack and brown wool is better than white. However, a white wool lure which is moved in an imitation of a robber's flight works as well as a brown lure held still. Any stickleback model evokes more aggression from a male on his territory if it is held in the head-down threat posture. Here position and red colour can be shown to be additive in their effects. The term 'heterogeneous summation' is used to describe this 'adding up' of diverse stimuli. It may be considered as an extension of the stimulus summation already described for reflexes and complex behaviour in Chapter 1.

Selective responsiveness to sign-stimuli is clearly of great adaptive significance in the lives of many animals, particularly those which rely primarily upon inherited behaviour. Sign-stimuli will usually be involved where it is important never to miss making a response to the stimulus, but a few false responses do not matter much. Driving away rival males from its territory is so important to a male stickleback that it must show extreme responsiveness to red. Nearly all the red objects which it sees will be rival males, but occasionally red petals from flowers may fall into the water and cause a male to waste some time in futile aggression. Tinbergen describes how a male he was observing made a threatening display towards a red post-office van which drove past the windows of the aquarium. Such rare false positives do not count for much when set against the advantages of being consistently aggressive towards other males.

Again animals must never fail to respond to the sign-stimuli provided by a predator, or by the alarm calls of other individuals. It is advantageous to respond to the alarm calls of other species too, so that no matter which first perceives danger, all are warned. Marler[329] discusses the characteristics of an 'ideal' alarm call which must carry as far as possible whilst giving the least chance of a predator being able to locate the calling bird. Among other things, this means the call should be of constant pitch and have a gradual beginning and ending. The alarm calls of many finches, buntings and thrushes are remarkably similar and it seems probable that they have evolved towards each other so as to provide a common sign-stimulus to which all respond (see further discussion in Chapter 6, p. 228 and Fig. 6.11). The strong selective forces which determine the response to preda-

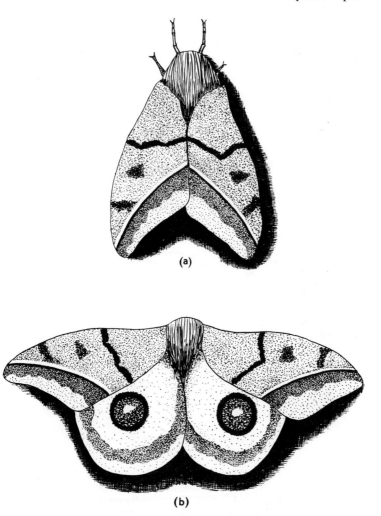

Fig. 3.4 The moth *Automeris coresus* (a) at rest, and (b) displaying the vivid eye-spots on its hind wings in response to a light touch. (Modified from Blest,[56] 1957, *Behaviour*, **11**, 257.)

tors have allowed some animals which are themselves the prey of small birds to gain some protection by mimicking the sign stimuli used by the latter. One way in which many small birds are protected against predators is by possessing a strong flight response from any animal having large staring eyes, such as are characteristic of most birds of prey and owls. Blest[55] has shown that small birds are far more frightened by a simple pattern resembling an eye which is flashed at them, than other patterns with equal amounts of contrast and outline. Many different types of butterfly and moth—potential prey—have developed eye patterns on their wings. Sometimes these are of amazing detail (see Fig. 3.4) and whilst ordinarily hidden they are suddenly flashed into view when the moth is touched. Small birds may be repelled and thereby miss a meal, but it is better to be frightened

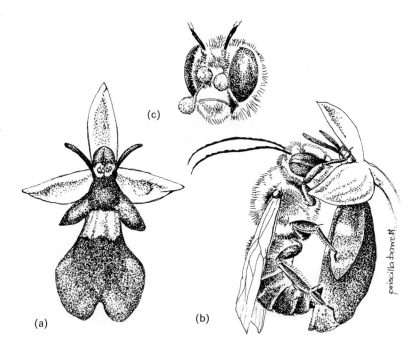

Fig. 3.5 The 'parasitic pollination' of an orchid. This is the fly orchis (*Ophrys insectifera*) whose flower (a) has an enlarged lower petal resembling a bee. In (b) a male long-horned bee attempts to copulate with the flower. In doing so his head comes into contact with the sticky pollen sacs which adhere to his head (c). They are now positioned so that they will make contact with the stigma of the next flower the male bee visits. (From Wickler, [497] 1968. *Mimicry in Plants and Animals*. World University Library. Weidenfeld and Nicholson, London.)

away by a harmless moth occasionally than ever to fail in the response to a predator. In an analogous way many species of orchids, whose flowers secrete no nectar and have no surplus pollen, nevertheless manage to attract visits from insects which thereby pollinate them. They are visited because the shape, colour and patterning of the petals mimic the bodies of the females of various types of Hymenoptera. It is the males of each corresponding species which visit the flowers and attempt to copulate with them, Wickler, [497] see Fig. 3.5.

In some contrast to dependence upon sign-stimuli, responses to other individuals or objects in the environment may be based on a complex of features. The features used by male and female ducks for recognition of a mate illustrate this contrast. The males of the various species have strikingly different and boldly patterned plumage. Conversely the females resemble one another much more closely because they are cryptically coloured in mottled shades of brown. Evidence is now accumulating that females have an inherited ability to recognize their own type of male, but males have to learn to recognize the subtle pattern of their females during the process of sexual imprinting. This is usually completed during a few weeks following hatching using their mother as a model (see p. 37 for more details of sexual imprinting). Seitz (see Baerends & Baerends-van Roon[22]) studied the tropical fish *Astatotilapia* which has brightly marked males and dull females. He found a situation somewhat analogous to that of the ducks. Females have an inherited response to sign-stimuli provided by the males, but males learn details of the female markings. This means that whilst females will readily respond to male models, only inexperienced males will court simple models of females. Once they have sexual experience they respond only to the whole complex of female markings.

Certainly recognition based upon complex features is common in higher mammals. For example, the members of a monkey group know one another individually, and a whole range of stimuli provided by—say—the dominant male will allow the others to recognize him. These will include not just size and any special markings he may show but posture, movement and smell, even the fact that he tends to sit in particular places.

We may also consider the complex range of features by which we recognize a particular human face. There are no sign-stimuli here, but almost every aspect combines to produce a unique pattern. It takes time to learn this and we are clearly influenced by our culture. We find it more difficult to distinguish Mongoloid faces than Caucasian ones and Mongoloid people apparently have a similar difficulty with Caucasian faces.

In our discussion of sign-stimuli we have constantly used examples of responsiveness to the particular sounds, scents or colours of another animal. It was Lorenz who first developed the idea that such characters are in fact specially evolved to evoke such responses. He called them **releasers**, and pointed out that the releaser and the response from the animal which 'receives' have become mutually adapted to each other. They form a signalling system—and 'social signal' is often used as an alternative term for releaser—which has often become elaborated into a form of language in which the effect of a releaser is enhanced by a display movement. Cockatoos erect their brightly coloured crest feathers whilst honey-bees open the Nasanoff gland on the abdomen in order that its scent may be dispersed. We shall refer to releasers in other parts of this book because their evolution is linked closely to the evolution of communication as in courtship and threat displays. Until their signal function was understood, many of the bright colours of such animals as birds and butterflies were hard to account for. Now it is often possible to deduce quite accurately the type of display a bird or a fish will use simply from seeing its colour pattern and shape. Fish with colour markings on the gill covers will tend to have elaborate frontal displays in which the covers are spread, and so on (see Fig. 5.3a). We shall discuss the role of releasers in communication more fully at the end of this chapter.

The recent work of ethologists has provided some beautiful analyses of the sign-stimuli which evoke particular responses. One of the most interesting findings is that different aspects of the same object may be effective according to the type of behaviour which is involved. For example the answer to the question, 'How does a gull recognize an egg?', depends on whether the bird in question is incubating eggs or feeding on them. Baerends & Kruijt[24] summarize an extensive series of their experiments which show that herring gulls respond most strongly to green eggs as objects to incubate, but this colour preference is dwarfed by their response to speckling. The more spots an egg has and the more they contrast with the background, the more strongly is the gull stimulated to roll it into the nest and incubate. Shape, on the other hand, is ignored; square or cone-shaped eggs are accepted perfectly well provided they are speckled. We

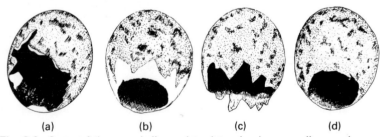

(a)   (b)   (c)   (d)

Fig. 3.6 Some of the egg-shells used to determine how a gull recognizes an empty shell. The real shell (a), 'painted rim' (b) and 'notched rim' (c) elicited more carrying than 'smooth rim' (d) which was rolled back into the nest. (Modified from Tinbergen et al.,[469] 1962, British Birds, **55**, 120.)

may contrast this with the responses of gulls which are robbing nests to eat the eggs. Now their preference is for blue or red background colour and green is least popular. It seems unlikely that this is a result of increased conspicuousness since the robbing gulls located green eggs just as easily. There appears to be a real switch of the preferred sign-stimulus. Supporting this conclusion are other tests which suggest that speckling is not nearly so attractive to feeding gulls as it was to those incubating eggs.

Tinbergen and his co-workers[469] have been able to analyse how black-headed gulls recognize egg-shells and distinguish them from eggs. They remove the egg-shells from the nest as soon as the chicks have hatched because empty shells attract predators. The gulls recognize a shell mainly because it shows a thin white edge which is serrated and 'hollowness' by itself is not effective. Fig. 3.6 illustrates some of the models used in their tests.

'Supernormal' stimuli

Ethologists usually investigate sign-stimulus situations by making models of the stimulus object and changing parts of it in rotation to see which are most important. Not uncommonly they have found that it is possible to produce a model which is 'supernormal', i.e. it evokes a stronger response than does the natural object. The most striking examples have again come from the incubation behaviour of birds. Tests have been made with the herring-gull, the grey-lag goose and the oyster-catcher and in all three the larger an egg is, within broad limits, the more it stimulates incubation. This results in the bizarre situation illustrated in Fig. 3.7 where an oyster-catcher

Fig. 3.7 An oyster-catcher attempting to brood a giant egg in preference to its own egg (foreground) or a herring-gull's egg (left foreground). The bird's original nest site was equidistant between the three 'test' eggs. See also Fig. 2.6 .(From Tinbergen,[460] 1951, *The Study of Instinct*. Oxford University Press, London.)

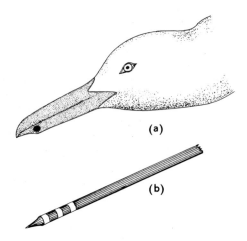

Fig. 3.8 (a) Accurate 3-dimensional model of a herring-gull's head, and (b) 'supernormal' bill. The latter received 26% more pecks from young chicks. (From Tinbergen and Perdeck,[470] 1950, *Behaviour*, **3**, 1.)

invariably chooses a giant egg in preference to its own as an object for brooding. The same species chooses an artificially large clutch of five or six eggs in preference to its own clutch of three.

Herring-gull chicks will peck at models of the parent's bill from which they are normally fed with regurgitated fish. The herring-gull's bill is clear yellow with a red circular patch on the lower mandible (see Fig. 3.8). The artificial bill, also shown in Fig. 3.8 is thinner than the real one, coloured red with three white bars at the tip; this attracts more pecks from naïve chicks than does a realistic copy of the natural bill and head.[470]

A further supernormal stimulus was revealed by the work of Magnus[320] with the silver-washed fritillary butterfly. Males are attracted to females by the flashing orange wing pattern as they fly by. Magnus could attract males to a revolving drum which 'flashed' a wing pattern at any required speed. The normal wing-beat frequency of a female fritillary is about 8 per second, but males show stronger responses the faster the wings of models are made to beat, up to as high as 75 per second.

In all these examples there is no difficulty in understanding why the natural stimulus has not evolved further towards the supernormal condition. Female fritillaries would attract more males if they beat their wings faster, but wings are not only for this purpose, they have to move insects through the air and there are severe mechanical limitations on their speed. Similarly a herring-gull's bill would probably be highly inefficient in all but attracting the pecks of its chicks, if it were as long and narrow as the supernormal model.

There remains the converse question of why the response of the receiver is not perfectly adjusted to the natural situation. We do not know the answer to this. Perhaps the mechanism which provides the selective responsiveness to a sign-stimulus has no upper limit. An increase in the 'quantity' of sign-stimulus inevitably results in an abnormally strong response. Natural selection can rarely have an opportunity to operate against this kind of system. On the other hand the important thing about supernormal stimuli may not be an increase to the 'sign' aspects but the fact that they are more conspicuous or arouse the animal more in a general way, so that its response is stronger. This seems likely to be the explanation for supernormal female models in the fritillary. Bees and other insects are known to approach flickering lights and within limits the faster the flicker, the more attractive. For the same reason bees prefer objects such as stars with broken outlines to circles or other plain shapes. As a bee flies over a broken shape the images of its margins will pass across the separate ommatidia of the bee's eye producing the effect of a flickering light source. The fast-moving flicker of the fritillary model will work in the same way.

STIMULUS FILTERING

In all our discussions of sign-stimuli it has been implicit that the animal selects one part of the environment to respond to and ignores the rest. This immediately raises the question of how far the ignored stimuli penetrate. Where is the 'filter' which separates them from the sign-stimuli and renders them ineffective?

Before we discuss this problem it is worth pointing out that no animal's brain can ever handle more than a fraction of the potential information that comes from its sense organs. Sense organs transform some physical feature of the environment—a colour, a sound, a smell—into a series of nerve impulses which travel along the sensory fibres up to the brain. Barlow[26] calculates that there are roughly 3 million such fibres entering the human brain. If each of them is regarded as a switch which can be either 'off' or 'on' then $2^{3 \text{ million}}$ different combinations of input are possible. To deal with them the brain has 10^{10} neurons and, vast as this latter number is, it is paltry in relation to the former which is far larger than Eddington's figure (10^{79}) for the number of particles in the entire universe!

Obviously not all of the $2^{3 \text{ million}}$ combinations of sensory fibre activity can occur, and anyway a large number of the combinations may be meaningless as far as the brain is concerned and can be ignored. Barlow suggests ways in which the brain can further economize on handling sensory information: some form of filtration is universal.

As applied to sign-stimulus situations filters can operate predominantly at the periphery or more centrally, and we can best explain these by examples.

Peripheral filtering

Quite obviously what an animal responds to is limited in the first place by its sense organs. We have already discussed the capacities of sense organs and often they act as filters. Our own eyes filter off all ultra-violet light and our ears transmit no sound with a frequency higher than 20 kilohertz. Sometimes selective transmission by sense organs produces the effect of a sign-stimulus. Narins and Capranica[363] have discovered a remarkable sex dependent filter mechanism in the ears of the tree frog, *Eleutherodactylus coqui*. The specific name of this frog refers to the call of the males—a double note 'co-qui'. By playing back tape recordings to the frogs, Narins and Capranica could show that males (who attack other males which approach them) respond only to the 'co' note, and females only to the 'qui'. Neurophysiological recordings made it clear that each sex heard only the note of relevance to itself because the neurons of the inner ear were tuned differently in males and females.

Frogs also provide a good example of a visual filter at the periphery. When feeding, they respond much more strongly to small, dark, circular objects moving close to them than to movements of large objects or the whole background. This makes sense in an animal whose chief diet is flies. Lettvin *et al.*[291] have shown that the light receptors of the frog's retina are connected together so as to form 'receptor fields'. Some of these fields are specialized to respond preferentially to the intermittent movement of small, dark, convex objects and Lettvin and his co-workers appropriately call these receptor fields 'bug detectors'. The responsiveness of the frog to such stimuli is produced by peripheral rather than by central filtering.

It is interesting to compare this example with that of the kittens studied by Blakemore and Cooper,[54] to which we referred in the previous chapter, p. 48. The cat's retina does not play such an important role in processing visual stimuli, it is the visual area of the cerebral cortex which is organized into receptor fields. As we discussed earlier the forms to which these receptor fields will respond are greatly affected by the early visual environment. Normally kittens will grow up capable of perceiving a complete range of visual features: there is probably much less filtering in mammals than in the frog.

On the other hand, there are numerous clear examples of peripheral filtering from the insects whose sense organs are often very highly specialized to respond to a particular range of stimuli and to this range only,

Roth[399] describes how male mosquitos respond with high selectivity to the sound of their females' wings which beat at a characteristic frequency, different from their own. It is the delicate hairs on the male's antennae which respond to the sound. Roth suggests that their physical structure is such that they vibrate most readily at the female's wing-beat frequency, just as a tuning fork will vibrate if its own frequency is sounded near by.

The sense organs also determine the amazing selective responsiveness which some male moths have for the scent produced by females. Males of the silk moth will gather from considerable distances when a virgin female is put out in a cage; once she has mated she ceases to produce the scent. It has been possible to isolate the female scent substance and define it chemically. A few synthetic substances which are closely related to female substance will attract males, but no others. Subsequently Schneider[423] has found that the scent receptors on the male's antenna are extremely sensitive to female substance, but show little or no response to any other chemicals.

Central filtering—the innate releasing mechanism

In contrast to these examples there are many where an animal's responsiveness to a sign-stimulus cannot easily be ascribed to its sense organs. For a good example we may look at the work of Burghardt[80,81,174] and his collaborators on the feeding responses of garter snakes and water snakes. These are aquatic animals living in rivers and ponds of the U.S.A. which feed largely on small fish and worms. These snakes, like many of their relatives, detect their prey by smell and taste and in particular use chemoreceptors located in pits on the roof of the mouth—Jacobson's organ. The snake's tongue is flicked out, picks up chemicals from the air or contact with the prey and carries them to Jacobson's organ where they are 'tasted'.[81]

Burghardt found that different species of garter snakes showed different levels of attack to cotton swabs dipped in solutions extracted from various potential prey—fish, frogs, salamanders, worms etc.—and that their preferences always related to the prey type which they most commonly took in the wild. He found that such preferences were present in newborn snakes which had never fed, and that they persisted no matter what diet had been fed to the mother snake during her 'pregnancy' (garter snakes develop inside the mother, not from eggs laid externally). The case for an inborn, genetically coded preference for certain chemicals seems very strong.

This could be another example of peripheral filtering—Jacobson's organ might be specialized like a male moth's antenna—but this is not a likely explanation. Burghardt[80] found that force feeding young snakes for months on an artificial food (liver extract) did not change their original preferences at all. However if the snakes themselves switch to another prey, their preferences for chemical extracts do change rapidly. Thus adults caught near a fish hatchery where goldfish were abundant, preferred goldfish extract to that of naturally occurring prey fish such as minnows. Young snakes fed on goldfish in the laboratory responded in just the same way.[174]

Each type of prey will present Jacobson's organ with an array of chemical stimuli of varying types and at varying concentrations. It seems probable that it passes on much of this to the snake's brain where some parts operate as a filter to determine whether it should attack or not. The filter has a genetically determined bias, modifiable according to experience.

We discussed earlier that sometimes animals respond to different aspects of the same situation according to their 'mood'. Gulls responding to eggs as food or objects to be incubated was a case in point. Here again we must assume that some part of the selection is taking place centrally; i.e. the brain is picking out certain fractions of the stream of information from the sense organs and using them to direct a certain type of response.

Lorenz[303] suggested that a special mechanism was responsible for stimulus filtering and proposed the term 'Innate Releasing Mechanism' (IRM) to describe it. An IRM is defined by Tinbergen[460] as a '. . . special neurosensory mechanism that releases the reaction and is responsible for its (the reaction's) selective susceptibility to a very special combination of sign stimuli'.

It will be seen how closely the IRM fits in with the concept of the releaser. IRMs and releasers are mutually adapted to one another. One animal gives a stereotyped signal by its releaser to which another animal is especially responsive via its equivalent IRM.

As defined by Tinbergen the IRM concept does not exclude the possibility of peripheral filtering, but there can be no doubt that ethologists used to think primarily in terms of central filtering. Perhaps some of the trouble lies in the 'mechanism' part of the term IRM. This might be taken to imply that there is one site in the nervous system where the filter operates and that this is similar in all cases. One might equally object to the word 'releasing', because this implies that the stimulus merely triggers off a response which is ready-primed and ignores the arousing effects of all stimuli.

However we may with reservations retain the term if it proves useful and provokes research into the exact nature of stimulus filters. In fact although we have several excellent analyses of the stimulus situation to which a particular IRM is adapted, we know next to nothing about how such a filter may work. Before this problem can be approached we need to define very exactly what are the characteristics of particular IRMs. Thus among other things, we must determine just how 'central' the filter is or whether some of the sensitivity to key stimuli is due to sense organs after

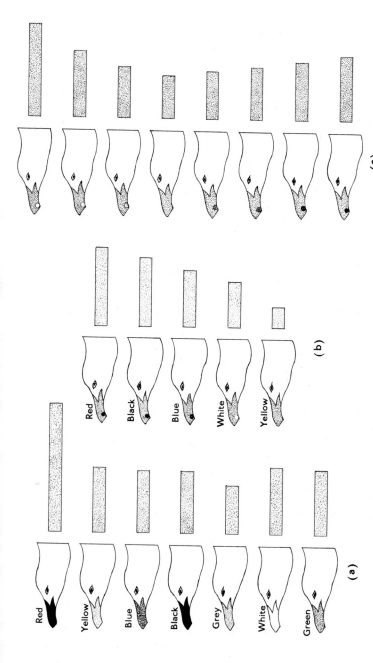

Fig. 3.9 Three series of model heads used in the pecking tests; (a) measures the effect of bill colour, (b) that of patch colour (all the bills were yellow), and (c) the effect of varying the contrast between patch and bill colour (all the bills were grey). The length of the bar beside each model is proportional to the number of pecks it received. (From Tinbergen and Perdeck,[470] 1950, *Behaviour*, **3**, 1.)

all. We will examine two examples where analysis of an IRM has gone some way.

The pecking response of gull chicks*

The young chicks of gulls and terns solicit feeding by pecking at their parents' bills. Tinbergen and Perdeck's[470] work on the stimuli which evoke the herring-gull chicks' pecking has already been mentioned. The adult's bill is yellow with a red patch on the lower mandible (see Fig. 3.8). Tinbergen and Perdeck made a number of cardboard models in which they varied (i) head colour, (ii) head shape, (iii) bill colour, (iv) colour of patch on lower mandible and (v) degree of contrast of patch (black through shades of grey to white) on a medium-grey bill. Fig. 3.9 shows the results of series (iii), (iv) and (v) diagrammatically. In all these tests they took care that the models were presented to the chicks in standard manner. Each type of model was presented first to equal numbers of completely naïve chicks, i.e. those which had never seen a model or a real gull's head before; thereafter the order of presentation was randomized.

In summary, Tinbergen and Perdeck found that the mandible patch was the most important aspect of the parent's head. It worked partly by contrast (grey bills with black or white patches got more responses than plain bills), and partly by virtue of its red colour. A plain red bill is more attractive than any other colour, and a red patch more effective than a black one even though its contrast may be lower. The colour or shape of the head beyond the bill are immaterial. There are, in addition, other features of bill shape and movement which we will ignore here.

We can now begin to specify some of the characters of the IRM which filters out the sign-stimulus in the normal situation. It selects redness and contrast but eliminates bill colour and stimuli from the parent's head.

Weidmann and Weidmann[490-1] have repeated these experiments using black-headed gull and Hailman[190] has used the laughing-gull. Both these species have a plain red bill and in both cases, as with the herring-gull, the naïve chicks' preference for red was strong. The same is true of kittiwakes and arctic terns which also have red on the bill or inside the mouth.

There is plenty of good evidence that birds have full colour vision similar to our own, therefore it is tempting to consider the red on the parent's bill as a releaser and the chick's corresponding responsiveness as being one property of the pecking IRM, the two being mutually adapted.

However this preference for red might not be due to a central filter. In the retinal light receptors of gulls and many other birds, there are abundant droplets of red, orange and yellow oil. Incoming light has to pass through these droplets before it reaches the light-sensitive pigment. Light at the

*There is a very full discussion of this topic in Hailman.[192]

blue end of the spectrum will tend to be filtered out whilst red light will pass through. This does not mean that birds cannot see blue, but merely that it will appear darker than an equivalent intensity of red light and therefore is less likely to seize their attention. Perhaps then part of the properties of the IRM are in the gull chick's retina.

Comparative evidence shows that whilst this may be a sufficient explanation in those species which prefer red, it is not the whole story. Two species of tern have been studied in which the young chicks respond more strongly to black bills than to red (sandwich tern[491]) or find black equally attractive (wideawake tern[114]). It cannot be simply coincidence that the adults of these two species have black bills; that of the wideawake is plain black, whilst the sandwich tern's bill has a yellow tip. There seems to be no possible property of the retina which could make black appear as bright red provided, as in these tests, precautions are taken to equalize contrast with background. The IRM in these two terns probably includes a truly central component which responds selectively to black.

The universality of oil droplets in the retina of birds makes it likely that the red bills of the great majority of gulls and terns are adapted to elicit the maximum strength of response from their chicks who see red as the brightest colour. For the same reason flowers which are pollinated by birds are nearly all a vivid red colour. A number of other factors in the feeding behaviour of the adult, such as the angle at which it holds the bill and the way it is moved back and forth, also serve to focus the chick's attention on the bill tip, where food will be presented.

With the majority of gull species we do not need to ascribe any filtering properties to the IRM beyond those of the retina itself. The chick pecks towards the most conspicuous object in front of its eyes. The irrelevance of the parent's head, both in colour and in shape, is probably because it is right at the edge of the visual field of a chick whose attention is focused on the bill tip. Weidmann[491] considers that black-headed gull chicks cannot see the parent's head at all clearly when they are pecking.

With the black-billed terns, the mutual adaption of releaser and IRM have proceeded one stage further. For some reason selection has not favoured the retention of red bills and the chick's responsiveness has followed the change to black.* Here there may be a central component selecting black in the chicks' IRM, but presumably because of the basic retinal characteristics red still remains very attractive to them and second best to black.

---

* At least this is the simplest explanation. When deducing evolutionary trends without fossil evidence one must always allow for a trend proceeding in the opposite direction. Ancestral gulls and terns may *all* have had black bills and only these two species have retained them. In which case the argument, somewhat tortuously, must run backwards too.

There is some other evidence that central filters which are not affected by experience may be involved in the colour preferences of young birds. Kear[262-3] placed newly-hatched chicks of a number of precocial birds (those which can run about and peck after hatching) in an arena where they could peck at coloured spots. Most species showed strong preferences though they were never rewarded for pecking at any particular colour. She confirmed the attractiveness of red for gull chicks. Moor-hen and coot chicks also preferred red and both these species are fed from the parent's bill at first. The correspondence is good in the moor-hen which has a red bill, but although the coot's bill is vivid white, young coots still prefer red. Pheasant chicks and ducklings of several species all pecked most strongly at green, although green was very unattractive to the other birds tested. It is possible that this preference for green may simply be a response to brightness because the eyes of ducks and pheasants are more sensitive to green wavelengths. Kear[263] did find that varying the brightness of the colour spots affected the relative popularity of some colours. However it seemed most likely that ducklings do select green by its hue and it may be relevant that pheasant chicks and ducklings have to feed themselves as soon as they leave the nest and mostly eat green plants of various kinds.

Oppenheim[374] has recently examined the colour preference of mallard ducklings in great detail. Like Kear he found a strong preference for pecking at green, and this seems to be because the ducklings are attracted by green objects which contrast with their background and not just because they move towards any area of green colour. Oppenheim found that the initial preference for green is unaffected if eggs are incubated and the ducklings hatch in total darkness, or if he opened a window in the shell and illuminated the eyes of the embryo with white or yellow light for many hours before hatching. He was unable to account for the green preference by the properties of the duck's retina alone.

All these observations suggest that in some cases colour preferences have an inherited basis which is 'built in' to the central nervous system. The next example examines selective responsiveness to other aspects of visual perception—shape and movement.

The alarm response to a flying predator

Many ground-nesting birds such as ducks, geese, pheasants and turkeys give alarm calls and crouch when a bird of prey passes overhead. This response is particularly strong in a female bird with chicks. Turkey hens spread their tails as they call and the chicks run to shelter beneath them.

Tinbergen[460] describes some experiments which Lorenz and he performed in 1937. They constructed a number of simple cardboard silhouettes of flying birds which they caused to 'fly' on a wire across a pen containing geese or various game-birds.

They recorded the presence and strength of alarm responses and found that the most effective models had the common feature of a 'short neck' which is characteristic of most birds of prey. The most striking experiment which has become something of a classic, albeit a controversial one, used the silhouette shown in Fig. 3.10. This evoked no alarm if flown to the left with the long neck leading (simulating a goose), but did evoke alarm when the short neck led (simulating a hawk). Here then is a sign-stimulus situation in which it is not merely shape which is effective, but the relationship of shape to movement. It is unlikely that such selective responsiveness can be due to properties of the bird's retina alone.

A number of critics have been sceptical of the suggestion that a response to such a complex stimulus configuration can be inherited. Schneirla[424] has suggested that there is a far more general explanation of such responses. All animals from Protozoa upwards, he argues, tend to approach any sources of moderate stimulation—sound, light or chemical—and to withdraw from strong and suddenly increasing stimulation. He suggests that the silhouette flown in the 'hawk' direction is more alarming simply because it has a broad leading edge. This will cause a more abrupt darkening of the bird's visual field than the 'goose' silhouette which broadens gradually from a tapered leading edge. Schneirla discounts any IRM specially adapted to the hawk silhouette, and predicts that one would get the same differential effect from a plain triangle depending on which way

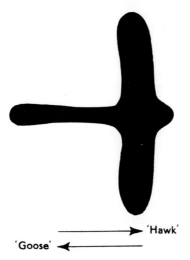

Fig. 3.10 The 'hawk-goose' silhouette used in experiments on the alarm response of geese and game-birds. (From Tinbergen,[460] 1951, *The Study of Instinct.* Oxford University Press, London.)

it was flown. Base leading, he suggests, will induce more alarm than apex leading.

Certainly we must examine critically any claims to have demonstrated inherited specific responsiveness to complex stimuli and the original experiments are open to a number of criticisms. In the first place nothing can be argued about the nature of an inherited response to a 'hawk shape' unless naïve birds are used. Lorenz and Tinbergen's birds could have learnt previously what hawks look like. Secondly, they scored only the reaction of the group, and individuals may vary and be influenced by the behaviour of others.

The first attempt to apply more strict criteria was that of Hirsch, Lindley and Tolman[226] who used newly-hatched white leghorn chicks and scored their responses individually. They found no greater alarm to the hawk model than the goose. Both proved moderately alarming when first flown overhead, but the chicks rapidly got used to them and took little notice.

Tinbergen[464] and others have quite reasonably suggested that it is difficult to generalize from white leghorns to wild geese. The former may have lost any inherited response to hawk shape from many generations of domesticity with no selection in its favour. More recently Melzack et al.[340] have repeated experiments of this type with wild mallard ducklings. They could confirm Lorenz and Tinbergen's original result to the extent that roughly twice as many naïve ducklings showed fear responses to the hawk when first flown over, as to the model reversed and flown as a goose. However, like Hirsch et al., they found that the ducklings rapidly ceased to show fear to either model, although they never relaxed their vigilance and always followed with their eyes any model that was flown. Green et al.[176-7] got the same result—the hawk elicited much more alarm than the goose from naïve ducklings which had no prior experience of overhead objects. In addition, they tested Schneirla's prediction concerning plain, triangular silhouettes and could not uphold it. The ducklings showed very low levels of alarm to the triangle whichever way it was flown.

However a further complication was added by the work of Schleidt[420] with turkeys. He found that he could release the alarm response with models of almost any shape—a plain circle for example—provided only that it flew over at the correct speed. This is rather slow, between 5 and 10 model-lengths per second and corresponds well to the apparent speed of birds of prey in flight. A tiny model only an inch long will evoke a response if it moves at this speed. Schleidt found that turkeys would rapidly cease to respond to any model with which they were familiar, but novel shapes, no matter how they moved usually frightened them. He suggests that Lorenz and Tinbergen's geese did not respond to long-necked models because they were already accustomed to the sight of geese flying over, but this cannot explain Melzack's or Green's results.

It is rather difficult to equate the results of all these experiments with the natural situation. Birds cease to respond to familiar models no matter how hawk-like, but in the wild they go on responding to birds of prey although few will ever have been attacked. Perhaps this is because the models always appear in the same place and in the same way. Also, as Schleidt points out, wild birds are rarely confronted with flying hawks at the high frequency with which experimenters present hawk models! The scarcity of hawks must be a powerful factor in maintaining responsiveness.

In summary, there is evidence that wild birds do possess an IRM which enables them to respond to birds of prey on the first occasion that they see them. This IRM probably has different properties in different species but short neck and relative speed of movement are among them. Peripheral filtering is unlikely to be involved here.

Any attempt to analyse the properties of an IRM is bound to involve work on both sensory physiology and behaviour. Clearly the IRM, if it has any use as a concept, must be taken to include the whole range of possible filters beginning at the sense organs themselves. The end result of the filtration process is to adapt the animal to respond selectively to some sign-stimulus from the environment, which may be a specially evolved releaser whose properties are mutually adapted to those of the IRM. The main un-solved problem concerns the nature of central filters. It seems unlikely that these rely on any single type of mechanism, although the properties of central filters which mediate learnt discriminations and those such as we have dealt with here, whose development is controlled by genes, probably have much in common.

The experiments of psychologists investigating how an animal recog-nizes objects, show that it does not respond to all their features simul-taneously. Rather it directs what may be called a central 'attention mech-anism', first upon one set of cues—say 'brightness', 'colour' or 'shape'—and then upon another.[168] Clearly such an attention mechanism shares many properties with the central filters we have been discussing. Etholo-gists have been impressed by the manner in which a change to an animal's 'mood' causes a switch in attention; one filter is inhibited and another is brought into operation. When he first emerges from the pupa a male grayling butterfly responds most strongly to blue and yellow which lead him to flowers. Later, using the same sense organs he responds preferen-tially to brown when courting a female. In some way the filter changes as his 'mood' changes from feeding to courting.[460]

COMMUNICATION

As we have been discussing in this chapter, many of the most important external stimuli to which animals respond are those provided by other animals—stimuli from predators or from prey, stimuli from the parents to which the young respond, courtship stimuli from one sexual partner to another, and so on. Further, there is abundant evidence that animals have frequently evolved special structures—releasers—and special behaviour patterns whose chief function is to send stimuli to another animal. In other words, animals communicate and we shall need to refer to various aspects of communication throughout this book. More detailed accounts of the whole topic are to be found in Hinde[218-19] and Sebeok.[429]

First it is necessary to define what we mean by 'communication' and this is not a completely straightforward task. Like many words used both technically and in ordinary speech it has gathered several different meanings through constant use. Communication always involves the passage of information and one of the widest definitions would be to accept this and define it as 'any transfer of information'. It is perfectly reasonable to say that the different parts of a computer communicate one with another and that a computer communicates with its human operators. However if we do use a definition as broad as this it will not help in making distinctions which are very important biologically. Thus if a male robin sees another robin intruding in his territory and flies across to alight on a nearby branch, we have to conclude that both the rival and the branch have communicated with the territory holder.

Most biologists would prefer to confine the term for information transfer between animals, but questions still remain. Should we restrict the definition to transfer *within* a species? If we do not then we must conclude that the wildebeeste communicates with the lion which is stalking it and, *vice-versa*, once the lion begins its charge causing the wildebeeste to turn and flee. Burghardt[79] in a most useful essay on the problems of definition, quotes another example which makes this point even more forcibly. Foraging ants commonly lay trails of scent which nest mates pick up and follow out to sources of food. A small species of snake (*Leptotyphlops*) also detects the scent trails and follows them back to the ants' nest where it devours the brood. Most of us would probably refer to the scent as a signal, evolved by the ants for communication, but we would not consider that the ants communicated with the snake.

We are approaching a restriction on our definition which involves the use of a concept of 'intent'—signaller and receiver are mutually adapted to one another. It must be admitted that intent is not itself easy to define rigorously but if employed with caution it will be useful. Clearly the intent to transfer information will often not be conscious but rather describes a relationship between communicating animals which has evolved to their mutual benefit. We have already discussed the mutual adaptation between releasers and IRMs which was postulated by ethologists, (p. 72 and the term 'social signal' certainly implies a communicatory function. In the example we have just described the ants intend to communicate with each other by their scent trails but one of the penalties they incur involves leaving information which the predatory snake can use. We can be sure that the advantages the ants derive outweigh the occasional disadvantages.

Our use of intent in this way certainly means that we cannot reasonably restrict communication to information transfer between members of the same species. The nectar-guide patterns on flowers (see Fig. 3.2 and p. 65) communicate with bees to the mutual benefit of both partners. The skunk communicates to potential predators by presenting its hindquarters and raising its tail. The rattlesnake's rattling serves a similar function.

Not all workers in animal communication have chosen to use the definition we have evolved. For example, both Altmann[8] and Hinde and Rowell[221] studied social communication in rhesus monkeys and attempted

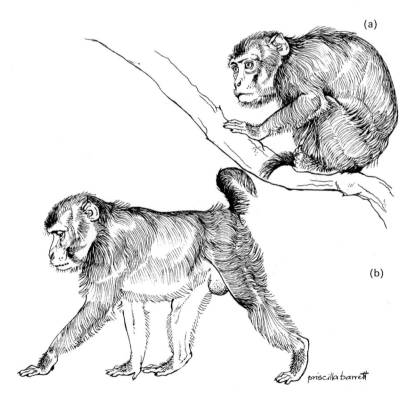

Fig. 3.11 Typical body postures assumed by (a) subordinate and (b) dominant rhesus monkeys. (From Hinde,[219] *Biological Bases of Human Social Behaviour.* © R. A. Hinde 1974. Used with permission of McGraw-Hill Book Company.)

to describe their visual signals—the former study lists about 50, the latter only 22. The reason for this discrepancy lies in their different uses of the term communication. Altmann's definition of social communication is, 'a process by which the behaviour of an individual affects the behaviour of others'. All kinds of movements or postures may do this—the sight of a monkey feeding for example. Hinde and Rowell, on the other hand, use a definition similar to that discussed above and they classified as visual signals only those which were likely to have evolved for this purpose. Even though we shall concentrate on specially evolved social signals in this brief discussion of communication, it is important to recognize that animals do receive a great deal of information from others in the way Altmann describes. We have already mentioned (p. 71) the diverse stimuli from a dominant male monkey to which others in his troop respond. His general body posture and manner of walking will convey information quite apart from any signals, such as threat movements (see p. 168) he may make. Almost every feature of his body when moving or at rest is in strong contrast to that of a subordinate male (see Fig. 3.11). In such cases it may be difficult to decide whether intent to communicate is involved or not; it is not always possible to identify a social signal with certainty, particularly in the primates.

We can organize our discussion of communication around three interrelated questions.

1. How are social signals adapted to their role?
2. What is communicated?
3. How can we investigate whether communication occurs?

1. How are social signals adapted to their role?

A survey of communication through the animal kingdom (and excellent ones are provided by the central chapters of Marler and Hamilton[333] and by Brown[73]) reveals that the different groups are more or less specialized in the types of signal they use depending on the development of their different sensory modalities. Further there is an overall correlation between the properties of a signal and the function it has to perform.

Touch is limited in its scope for transmitting information but it is in many ways the most basic of the channels of communication for almost all living material responds to physical contact. Tactile communication certainly dominates the social interactions of many invertebrates; for example, the blind workers of some termite colonies which never leave their subterranean tunnels or earthworms which emerge from their burrow at night to mate. Amongst invertebrates touch is closely associated with the chemical senses for specialized tactile organs like the antennae or palps of insects often carry chemoreceptors as well. The social insects transfer a great deal of information through their colonies by a combination of tactile and chemical signals (see Chapter 8, p. 269). Tactile communication remains important in many of the vertebrates, particularly so in the mammals where some of the more social species spend a good deal of time in physical contact with each other. It is probably not appropriate to think of specially evolved social signals in the majority of such cases; tactile stimuli from another member of the same group convey more general information on the social relationship between the two animals as when one monkey grooms another (see p. 291 and Fig. 8.9, p. 290).

Tactile communication can by its very nature operate only at very close range. The long antennae of cockroaches and lobsters act as 'feelers' which enable them to explore the world over one body's length ahead, but this is about the limit for touch. The other sensory modalities, vision, hearing and smell can allow communication over a considerable distance. Sound and smell have the added advantage that they can travel past, and to some extent through, natural obstacles such as dense vegetation and hence long distance signals are usually calls or specially produced scents.

Sound signals and the way they are delivered are closely adapted to their function. Low frequency calls penetrate dense vegetation best and the call notes of tropical forest birds are usually of this type. Hooker and Hooker[231] describe the low flute-like calls of the African bou-bou shrike where male and female call alternately to one another in the form of a 'duet' whilst keeping concealed in thick forest or scrub. Marler[330] has pointed out that the calls of many primates living in tropical forests are similarly adapted to carry long distances. Again we have already discussed the characteristics of an ideal alarm call and noted the way in which several bird species have converged in their evolution of the optimal sound signal for this function. We return to this example in a discussion of evolution, Chapter 6, p. 228.

The manner in which a sound signal is delivered can help its dispersal too. The territorial songs of birds are usually delivered from a raised song post which increases the effective range. Grassland birds such as larks and pipits sing whilst in flight above their territory.

Sound travels better with less attenuation in water than in air and hence aquatic animals use sound extensively for communication. The development of the underwater microphone allowed us to discover the amazing range of sounds produced by fish and whales. Payne and McBay's[379] remarkable study of humpback whales suggests that their 'song' could be picked up by other whales several hundred miles away! This is certainly a long-distance record for animal communication.

Chemical signals are particularly well-developed in insects and mammals (there are good reviews by Schneider[423] and Wilson[505] for insects and Ralls[388] for mammals). One of the drawbacks of chemical communication is the difficulty of changing the signal quickly—it is normally not possible to produce a pattern of scent, whereas patterning is readily achieved with sound and visual signals. Consequently most chemical signals are used to pass a single, relatively stable message. Many mammals mark territory by smell, often concentrating it at particular scent posts. Smell is also used to communicate the breeding condition of female mammals which secrete special chemicals when they are in oestrous, i.e. fertile and receptive to the male. This type of signal is analogous to the scents produced by virgin female moths (see p. 77) and such chemicals, secreted to the exterior and acting to stimulate reproduction are called *pheromones.*

Wilson[505] and others have discussed the way in which the chemical structure of scents can be adapted more specifically to their function. Sex pheromones and territory marks need to be persistent and therefore their constituents must have a fairly high molecular weight. It cannot be too high, or it will be difficult to secrete and may not disperse well. Moth pheromones strike a compromise between persistence and good dispersal. In favourable wind conditions males can detect them from 4–5 kilometres downwind. On the other hand the chemicals used by some ants to signal alarm must not persist—if they did then precise localization of the immediate source of danger would be impossible. These chemicals are volatile and disperse well over a short range of 3–5 cm but they have usually faded below detection level within a minute or even less.

Visual signals can operate only at relatively short range, at least if they are to convey much detailed information. Simple alarm signals often involve flashes of white—the scut of deer, the tails of rabbits—which may be effective over longer distances. Visual communication is especially characteristic of the vertebrates and the cephalopod molluscs both groups which have good eyes. It is interesting to note that colour vision is virtually universal except in most of the mammals. Their early history involved many millions of years as nocturnal insectivores when colour vision was of little advantage and was effectively lost. The brilliant colour patterns on some fish, reptiles and birds are in striking contrast to the universal greys, blacks and browns of most mammals. The exceptions are some squirrels and the primates, the two groups of mammals which have regained colour vision and here we find blues, reds, greens and yellows returning.

The arthropods have excellent colour vision but, perhaps because they are small and their compound eyes do not permit good resolution of form, visual communication is not very common although colour signals are used in the courtship displays of butterflies and fiddler crabs, for example. The fireflies have developed a remarkable form of visual communication which can operate over long distances, because they manufacture their own light and signal only at night.

In natural situations it is not very realistic to separate the different

Fig. 3.12 A male frigate bird performs the advertising display on its nest. A female sits to the left. Further description in text. (From Nelson,[364] 1976, *The Living Bird*, **4**, 113–156. Cornell University Press.)

sensory modalities as we have just been doing. Animals have evolved effective combinations of signals involving, for example, both sound and vision. The following description from Nelson[364] (see also Fig. 3.12) gives a vivid impression of how male frigate birds catch the attention of females and entice them to alight at their nest sites.

'The sexually motivated advertising display of males in (*Fregata*) *minor* is a Gular Presentation and has three main elements: (1) an upward presentation of the inflated crimson pouch with the head thrown back and turned from side to side; (2) trembling of the spread wings, the silvery under-sides facing upwards; and (3) accompanying vocalisation, described as warbling or whinnying, a rapid falsetto *whoo-whoo-whoo*. The wing-trembling is striking, for the males swivel on their perches as they orient their display directly toward the female, giving the effect of two enormous beseeching black arms with the crimson sac couched between. If the female descends, the male's head-waving becomes more exaggerated and disjointed. A fine-drawn reeling sound and mandible-clapping, sometimes interspersed with a peculiar quavering sound, accompanies the display. Any one of three such conspicuous components would be dramatic; they produce an astounding performance when delivered by a group of birds'.

2. What is communicated?

We have already mentioned some of the types of information which animals communicate; alarm calls and sex pheromones are examples of signals which transmit fairly simple messages. In a valuable discussion of communication in vertebrates Smith[441] suggests that we need to distinguish between the information the 'transmitter' animal puts into a signal and that which the receiver interprets from it. He calls these 'message' and 'meaning' respectively and it is important to recognize that they will not necessarily be the same. Very often the message arises from a shift in the internal state of the signaller. A male chaffinch wakes up in his territory and begins to sing, and his message is probably no more than a measure of his reproductive condition. However the receiver can extract more information for there is also contextual evidence for him to go on. The signaller will be a male chaffinch in reproductive condition and on his territory, the receiver will also know precisely where he is and be able to respond accordingly. Depending on the time of year it may be possible to estimate whether the male has a mate or not (males whose females are incubating on a nest sing much less) and finally the detailed characteristics of the song will enable a neighbouring territory holder to identify the singer as A, not B or C. This is of course a human interpretation of the male's song, the meaning it has for other chaffinches is probably not so elaborate, but it gives an impression of the information that is potentially available from some simple messages.

In Chapter 5 which deals with conflict, we shall be able to discuss other examples of signals which convey information about the internal state of an animal, especially its likelihood of attacking or fleeing. However signals can convey a whole range of messages. Many are purely sexual, as in the pheromones we described earlier, or the flashing of fireflies. A whole range of signals exchange information between parents and offspring, alarm calls and calls to food for example. Animals which live in groups often have signals which may help them to keep in contact, perhaps without any specific message beyond this. Most birds which form flocks have contact signals of this type—the calls made by geese and finches in flight, or when about to fly are familiar examples. Such calls are often made by single birds and are probably linked causally in some way to flying. In *King Solomon's Ring* Lorenz describes the 'decision making' of a flock of geese on the ground. One bird, then perhaps a few begin flight calling. If this is taken up generally the flock usually takes off, but if too few respond the original callers subside into silence for the present and the flock remains on the ground.

There are a few interesting examples of a rather distinct type of signal whose message is to qualify other signals which follow it. This phenomenon has been called **meta-communication** and is best known from play situations in carnivores and monkeys. Fig. 3.13 shows an adult male lion inviting a cub to play. His posture with forequarters lowered is not seen in any other context and its message is that all aggressive movements which follow are play. Dogs use almost exactly the same posture[45] and may also

priscilla barrett

Fig. 3.13 Meta-communication in lions. The male's posture with lowered fore-quarters is only seen as a preliminary to play. He invites the cub to play and cuffs it gently as the cub joins him. (From Schaller, 1972, *The Serengeti Lion : a Study of the Predator-Prey Relations.* © by the University of Chicago Press, Chicago.)

wag their tails during play fights (as Darwin noted in *The Expressions of the Emotions in Man and Animals*). Monkeys adopt a 'play face' in similar situations. Playful aggression is an important part of behavioural development in carnivores and it has obviously been necessary to evolve some convention which allows stalking and attacking to take place without the disruption and damage of real fights.

Graded and fixed signals

Fig. 3.14 illustrates different forms of a visual display given by Steller's jay during territorial disputes. It is obviously graded and Brown[72] found that the higher the angle of the crest the more the bird resisted in aggressive interactions. The crest is held flat if the bird flees and also during courtship. Clearly there may be many advantages in evolving graded signals for they could allow the message to be similarly graded in its intensity. The various calls made by the rhesus monkey during aggressive encounters can be arranged in a graded network which can be correlated fairly successfully

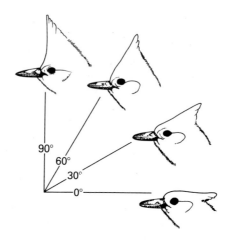

Fig. 3.14 The range of crest elevation shown by Steller's jay. Brown used angle of elevation in this graded signal as an index of aggressive tendencies. (From Brown,[72] 1964, *Univ. Calif. Publ. Zool.*, **60**, 223–328.)

with the likelihood that the caller will attack (Rowell[403]). Figure 5.8, p. 177 shows a similar attempt to interpret the graded facial expressions of cats. The food dance communication of honey-bees, which we shall discuss in detail below, is a further example of a graded signalling system whose message is very exactly related to its form.

However there are some drawbacks to signals whose form varies according to circumstances. They may be ambiguous to the receiver, particularly if they are seen only briefly. Many signals are fixed in form, or vary only within narrow limits (see the discussion on 'typical intensity', p. 221). Many alarm calls are of this type and many sexual signals which function to bring the sexes together, such as female pheromones or the songs of male crickets and cicadas. In such cases it makes sense to have a stereotyped signal—it is only produced when the signaller is in a fully-receptive condition. By contrast many of the close-range courtship displays in fish and birds are graded since relatively detailed information about readiness to mate, aggressiveness and fear has to be conveyed to the potential mate (see the discussion in Chapter 5, p. 187).

The distinction between graded and fixed signals inevitably becomes less precise when we look at the details of communication. We discussed earlier (p. 64) the way in which stimuli can act as 'pumps' or 'triggers' and communication obviously can operate in a 'pump' fashion, where the effects of signals gradually accumulate and change the probability that the receiver will respond. A fixed sexual signal like the courtship bow of a male pigeon may yet have a graded effect on the female because it must be repeated many times over a matter of days if it is to produce its full effect. Schleidt[421] calls this 'tonic communication' and discusses other examples. Much depends on the function of the signals involved. Other types of fixed signal, e.g. alarm calls, have to operate as triggers and change the behaviour of the receiver abruptly if they are to be of any use.

Animal communication systems lack the unlimited resources of human symbolic language. They have to rely heavily on context and on combinations of signals to extend the range of their messages. Even amongst the primates—our closest relatives—the number of messages is quite limited although as we have seen (Fig. 3.11 and p. 88) their great powers of learning and long experience in each other's company mean that their social interactions are often very elaborate. In Chapter 7, p. 257 we discuss some of the experiments on teaching symbolic language to chimpanzees. It is a remarkable fact that although they can learn the complex gestures of a human sign language, wild chimpanzees exhibit very few. Begging with outstretched hand is quite common but in many hours of observation Menzel[341] never saw chimps use a beckoning gesture in order to invite approach. This proved one of the first signs learnt by a young chimpanzee during deliberate training (Gardner & Gardner).[164]

One distinction between human and non-human communication which is often given great attention concerns deception. Humans can deceive, but animal communication is so much a direct product of their own internal states that deception seems impossible. We cannot imagine a bird giving a false alarm call, for example. However it is at least arguable that the eye-spots on moth's wings deceive. They do not represent any internal state of the moth, they are simply a visual releaser, but one which has evolved to mislead predators. The numerous insects (and other groups) which live parasitically in ants nests (myrmecophiles) all give false signals (often super-normal signals) to their ant hosts, which lead them to be accepted into the nest (see Hölldobler[229]). The 'broken-wing trick', by which some ground-nesting birds such as plovers lure predators away from their nest is more difficult to explain. We know too little about the internal state of the bird when it displays but there is unlikely to be any degree of conscious deception. Again it is evolutionary deceit.

The most remarkable example of evolved deception concerns fireflies. Virgin females of *Photuris versicolor* emit their species-specific pattern of light flashes in response to their males' flashes and the latter approach them and copulate. Once mated a female stops flashing and over the following two nights her behaviour changes. She takes up a predatory stance with fore legs raised and mandibles open. Now she begins to flash again but this time it is not the code of her own species. She flashes a pattern which is typical of a related smaller species of the genus *Photinus*. When the

Photinus males approach in response, she kills and eats them! Lloyd[301-2] who discovered this extraordinary piece of deception rightly called the *Photuris* females, 'firefly femmes fatales'.

3. How can we investigate whether communication occurs?

So far in all our discussion of communication we have assumed that social signals are effective in evoking appropriate responses. We may feel confident on evolutionary grounds that animal communication systems work, but we can never learn much about their operation and their limits unless we can begin to make measurements.

Many interactions between members of the same species include a diversity of potential signals. Experiments which employ models or interfere with the normal range of stimuli often help to identify which are the key features of a communication system and this is a first stage in a proper analysis. We have already described and illustrated (Figs. 3.3 and 3.9) some model experiments which prove that the red throat of a male stickleback and the bill spot of a herring gull are major components of their communication in certain situations. In an exactly analogous way early experiments which showed that female crickets would walk into the horn of a loudspeaker which was emitting the male song, first proved the efficacy of sound communication in insects. Many of the experiments with the sex pheromones of moths also give unequivocal results—no other stimuli from females could possibly be acting over such distances.

Some communication has been proved to involve more detailed features. Amongst monomorphic bird species—those in which males and females are alike—sexual identification presents problems. The only obvious difference between male and female yellow-shafted flickers—a type of American woodpecker—is the small black moustache of the male (Fig. 3.15). Noble[367] captured the female of a mated pair of flickers and stuck a moustache of black feathers on to her. The male promptly attacked her and drove her out of their territory. When the moustache was removed she was once more accepted by her mate. The fact that the male's response could be switched on and off in this way convincingly demonstrates that the signal value of the moustache outweighs any other aspect of recognition. Smith[440] has investigated a most remarkable example of interspecies communication of this type between four species of arctic nesting gulls. These are all predominantly white birds with grey backs. For the human observer they are hard to tell apart—the pattern of their wing-tips is perhaps the easiest distinguishing mark—but close examination reveals that they also differ in their eye colours. The colour of the iris varies between the species and more conspicuously, the narrow ring of pigmented skin surrounding the eye and visible beneath the lids when they are open is differently coloured.

Yellow Shafted Flicker

Fig. 3.15 The heads of male (left) and female yellow-shafted flickers. The male's black moustache is a key feature in sexual identification

Smith captured gulls and managed to paint their eye rings to resemble the colour characteristic of other species. If this was done early in the season it proved sufficient to break down the normally complete barriers to inter-specific pairing. Further, pairs which had formed naturally often broke up if eye ring colour was changed early in the season, though they usually persisted if the change was delayed. The gulls identification of mates of their own species depends crucially on the eye rings.

Identifying the stimuli used in communication is a first step, but it is then important to try to get some measure of its effectiveness. This is not always a straightforward matter. In the first place it is often not possible to decide whether no response means that the 'receiver' failed to perceive the 'signaller's' message or that he received it but simply failed to respond for some reason. This is a difficulty even with social signals which—if they work—usually produce an immediate response; it is far more of a problem with signals which act tonically or as pumps rather than triggers.

There is no substitute for careful and repeated observation of social encounters, testing the efficacy of a signal by recording the response of 'receivers' when a certain signal is given and when it is not. In Chapter 5 we shall describe some analyses of this type for hostile encounters between birds (see p. 178); here we can use Chalmers'[88] work with the mangabey as a useful example.

Mangabeys are large forest-living monkeys from Africa. Chalmers de-scribed a number of their vocal and visual displays and tested their effec-tiveness in communication by the method just outlined. He had to pool data over a large number of encounters between different individuals in order to get an adequate number of encounters with and without the dis-play for comparison. One common visual display is 'presentation' when a subordinate monkey—of either sex—turns its back on a more dominant animal and 'invites' it to mount, (see Fig. 8.10 and discussion on pp. 179 and 290). Presentation has often been interpreted as an appeasement display which inhibits attack. Table 3.1 shows the consequences when dominant animals approach subordinates comparing occasions when the latter presented or did not present.

Table 3.1 Behaviour of dominant mangabey towards subordinate after the subordinate (a) presents and (b) does not present.

Behaviour	Dominant does not attack	Dominant attacks
(a) Dominant approaches subordinate who presents	53	0
(b) Dominant approaches subordinate who does not present, and gives no other gesture or call	21	9

A far smaller proportion of attacks followed when presentation occurred (statistically significant at the 0·1% level on this data) and thus the subordinate's signal appears to work. Chalmers also studied the effects of one of the mangabey's calls, a low repeated grunting often made when monkeys approached one another. It seemed to reduce the probability of the encounter being a hostile one, a suggestion which is borne out by the data of Table 3.2.

Table 3.2 Number of hostile and peaceful encounters following the approach of silent or grunting mangabeys.

Behaviour	Hostile	Peaceful
Silent approach	19	25
Grunting approach	3	30

These results are convincing but—as Chalmers himself points out—on their own they cannot be conclusive proof of communication by presenting or grunting. It could be that a monkey can detect by some other means whether a dominant is going to attack or not and presents only to the latter. Again monkeys who are not going to attack may grunt as they approach one another, whilst hostile monkeys approach silently. These are certainly not probable explanations but we can eliminate them only by further study of all the contexts in which displays occur.

The honey-bee dance

There is no better way to conclude this section on communication than by some account of the honey-bee dance. Not only is it one of the most remarkable of all systems of animal communication, but also it has recently been a source of a continuing controversy arising from just those problems of measurement and interpretation which we have been discussing.

The fact that honey-bees must communicate about flower crops has been known for centuries, but it will now always be associated with the name of Karl von Frisch who was the first to unravel the nature of this communication and whose book[156] provides a full survey of the whole dance system. Like others before him von Frisch had noted that if a source of sugar solution was put out to attract bees it was often many hours before the first one found it, alighted and drank. However once a single bee had located the source it was usually only a matter of minutes before many other foragers arrived—somehow the information had been passed on. It took von Frisch some twenty years of painstaking observation and experiment before he had worked out the bee's communication system to his own satisfaction. The conclusions he came to were so extraordinary and unparalleled that he himself declared that no good scientist should accept them without confirmation. Following World War II, other zoologists did confirm von Frisch's results and even worked with him on some final experiments. There is now no doubt about the nature of the bee dance itself.

Von Frisch marked foragers as they drank at a dish of sugar syrup and then watched their behaviour when they returned to the hive, using glass-sided observation hives for this purpose. The forager usually contacts a number of other bees on the vertical surface of the comb and gives up her cropful of sugar solution to them. She then begins to dance and we may first consider the case where the food dish which she has just visited is close to the hive, within 50 metres. Her dance then is rapid in tempo and forms a roughly circular path just over her body's length in diameter. The bee moves in circles alternately to the left and to the right. She stays approximately in the same place on the comb and may dance for up to 30 seconds before moving on. Other foragers face the dancer, often with their antennae in contact with her body and follow her movements closely, being themselves carried through her circular path (Fig. 3.16). The 'round dance' as this is called, stimulates other workers to leave the hive and search nearby. It appears to convey the information, 'search within 50 metres'. It also may convey some olfactory cues because if the food source is scented the dancer will carry this scent on her body and perhaps in the sugar solution itself. If the sugar dish is not scented the forager may 'mark'

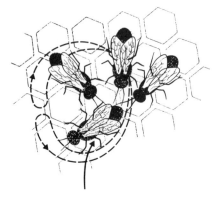

Fig. 3.16 The 'round dance' of the honey bee worker on the vertical face of the comb : her path is indicated by dotted lines. Note how she is closely attended by other workers.

it to some degree by opening the Nasanoff scent gland on her abdomen as she drinks.

Thus far the bee dance is not very exceptional because many ants and termites have similar 'alerting' displays and pheromones which help to organize foraging activity when a new food source has been found near the nest.[505] The extent of the honey-bees' communication system is not revealed until the food source discovered by the forager is further from the hive, beyond 100 metres.

Von Frisch observed that as his food dishes were moved beyond 50 metres the forager's round dances gradually changed in form. A short straight run became incorporated between the turns and on this run the dancer wagged its abdomen rapidly from side to side. At about 100 metres distant the dance had become the typical 'waggle-dance' illustrated in Fig. 3.17 and this form remains the same as the dish was moved further, to 5 kilometres or even beyond. More recent work has revealed that during the waggle run, the bee produces bursts of high pitched sound (Esch, Esch and Kerr[146]). It is this waggle dance which von Frisch claimed transmits so much more information and is 'read back' by the dance followers as they follow every move the dancer makes.

The waggle-dance certainly includes information about both the distance and the direction of the food source. Distance is correlated with several features of the dance. Von Frisch concentrated on measuring its tempo and this falls off with distance, steeply at first and then more gradually. Thus there are 9–10 complete cycles per 15 seconds with the food at 100 metres, but only 2 when the food is 6 kilometres away. The number of waggles and the duration of the waggle run also correlate with distance,

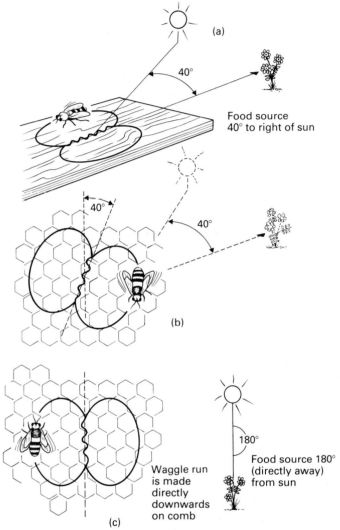

Fig. 3.17 The 'waggle dance' of the honey bee : further details are given in the text. (a) As the dance is occasionally performed on the horizontal entrance board to the hive. The waggle run 'points' directly towards the food source. (b) on the vertical comb the angle of the waggle-run to the vertical is equal to the angle the sun makes with the food source. (c) Shows the honey-bee's convention that directly downwards on the comb represents directly away from the sun. In the same way upwards represents towards the sun. (Modified from Curtis, 1968, *Biology*. Worth, New York.)

both *increasing* with it. Finally the duration of the sound bursts during the waggle run also increases with distance. Von Frisch had no evidence which enabled him to identify which of these distances cues was important for the dance followers.

It is its relation to the *direction* of the food source which is perhaps the most remarkable feature of the waggle dance. The little south Asian bee *Apis florea* is a close relative of the honey-bee. It builds a single vertical comb in the open with a flattish platform on top. Lindauer[297] describes how they perform their waggle dances on this platform. The waggle run is made *towards* the direction of the food source, i.e. it operates like a pointer. Now very occasionally honey-bees will dance on a flat platform by the hive entrance. If they do so, then like *Apis florea*, their dances also point directly to the food source (Fig. 3.17a). However this observation did not help von Frisch because he watched the dances at their normal site on the vertical face of the combs inside the hive. There he noted that the direction of the waggle run was consistent within a dance and that it was the same for all the foragers who danced after feeding at the same dish. Given that other bees foraging at different dishes made waggle-runs at other angles even when the distances were comparable, this was strong circumstantial evidence that the angle related to direction in some way. Now came the crucial observation. Von Frisch recorded dance after dance throughout the day as foragers returned from the same food source and he found that the direction of the waggle-run gradually changed. Its mean direction shifted by $15°$ per hour and this could mean only one thing; that it relates to the apparent movement of the sun. Fig. 13.17b & c show the way it does so. The foraging bee, like many other insects, uses the sun as a compass and records the position of the food source with respect to it. To get to the foods it steers—say—$40°$ to the right of the sun. When dancing on the vertical comb the sun is not visible and the bee transposes the angle to the sun into the same angle with respect to gravity. The honey bees' 'convention' takes vertically upwards to represent directly towards the sun. Thus the forager dances with her waggle run $40°$ to the right of vertical. She will change this angle to match the sun's apparent movement through the sky. Von Frisch had at last understood the honey-bee's dance language and the world was forced to accept that another animal apart from man— and a humble insect at that—could convey information in a symbolic fashion.

Von Frisch and his co-workers had no doubt that the dance did communicate. Other foragers picked up the dance's rhythm and orientation as they followed through the dancing bee's movements on the comb and they then transcribed back from an angle with respect to gravity to an angle with respect to the sun. This assertion was based on numerous experiments in which an array of food dishes was offered, so arranged as to test

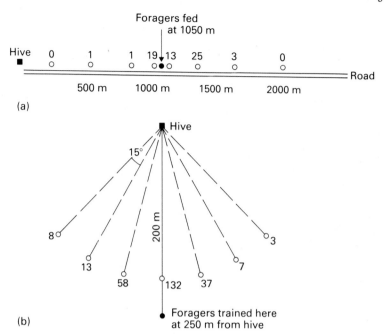

(a)

(b)

○ Scent plates at which recruits were recorded, the number of visits is given by each.

Fig. 3.18 Experiments carried out by von Frisch and his co-workers to test the communication of distance and direction by the honey-bee waggle dance. (a) A distance test. Foragers were trained to a dish of dilute scented food 1050 m east from the hive. Then a series of scented plates without food were put out at distances varying in the same direction from 100 m to 2000 m. Dancing was induced by suddenly increasing the sugar concentration at the feeding dish. Recruits were counted, but not captured, as they approached the different scent plates. The numbers above them record the number of visits made to each scent plate during the test. Most recruits appeared at plates close to the feeding station. (b) A direction test. The procedure is much as for the distance test, but here the scent plates are put in an array at the same distance but different directions from the hive. The majority of visits were made to dishes close to the bearing of the training station. (Modified from von Frisch,[156] 1967. *The Dance Language and Orientation of Bees*. The Belknap Press of Harvard University Press, Cambridge, Mass.)

the accuracy with which foragers recruited by the dance interpreted its information on distance and direction. Fig. 3.18 shows the two commonest types of configuration; a line of dishes at different distances on the same bearing from the hive tested distance communication, a fan pattern of dishes in an arc equidistant from the hive tested for direction. The figures also indicate the results of typical experiments showing that the number of recruits is highest at the dish closest to that at which the original forager fed and recruiting falls off rapidly to each side. (Further details of the experiments are given in the captions.)

This type of evidence was very generally accepted as convincing proof that the dance was the effective communication system. However following a series of experiments (best summarized by Wells and Wenner[495]) Wenner and his co-workers suggested an alternative hypothesis: that the dance merely stimulated foragers to go out and search and that they then located the food by olfaction.

Before discussing this hypothesis it is important to note that Wenner's group do not disagree with von Frisch's finding on the form of the dance and its relation to direction and distance, but they claim that this information is not communicated. One obvious riposte is to ask why then has such a remarkable relationship evolved? However intuitively reasonable this reply seems it does not really supply a secure argument for it assumes that every biological phenomenon must be functional. Wells and Wenner can point to several examples of behaviour which certainly contains information but which just as certainly is not used by con-specifics. Dethier[131] has described the searching movements made by flies after they have located and then exhausted a small source of food—e.g. a drop of sugar solution. Their 'dance' can convey information to a human observer about the

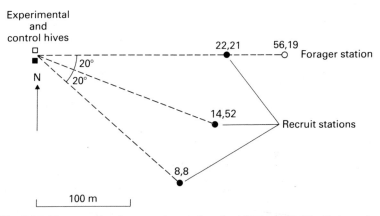

Fig. 3.19 The two-hive fan experiment described in the text. The first number beside each station records the percentage of recruits from the experimental hive when the control hive was shut up. The second number gives the same data when the control hive also was open and its recruits active. Notice how the experimental hive recruits are strongly biased towards the middle of the three recruit stations when the control hive is open : this happened to be the most popular station with the control bees. The wind was from 140° at 8 knots throughout the experiment. (Redrawn from Johnson, D. L., 1967, *Science, N.Y.*, **155**, 844–847.)

'shape' and concentration of the source, but other flies do not respond. Certainly other bees do respond to the waggle dance but then Wenner's hypothesis requires that they are aroused to search thereby, just as ant foragers are. Further, Wenner agrees with von Frisch that recruits pick up olfactory cues about the food source from the dancer—its scent will adhere to her body.

The experiments of Wenner's group were similar to those of von Frisch in their general form. Arrays of dishes were presented to recruits to test distance and direction finding. More attention was paid to the olfactory cues provided by the food source and to the wind direction during the period that the recruits were searching. In some of their experiments recruits were given a choice between dishes at sites which had previously been scented and indicated by dances but were now without scent, and dishes which had never before been visited but which were at the same distance and were scented. Large majorities favoured the latter, although its position cannot have been indicated by dancers. A result of this type appears to be quite inconsistent with the dance-language hypothesis, but it is known that strong olfactory stimuli in the hive will sometimes cause foragers to leave and search even in the absence of dancing. It is possible that the high levels of scent used in such experiments worked in this way and since little dancing was involved recruits homed in on the scented dishes. Certainly bees can detect very small olfactory cues and may be further stimulated to alight at dishes by the sight of bees already feeding there.

Whilst we can produce an alternative to the olfaction hypothesis in the experiment just described there are numerous experiments by Wenner's group which show convincingly that olfaction must play some role in determining which dishes of an array bees visit. The direction of the wind proves to be important for one thing, and this was rarely noted in von Frisch's original experiments. In many of his experiments all the dishes were provided with scent, but even when they were not we must remember that bees mark food sources with their own scent from the Nasanoff gland on the abdomen.

Fig. 3.19 illustrates one of the most convincing demonstrations of the effects of the sight and smell of bees already feeding at dishes. Two hives side by side were used and their inmates had different body colours so that they could easily be distinguished. A fan array of feeding stations was set out about 200 metres from the hives as in von Frisch's experiments to test communication of direction; all the stations provided scented sugar solution. The foragers from one hive, which was called the experimental hive, were trained to the forager station only, those from the other hive (the control) were similarly trained to all the other three stations but not to the forager station. The experiment compared the distribution of recruits from the experimental hive (a) when the control hive was closed and (b) when it was open. All recruits were captured before they could return to their hives. During (a) the recruits from the experimental hive had only the dances of their foragers to direct them together with their sight and smell at the forager station. During (b) there was added to all the recruit stations the sight and smell of foragers and recruits from the control hive. Fig. 3.19 shows that experimental hive recruits favoured the direction of the forager station when only their own hive was open but distributed themselves much more evenly when the control bees were also active. This result certainly suggests that the sight and smell of other bees can outweigh the dance information in some circumstances.

Supporting this conclusion Wells and Wenner's critique of the language hypothesis draws attention to the fact that the recruits visiting an array of dishes do show many errors (often far more than in the examples illustrated by Fig. 3.18). Further, they often take a long time to find dishes—longer than would be predicted by a direct flight following an exact fix on a dish's position as indicated by the dance. Such delays and errors are readily compatible with the olfaction hypothesis.

Von Frisch himself recognized that communication by language was prone to error. A most useful early paper by Haldane and Spurway[193] had analysed its accuracy quantitatively. Even so many other accounts of von Frisch's work refer to 'precise information' or 'exact location'. In fact if one watches a waggle dance in progress it is obvious that the angle of the straight run varies considerably between turns. The dancer is often performing on a densely packed comb and is jostled by other bees. The human eye can compare the line over successive runs and compute a reasonable average angle. Perhaps the dance followers can do the same, but we cannot expect the direction indication to be absolutely precise. Similar errors are likely to occur in tempo or sound production which indicate distance. Clearly such errors do not in any way invalidate the language hypothesis or affect the way the dance functions. Experiments with small, discrete food sources are quite artificial. In nature the source indicated by the dancer will usually be some part of a much larger diffuse crop of flowers —a grove of lime trees or a field of clover for example. Recruits will find such crops quite easily even if the dance information is less than exact.

After several years of cross fire in the literature (very clearly and attractively reviewed by Gould[173]) it would be fair to say that conclusive proof of either hypothesis was still lacking. A majority favoured von Frisch but probably because they could not believe that the proven relationship between the form of the dance and the position of the food source could be without function. It should have been relatively simple to clinch the matter but in fact almost all the experiments allowed some uncertainty to remain, if only because the olfaction hypothesis is difficult to disprove totally if bees are visiting food dishes. Some traces of their scent might remain behind. It was necessary to have a bee performing a waggle-dance which in-dicated the direction and distance of a food dish which it had never visited. A 'model' of a bee made to perform a dance under human control would be the ideal solution. Unfortunately, although there have been several attempts, nobody has ever got bees to respond adequately to a model and follow its dances. However an extremely elegant solution to the prob-lem has recently been devised by Gould.[173] He made use of the fact dis-covered some years ago that if a small point source of light is provided at the side of the vertical comb, bees will often treat this as if it is the sun. This means that when dancing a bee treats this point source just as the bees dancing at the nest entrance treat the sun (Fig. 3.17a) and angle their waggle run 'directly' to the food source. Hence 40° to the right of the light source means 40° to the right of the sun, even though the dance is on the vertical comb. Now when the three ocelli or simple eyes on the dorsal surface of the head are blacked out, a bee's threshold to light is greatly raised. Such bees require much stronger light levels if they are to forage, they begin much later in the morning and cease flying earlier in the evening. They also re-quire a much stronger light on the comb to redirect their dances.

Using this threshold shift, Gould was able to set up a crucial test of the language hypothesis. He had ocelli-blackened bees perform dances on a comb with a dim light source. The dancer did not respond to this light and orientated her waggle run as normally with respect to gravity. The other bees with normal ocelli which followed her dances *would* treat the light as if it were the sun. Consequently they should be misled on direction and appear at dishes in a fan array displaced by the corresponding angle.

In Gould's experiment the ocelli-blackened dancers were kept foraging at the same dish (whose smell should have built up as time went on) and recruits were anaesthetized as they arrived at dishes. Every 30 minutes the light on the comb was moved so the angle apparently being indicated also moved. The results were clear cut; the new recruits *were* shifted by an equivalent angle across the array of dishes. This result is completely counter to the olfaction hypothesis and vindicates von Frisch's original ideas.

Many questions on the exact details of the honey-bee's remarkable communication system remain to be answered. At least we can now formu-late them more clearly. Certainly olfaction is an important stimulus, but we do know that information is passed by the dance being followed through by recruits. We shall discuss some aspects of the evolution of the dance in Chapter 6, p. 224.

4

Motivation

It is a common observation that the same stimulus given to the same animal at different times does not always evoke the same response. Something inside the animal must have changed and we invoke an 'intervening variable'. This is something which comes between two things we can measure—in this case the stimulus we give and the response we get out— and affects the relationship between them. We cannot observe such variables directly. Physiological studies of the brain are certainly providing information as we shall see, but the picture is a very complex one and it remains true that in some cases we know next to nothing about the real nature of such variables. Some groups of behaviour workers refuse to use them and concentrate entirely on directly observable aspects of behaviour. However, most people who have worked with animals under fairly natural conditions recognize the necessity to invoke intervening variables in behaviour. At this stage of our knowledge it is profitable to define them carefully and investigate their properties. Already in this book we have mentioned two factors with different characteristics which alter the relationship between a stimulus and the response it evokes. These were 'fatigue' (p. 9) and 'maturation' (p. 41). To these we may add two others: 'learning', which will be dealt with in Chapter 7 and 'motivation' which concerns us here.

Changes in 'motivation' are deduced when we can eliminate the other factors just listed, but still observe that an animal spontaneously changes its behaviour or shows a changed threshold to particular types of stimuli, particularly if such changes are reversible and behaviour fluctuates in a regular way. Sometimes a very slight stimulus may be adequate to evoke a powerful response, at other times far stronger stimuli are ineffective. As mentioned in Chapter 1, this is one characteristic which separates reflexes from more complex behaviour, because the former tend to have a consistently low threshold. With complex behaviour it is common to find that it is not just one stimulus which is effective or ineffective, but a whole range of stimuli which are related functionally to one another. Thus an animal's thresholds of response to all stimuli connected with food and feeding behaviour will rise and fall together; so will those connected with sexual stimuli, and so on. Because of this, many workers and particularly ethologists have regarded motivational changes as highly specific, tending to make the animal respond in a particular way. The result of such a tendency is that an animal's behaviour becomes organized so that it achieves a particular goal. A dog kept for many hours without food moves about restlessly; it is highly responsive to the smell or sight of food or any other stimulus, such as the sight of a food bowl or the sound of a knife being sharpened, which it has learnt to associate with food. It is not to be diverted by a bowl of water or a ball to play with; its restlessness continues until it has eaten.

However it may be misleading to think entirely in terms of specific motivational states. As we mentioned in the last chapter, any stimulus may produce a non-specific arousal via the reticular formation of the brain. This will render the animal more responsive to a wide range of stimuli and we might describe this change as a rise in 'general motivation' or, as it is called by many psychologists, 'general drive'. Although it might seem a straightforward question, it is in fact very difficult to collect really conclusive evidence as to whether motivation is general or specific; there is a good discussion of the problem in Chapter 9 of Hinde,[217] (see also Grossman[183]). In what follows, we shall assume that there is a considerable degree of specificity, although it is certainly wrong to think of motivation as being rigidly compartmentalized. Different motivational systems interact with one another, and in the next chapter we must consider how such interactions affect behaviour.

A specific motivation is often called a 'drive', thus the food-deprived dog in the example given above might be described as having a high feeding drive. This term needs to be used with care because we cannot measure drive directly. Usually we measure only the animal's response to various types of stimuli. If we find that the dog is highly responsive to food stimuli it might be preferable to say that it shows a high feeding tendency, or a low feeding threshold because this is what we actually observe.

In ourselves, specific motivational states are often associated with strong subjective feelings or emotions but we cannot tell whether animals feel emotions in the same way. All we can say with confidence is that there are various physiological changes in the body correlated with emotion and that animals show these. The mouth becomes dry, sweating starts, heartbeat accelerates and so on. These changes mostly follow the secretion of adrenalin into the bloodstream and are discussed more fully in Chapter 5. Although the emotions of anger, fear and lust feel subjectively very different, they share many features of this physiological arousal which prepares the body for violent action of any type.

Since we cannot measure emotion in animals, we usually have to deduce how an animal is motivated by observing how it behaves. Here we inevitably introduce a subjective element because we have to identify motivation with reference to a function that seems reasonable to us. If one animal attacks another we ascribe this behaviour to aggressive motivation; if it eats we ascribe this to feeding motivation and so on. Clear-cut cases such as this offer little difficulty if treated with common sense, but it must be admitted that not all behaviour is so easily interpreted.

SOME CHARACTERISTICS OF GOAL-ORIENTATED BEHAVIOUR

We have already introduced the idea that a specific motivational state leads an animal to organize its behaviour towards a specific goal. As when discussing 'intent to communicate' in the last chapter, we must recognize that there are problems in using the concept of 'a goal' with animals. It might be taken to imply a conscious purpose in their behaviour which some are reluctant to consider. We might try to use goal objectively by defining it as that situation which brings the behaviour in question to an end. Having fed, the dog ceases to search for food or eat; having completed building its nest tunnel the digger wasp, *Philanthus* ceases digging and so on. This is sometimes satisfactory, but we often find that other situations apart from the achievement of the normal goal bring activities to an end. The dog ceases feeding because he sees another dog approaching; the digger wasp stops building as darkness falls.

In some cases we may wish to use a definition of goal which involves some kind of mental image of the desired situation. The American schools of experimental psychology in the 1930's faced this question when they studied their rats learning to run mazes. A rat can learn to go to one part of a maze to get water and another part for food. Put into the maze when deprived of water it moves off, making the appropriate turns to bring it to the water. One school of thought interpreted such behaviour as a chain of responses to the stimuli provided by the successive turns of the maze. Others thought in terms of a 'cognitive map', i.e. some kind of concept of the goal which guided the rat's movements towards the water.

We may be more ready to accept purposive behaviour in mammals than in digger wasps. Certainly it is possible to programme a computer to behave 'as if' it was purposive, perhaps the wasp's responses are no more complex. We are unlikely ever to be able to prove the existence of a conscious goal but the word can still be useful. Hinde and Stevenson[223] and Griffin[180] provide a further discussion of these and other questions relating to the possibility of consciousness in animals.

When studying different motivational states we can quite often observe three stages in an animal's behaviour.

1. A phase of searching for the goal.
2. Behaviour orientated around the goal once it is found.
3. A phase of quiescence following the achievement of the goal.

1. The searching phase is usually called the phase of '*appetitive behaviour*'. This is best described with relation to feeding from which it probably took its name. A hungry animal can rarely just get up and eat; it has to seek out food and the behaviour patterns it employs may be many and diverse. This is particularly conspicuous with predatory animals which actively hunt their prey. Their hunting behaviour is variable and greatly modified by their previous experience.

It is often impossible to identify the nature of appetitive behaviour unless we observe the actual goal. Animals searching for food, water or a mate may behave very similarly but in each case the stimuli required to bring the search to an end are highly specific. As mentioned above we shall have to be cautious in our interpretation at times for situations irrelevant to the goal may cause the animal to change its behaviour.

2. The second phase is more clearly identified because once the appropriate goal stimuli are located, the animal's behaviour changes. The variable searching patterns now give way to a series of responses directed at the goal which are often stereotyped fixed action patterns. They are called *consummatory acts*; eating is the consummatory act of feeding behaviour, drinking relates to thirst, copulation to sexual behaviour, and so on.

3. Consummatory acts are normally followed by a period of quiescence when the animal is no longer responsive to stimuli from the goal and shows no further appetitive behaviour. This quiescence relates, of course, only to one type of behaviour; the animal may be actively pursuing some other goal. In some cases responsiveness slowly builds up again so that there is a fairly direct relationship between threshold to goal stimuli and time since the last performance of the consummatory act.

Whilst appetitive behaviour and consummatory act are still useful as descriptive terms, there is no question of their being rigid categories which can be applied universally.

For example, it is not easy to classify the nest-building behaviour of, say, a blackbird in this way. The blackbird may begin by searching for large twigs to form a foundation. When this is complete the sides are made from finer material; then mud is collected and shaped to form the cup and finally this is lined with fine grass and hair. We might classify this behaviour by postulating that, whilst the completed nest is the ultimate goal stimulus, there are a series of 'sub-goals'—nest foundation, sides, cup, etc.—along the way, each with its own appetitive behaviour and consummatory act.

Even in feeding behaviour, which in some cases fits the scheme very well, the pattern varies. A horse grazing in a field literally has its food at its feet. The appetitive behaviour phase is short or non-existent and the consummatory act of feeding may continue without a break for an hour or so until the animal is full. This situation is totally different from the feeding behaviour of a titmouse picking minute insects off leaves. Here each brief consummatory act is not followed by quiescence, but by a further phase of appetitive searching. Eventually after some hundreds of such sequences, appetitive behaviour ceases.

The existence of an identifiable goal and appetitive behaviour directed towards locating it has sometimes been used as evidence for postulating specific motivational states or drives. Some drives are labelled 'biogenic', or related to an urgent biological need. Everybody would agree that feeding and drinking come into this category and, as we have seen, they best fit the appetitive behaviour/consummatory act/quiescence scheme outlined above. Many would call sex a biogenic drive too, and it also fits the scheme fairly well, at least in vertebrates, although it is remarkable how much sexual arousal can be affected by the external stimulus situation in some cases— far more than is possible with feeding or drinking. For example Hale[194] has shown that even apparently trivial changes to the female test animal were sufficient to arouse a sexually exhausted bull to mount once more. Ethologists commonly postulate attack and escape drives, and in the next chapter we shall discuss these in relation to territorial behaviour. If we exclude predatory animals which attack their prey as part of their feeding behaviour, animals do not normally move around actively searching for things to attack or to escape from and both types of behaviour are far more dependent on external stimuli than are feeding and drinking. The tendency to attack does fluctuate—testosterone is one of the factors that affects this— but in addition it is rapidly aroused by certain types of stimuli and does not show a simple relationship between threshold and length of time since the last attack. Escape certainly does not fit into the standard scheme; it is almost exclusively under the control of external stimuli. Apart from this control, there is no evidence that the tendency to escape fluctuates save as a result of learning. Animals quickly cease to respond to stimuli which, though alarming at first, are not associated with punishment (e.g. scarecrows) but novel stimuli may once more evoke strong escape. It would be highly inadaptive if this were not the case.

At one time or another, sleep, parental behaviour and exploratory behaviour have all been ascribed to specific drives. Sleep, which we now know to be associated with a particular pattern of neural activity, not inactivity, superficially fits the appetitive behaviour/consummatory act/quiescence scheme rather well. However, little critical analysis in these terms has been made for sleep or any of the other suggested drives.

Experimental psychologists often use the term 'drive' in a rather different sense when they refer to 'secondary' or 'learnt drives'. (A 'primary drive' in their terminology is equivalent to those we have just been discussing.) Miller[346] says, 'Thus, if a child that has not previously feared dogs learns to fear them after having been bitten, it shows that fear is learnable.' From the biological point of view it is more appropriate to concentrate on the fact that it is the *stimulus* which is learnt, and thereby becomes associated with an escape system which already exists.

WHY POSTULATE DRIVE?

So far we have been using the drive concept in a descriptive way without comment. We have seen that what might be called the 'classical picture' of a drive in operation is rarer than might be supposed from the frequency with which the term crops up in ethological writings. In particular, it is difficult to find good examples from the behaviour of invertebrates. We must now consider in more detail exactly when and why we need such a concept.

The trouble with a term like 'drive' is that it can take on a substance which we cannot really justify. It may be stretched to account for a number of different behavioural phenomena which are thus given a kind of false unity. Hinde[213] discusses this problem and lists no fewer than six phenomena which drive has been used to explain. We can consider three of these in more detail.

Fluctuations in responsiveness

We have already dealt with the general nature of this phenomenon. In some cases it is easy to relate the fluctuating responsiveness to regular changes in an animal's physiological state. One only needs to know how long it is since an animal fed or drank to get a good idea of how responsive it will be to food or water respectively. Feeding and drinking provide the clearest examples of behaviour forming part of a homeostatic system. Some part of the system detects shortage of water, and appetitive behaviour and responsiveness to water increase until the animal drinks. Thereafter responsiveness falls until water begins to run short once more.

Sexual responsiveness also shows great fluctuations, especially in animals with a well-marked breeding season which show no sexual behaviour at all for much of the year. It is often possible to demonstrate a good correspondence between the level of sex hormones in its bloodstream and an animal's sexual responsiveness, see p. 158. Aggressive behaviour also increases under the influence of male sex hormones and males become more responsive to stimuli which provoke attack.

Sometimes the threshold for response becomes so lowered that consummatory acts can be evoked by very minimal stimuli. Indeed Lorenz used the term 'vacuum activities' for those cases where the behaviour is produced in the absence of any external stimulus. Tinbergen[460] describes a number of such cases; for example, a starling is seen to go through the motions of catching and eating a fly when none is there. When deprived of nest material, a male Bengalese finch will go into its nest box and perform all the movements of carrying and placing material although it has nothing in its beak. Clearly in a strict sense it is impossible ever to designate an activity as 'vacuum'; some external stimulus, no matter how minute, may be there. Nevertheless the term does draw attention to the extreme lowering of threshold which can occur.

To ascribe all such examples of changing responsiveness to a change in drive is no kind of explanation. The drawback, as mentioned earlier, is that by doing so we may give the impression that such changes are all produced in a similar fashion. We have little idea of the mechanism in most cases, but enough to indicate that it is not always the same. Sex hormones, for example, have often been described as increasing the sex drive and thereby changing responsiveness to sexual stimuli. We know that these hormones can act centrally on brain mechanisms (see p. 155) but they also affect responsiveness by peripheral action. Beach and Levinson[43] found that one action of testosterone in the male rat was to cause thinning of the epithelium of the glans penis. This increases the sensitivity of its tactile sense organs and may increase the rat's sexual responsiveness accordingly.

In most cases we assume that the incoming sensory information which results from a standard stimulus remains constant, but the brain's responsiveness alters. Only rarely do we have direct proof of this, but the elegant experiments of Dethier and his collaborators have shown that this does in fact hold during feeding in the blow-fly, *Phormia* (see Dethier[109]). When the chemoreceptors on its feet encounter sugar they can trigger off extension of the fly's proboscis onto the food where further sense organs determine whether it is sucked in. Direct recording from the sensory nerves originating in these sense organs shows that they always perform in the same way. They produce a burst of impulses when they first come into contact with sugar, but this decelerates back to the normal background level if the hair remains in the sugar—in physiological terms the sense organ adapts. The time it takes to adapt and the initial rate of firing depend on the sugar concentration. Low concentrations cause a very brief initial burst followed by rapid adaptation; high concentrations produce a prolonged, rapid burst and slow adaptation. These characteristics are the same whether the fly is fully fed or starving, but a hungry fly will respond to a more dilute sugar solution than one which has fed recently and it will go on feeding for longer. The fly's 'acceptance threshold', as Dethier calls it, fluctuates with food deprivation although its sensory threshold does not.

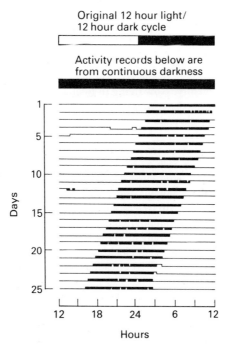

Original 12 hour light/
12 hour dark cycle

Activity records below are
from continuous darkness

Days

Hours

Fig. 4.1 The spontaneous wheel running activity of a flying squirrel in total darkness. Each line is the trace from an event recorder over 24 hours. Active periods appear as broad dark lines as the pen of the event recorder moves frequently up and down on the paper. Before these records were taken the squirrel had been kept on a regular cycle of 12 hrs light/12 hrs dark—indicated at the top. In total darkness it retains its rhythm and is active only in the period which had previously been dark. In the absence of external cues however its natural circadian rhythm which is slightly less than 24 hrs asserts itself and the period of activity begins a little earlier each night. (From DeCoursey, P., 1960, *Cold Spr. Harb. Symp. Quant. Biol.*, **25**, 49–55.)

The spontaneity of behaviour

Animals often start to behave in a particular way quite spontaneously. A dog wakes up on its own, stretches itself and then moves off in search of food. Of course the term 'spontaneous' can never be used very precisely. The dog may have been woken by hunger pangs transmitted by nerves from its stomach. Nevertheless some behaviour does show a very striking degree of spontaneity and it is for this reason more than any other, that the concept of an internal drive which in some way 'energizes' the relevant behaviour mechanisms, seems indispensable to many behaviour workers. Sometimes the spontaneous changes of behaviour have a particular rhythmic form.

Figure 4.1 records the activity of a flying squirrel living in a rotating cage. Each row represents 24 hours and the dark bands show when it was actively moving around. The squirrel is a nocturnal animal and it was normally active from just after 'dusk' until the lights came on again at 'dawn'. At the beginning of the records shown in Fig. 4.1 the lights were switched off permanently. Now living in total darkness the squirrel nevertheless maintains a fixed period of activity. This lasts almost exactly the same time as the original dark periods and occurs at extremely regular intervals. Interestingly enough these intervals are not 24 hours but just over 23 hours in this case, so that in real time the squirrel begins its activity a little earlier in each 24 hour period. A rhythm of activity of this type is called *circadian* from circa diem, about a day. Circadian rhythms are widespread amongst animals and plants and affect not just behaviour but all aspects of metabolism (see Saunders[416]). They certainly provide an extreme example of the spontaneity of behavioural change. In the absence of any change in the environment the squirrel patterns its behaviour in response to some kind of internal 'clock'.

Cycles can be annual as well as circadian, such as the familiar onset of the breeding season in many birds and mammals. In spring, a male chaffinch alone in an aviary, will begin to sing, and will continue singing for some minutes at a regular rate of some three songs per minute. It is very difficult to relate this song rhythm to any changes outside the bird's own nervous system. Nice[365] in her study of the American song sparrow observed that the tendency for males to sing gradually increased during early spring. Cold weather inhibits singing, and, over a number of years, she was able to plot how low the temperature had to be before singing was stopped on different dates. Fig. 4.2 shows a graph of her results; they strongly suggest that a spontaneous 'urge' to sing is increasing through January and February and accordingly cold has to be increasingly severe to suppress it.

At one time experimental psychologists viewed the concept of 'spon-

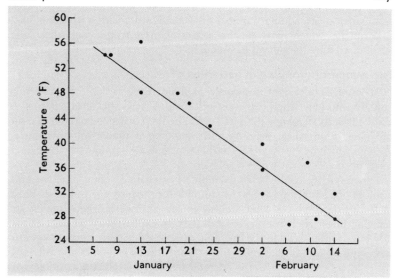

Fig. 4.2 The level to which temperature must fall in order to suppress singing by male song sparrows. As the season progresses the fall must be to lower and lower temperatures. (From Nice,[365] 1937, 1943, *Studies in the Life History of the Song Sparrow*. By permission of Dover Publications Inc., New York.)

taneous behaviour' with scepticism. They favoured a more rigid stimulus-in/response-out type of organization, but this has proved quite inadequate to account for complex behaviour. In fact, neurophysiology has overtaken behaviour theory and shown that neurons themselves are spontaneously active and go through regular cycles of discharge in the absence of any input; see Davis[121] for a modern review. In Chapter 1, Roeder's[396] work on mantids was used to exemplify how the neurons of the thoracic and abdominal ganglia of this insect are spontaneously active unless held in check by inhibitory impulses from the head ganglia. In the latter's absence they initiate all the movements associated with copulatory behaviour, whether or not any external stimuli are provided.

There are no physiological objections to the suggestion that specific appetitive behaviour arises spontaneously as a result of increased activity in specific parts of the central nervous system. Such activity may indeed be independent of external stimuli in any direct sense, but may result from internal changes such as those which follow many hours of food or water deprivation.

If we call such spontaneous activity 'drive', we are using the term according to its original derivation, i.e. that which 'drives' or 'urges' the animal to do something. Drive in this sense may also be used to describe how, for example, a hungry animal will actually work harder to obtain food. Experiments which illustrate this aspect of drive will be described later when discussing measuring techniques.

The temporal grouping of activities

An animal's behaviour is normally well organized so that each pattern is brought into play when it is most effective. A male stickleback begins to build a nest by digging a pit in the sand; he then brings material to the pit and glues it together, pressing and moulding it into a solid structure, through which he eventually forces his way to make a tunnel which he then keeps open.

This sequence could be an example of a 'chain reaction', i.e. the end result of one pattern provides the stimulus for the next one to begin. Thus sand-digging produces a pit and an empty pit may be the stimulus for bringing material. However, we still have to explain why the stickleback only builds nests early in the breeding season and is quite unresponsive to nest-building stimuli at all other times. The thresholds for all these different but related behaviour patterns rise and fall together although we are not dealing with a rigidly fixed sequence which is predetermined.

Such examples have been explained by proposing that all the patterns share a common drive. This is taken to mean that they are all activated simultaneously and other factors, partly internal organization and partly external stimuli, determine which particular pattern appears at any given time. Fig. 4.3 is reproduced from a detailed ethological analysis of this type by Baerends[21] and his collaborators. They have analysed the incubation, preening and escape systems of the herring gull and the way in which they interact with each other. The diagram is complex and its details need not concern us here. It represents a model (see p. 122) of how the gull's behaviour is organized and note its basically hierarchical structure. Each control system (which might be considered to generate a drive) controls the activity of subordinate systems, these latter in turn control other systems which are effectively fixed action patterns in this case. Thus top control system N (nesting) controls incubating, settling and building. Settling controls in its turn a set of behaviour patterns, shifting, ruffling, quivering and so on, which came into operation when the gull settles down on its clutch. Such a model of behaviour accounts for the co-ordinated rise and fall of threshold in sets of related activities. As the activity of N increases, so will the activity of all its subordinate systems. The model also allows that each activity might have some external stimuli which affect it more than others are affected. The different preening patterns, for example, almost certainly have some stimuli specific to each, but also some factors (like rain) which are common to all.

Sometimes ethologists have ranged such a series of linked behaviour

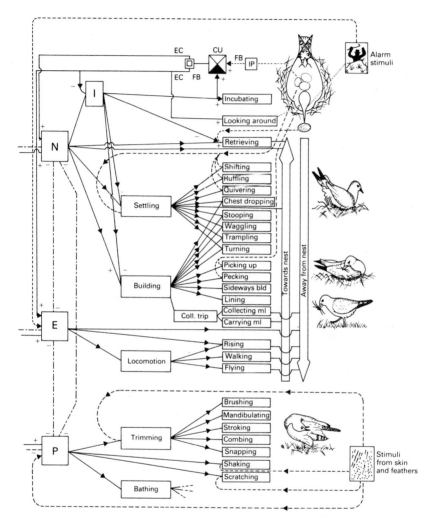

Fig. 4.3 Baerend's model of how the behaviour of a herring gull at the nest is controlled. There are three main control centres, nesting (**N**), escape (**E**), and (**P**) preening. Each receives some specific inputs and has outputs to a series of sub-centres controlling clusters of motor patterns denoted in the column of rectangles to the right and whose names are largely descriptive. (Coll.=collecting, ml= material). Further details in text. (From Baerends, [21] 1976, *Animal Behaviour*, **24**, 726.)

patterns on an 'intensity scale'. This implies that the strength of the drive
plays a large part in determining which pattern of the series is performed,
because the patterns have successively higher thresholds. If drive is
low only the pattern with the lowest threshold is activated; if drive
increases this may be replaced by the next highest, and so on. Soon
after settling on his territory, a male stickleback begins to dig the pit
which will receive his nest. At first we may observe no more than his
swimming low over an area of sandy bottom, later he dips his head repeatedly towards one part of the sand. Later still he touches the bottom
and sucks sand into his mouth only to drop it immediately. Finally he
sucks up sand, turns and carries it away before returning to collect more.
The whole sequence, from the first *intention movements* of sand digging—
as they are described—to the commencement of the pit in earnest,
may take a few hours. Intensity changes of this type are commonly observed early in an animal's reproductive cycle at a time when we might
expect drive to be increasing. To suggest that motivational changes underly
such behaviour changes is not to deny the part which external stimuli may
play in the changeover from one pattern to the next. Stimuli resulting from
the performance of an early pattern in the series may be one factor which
causes the drive to increase and so activate the next pattern.

One example where the concept of an intensity scale seems to be useful
comes from the work of Gardner[162] on the feeding behaviour of jumping
spiders. These spiders do not spin webs, but stalk their prey by sight.
When a hungry spider sees a fly it turns towards it ('orientation'); it then
pursues the prey, running rapidly at first but slowing down to a cautious
stalk as it gets closer ('pursuit'). When the spider is within jumping range
it crouches, and after a brief pause jumps, so as to land on top of the fly.

Orientation, pursuit, crouching and jumping form the normal sequence
of hunting. Spiders will catch and eat many flies in succession, and Gardner
showed that as a hungry spider becomes full the sequence begins to be
truncated. More and more sequences get no further than orientation,
although if the spider begins pursuit he usually goes on to crouch and jump.
Eventually the spider no longer even orientates to flies. There appear to be
three stages on the intensity scale.

1. At the lowest level of hunger no response can be elicited.
2. At the next level the spider orientates towards prey.
3. At the highest level it goes on to pursue, crouch and jump.

These last three patterns are apparently closely linked and Gardner can
find no evidence that crouching or jumping requires a higher drive than
pursuit.

This is a good example of an adaptive sequence of behaviour patterns
where an intensity scale can be used with some confidence and matched

against another, quite independent, assessment of drive; i.e. how long it is since the spider last fed. Not all sequences can be treated in this way. Hinde[212] has studied the nest-building behaviour of canaries. Building begins quite soon after a male and female are put together and gradually increases to a peak 3 or 4 days before the eggs are laid; female canaries do nearly all the building. The building behaviour can be classified into a number of categories such as gathering material, carrying it to the nest, placing it on the nest, and sitting, weaving and shaping the material into the nest. Each of these categories, in turn, consists of a series of more or less stereotyped patterns; sitting, for example, involves four or five different types of pattern.

Hinde used a standard observation period of 30 minutes and scored the number and duration of all these activities. There are some parallels with the spider hunting behaviour. Thus the more time spent in building, the more sequences, gathering, carrying and weaving, are completed. The transition from one activity to the next is only partly determined by the external stimuli from the nest, so there must be some internal factors which promote the change. Further, it is known that all the nest-building activities do share some common causal factors; oestrogen causes them all to increase in frequency, for example.

However, although it might seem reasonable to use total time spent building as a measure of a 'nest-building drive', this does not correlate with the proportion of time spent in the various constituent patterns. In other words, it is not possible to arrange the patterns on an intensity scale, nor is it possible to define the drive except as a tendency to nest-build.

To recapitulate this section, the concept of drive has been invoked to explain among other things, fluctuations in responsiveness to particular stimuli, the spontaneity of behaviour and the way in which a series of patterns appear grouped together. It is quite certain that no single entity with fixed characteristics can be involved. The concept must stand or fall on its usefulness; sometimes, as with feeding and drinking behaviour, 'drive' is a useful term whose physiological basis can be specified to some extent. In other cases, as with canary nest-building, it has no use save perhaps as a shorthand description of what we observe.

MODELS OF MOTIVATION

Having outlined some of the characteristics of goal-orientated behaviour with which the concept of drive has been associated, we should now turn to consider briefly whether we can offer any explanation for them in behavioural terms.

Our discussion will centre around two of the many 'models' of how behaviour is controlled—we have already discussed some aspects of another, hierarchical type model (p. 118). The purpose of any behaviour model is to devise a system of hypothetical components to which particular properties are given and which are so connected together that their 'behaviour' reproduces that which we observe. This is not such a useless operation as it may seem, because good models help to organize our thinking and suggest experiments which can test their own adequacy. If we find that one model consistently explains an animal's behaviour under a wide range of conditions, it may tell us a great deal about the *principles* upon which the nervous system is operating. It can tell us little or nothing about the *means* of operation. The nervous system uses large numbers of interconnected neurons for its operation and the interpretation of how the principles are put into practice remains a neurophysiological problem. A model can use components whose properties are borrowed from telephone exchanges, hydraulic systems or computers, so long as they can reproduce the result we are trying to interpret. Hinde[215] and the opening chapters of Deutsch's[134] book include excellent discussions of the principles and uses of behaviour models.

Sometimes ethologists have interpreted their observations with reference to a particular behaviour model which, in its final form (illustrated in Fig. 4.4) we owe to Lorenz.[305] He uses components which are borrowed from a hydraulic system, and this model is often called a 'psychohydraulic' one. Lorenz envisages that when motivation is increasing, as when an animal is deprived of food, there is an accumulation of 'action specific energy', i.e. energy which is earmarked for feeding alone and does not affect other types of behaviour. In the model this is represented by the gradual accumulation of water in a reservoir (R) supplied from a tap (T). Outflow from the reservoir represents the motor activity of behaviour, but this is normally held in check by a valve (V) held shut by a spring (S). There are two ways in which the valve can be opened. Weights on the scale pan (Sp.) can pull it open and these represent various strengths of stimuli. The gradually increasing pressure of water in the reservoir and the weights on the pan both act in the same direction—to open the valve. The higher the water level, the smaller the weight required and eventually the water pressure alone may push open the valve—a vacuum activity. Lorenz represents the different types of motor output by a graded trough (G and Tr.). If the valve is opened slightly a little water trickles through which reaches up only to the first and lowest hole in the trough. This represents the motor activity with the lowest threshold—often some form of appetitive behaviour. As the valve is opened wider, the trough discharges through other holes, which represent activities with higher thresholds and higher up the intensity scale. Once the reservoir is empty, the behaviour

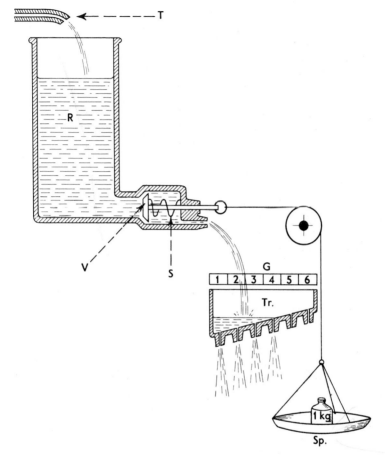

Fig. 4.4 Lorenz's 'psycho-hydraulic' model of behaviour; see explanation in text. (From Lorenz,[305] 1950, *Symp. Soc. Exp. Biol.*, **4**, 256.)

can no longer be elicited no matter how strong the stimulus; Lorenz talks about the 'exhaustion' of a behaviour pattern with this model in mind.

It is obvious that this psycho-hydraulic model accounts satisfactorily for cyclical changes in responsiveness. Note how quiescence following the consummatory acts is dependent on the *performance* of these acts, because this is the only way in which the reservoir can be emptied.

Now let us look at another model to explain the same set of facts. This one (Fig. 4.5) is put forward by Deutsch.[134] It forms part of a more comprehensive behaviour model which covers learning; here we illustrate only

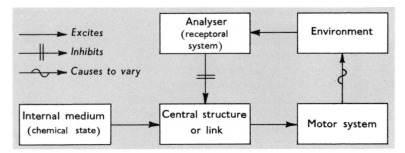

Fig. 4.5 Deutsch's model of behaviour ; see explanation in text. (From Deutsch,[134] 1960, *The Structural Basis of Behavior.* © by the University of Chicago Press, Chicago.)

the section relevant to motivation. Deutsch uses a series of hypothetical components called 'analyser', 'link', etc., whose operation he defines. We may not immediately 'see' how the model operates, as we can that of Lorenz, because it does not use 'working parts', but its operation is very simple. Some deficit or imbalance in the internal medium, shortage of blood glucose or water for example, is detected and excites a central structure or link. The persistence and strength of this excitation depends on the size of the deficit. The link, in turn, activates the motor system which produces behaviour. (This simplified portion of the model does not include any representation of external stimuli which release the behaviour or any way of 'grading' the behaviour in intensity, but these are easily developed.) As a result of its behaviour some aspect of the animal's environment changes. It has eaten or drunk and its stomach is now full of food or water, for example. The change in the environment is signalled to an analyser component which switches off the activity of the link so that it is no longer responsive to excitation from the internal medium. This inhibition slowly decays until the link is once more sensitive to excitation. We have already discussed systems which operate in this fashion in Chapter 1, p. 16 for Deutsch's model is a typical form of negative feed-back control.

To compare the merits of these two models we must see how well each accounts for the experimental data of behaviour. The psycho-hydraulic model is the more detailed of the two and does fit the facts derived from simple observation extremely well. It has been criticized because it uses hydraulic components which some critics say are too far removed from the actual operation of the nervous system. This type of criticism is quite invalid. There are no reservoirs filling up with water in the nervous system, nor are there black boxes labelled 'Link'; it is only the operating principles that count and the crucial test is which model fits the facts best.

Janowitz and Grossman[254] operated on dogs so that the oesophagus opened by a fistula to the outside of the throat. This meant that when such a dog ate the food fell to the outside and did not enter the stomach ('sham-eating' as it is called). The stomach could be filled with food from the outside without the dog having eaten anything. By this means it is possible to dissociate the behavioural act of eating from the normal results of having eaten. Suppose a dog with an oesophageal fistula of this type is kept for hours without food, but then has food placed directly into its stomach so as to fill it. Will the dog behave as if it is hungry or not? The psycho-hydraulic model would predict that it *would* still eat food because the reservoir, having filled with water (action-specific feeding energy in this case), has not had its valve opened and the water released through the trough (feeding behaviour); the reservoir remains full. Deutsch's model predicts that the dog will not eat. The link has been excited but nothing is 'stored' in this system, and although no motor activity has been performed, as soon as the environment changes (stomach full, in this case) the analyser is activated and this switches off the link so that no further feeding activity occurs. Janowitz and Grossman found that their fistulated dogs do not eat after their stomachs have filled and thus, in this respect, Deutsch's model is better than the psycho-hydraulic one.

It appears that stimuli resulting from a distended stomach are the most important factor in switching off eating behaviour. Receptors in the throat may play a part but a hungry, fistulated dog will continue sham-eating long after it would normally have stopped if none of the food reaches its stomach.

With drinking it appears that the mouth and throat receptors are more important than stomach distension. Bellows[46] found that thirsty dogs sham-drink little more than they would normally drink but of course, since they did not replace their water deficit, they soon began sham-drinking again. These observations taken on their own appear to fit the psycho-hydraulic model fairly well, but dogs which had their stomachs pre-filled with water soon ceased to respond to it although they had not performed any drinking behaviour.

It might be argued that feeding and drinking behaviour, because of their close relationship to an animal's physiological state, are not typical of other types of behaviour where the goal may be a more purely 'behavioural' one. Perhaps in sexual behaviour, for example, it *is* the performance of the consummatory act which reduces the drive. However in one instance, where it has been possible to separate the performance of sexual behaviour from its normal sequel, this has not proved to be the case. In the male stickleback there is a sudden drop in the sexual tendency immediately following the act of fertilization. Tinbergen[460] describes the stickleback's sexual behaviour in more detail; here we need only note that just after a female has laid eggs in his nest the male enters it and fertilizes them. Sevenster-Bol[443] has shown that the performance of fertilization is not necessary for the reduction in sexual tendency; it is also reduced if a male is allowed only to approach his nest and perceive the newly-laid clutch of eggs. This result again would not be predicted by the psycho-hydraulic model, but would by that of Deutsch.

Most recent evidence confirms that the period of quiescence following goal-orientated behaviour is the result of a sensory feed-back from some aspect of the situation which normally signals 'goal achieved'. This makes good behavioural sense. It is maybe some time before the circulating food reserves of a starved animal are restored to normal following digestion of the food it has eaten, but its stomach is filled within a few minutes. Feeding behaviour must be stopped long before the physiological goal is attained.

One reason why Deutsch's behaviour model is more satisfactory than the psycho-hydraulic one is because it incorporates such a feed-back from the environment via the analyser to the link which is thereby switched off. As we shall see in a later section of this chapter, we can now identify parts of the brain whose properties resemble those of the components in Deutsch's model.

Another fundamental objection to the psycho-hydraulic model is its use of stored energy to represent motivation. This is most unlikely to have any counterpart in the functioning of the nervous system. An energy model of this type inevitably represents the initiation and termination of behaviour as two aspects of the same thing—energy starts to flow and energy runs out and is exhausted. In fact as we have just seen, what switches on appetitive behaviour and what switches off the consummatory act and produces quiescence, may be completely different processes. Hinde[215] discusses the drawbacks of energy models of motivation in more detail.

MEASURING MOTIVATION—FEEDING AND DRINKING

Any study of behaviour in which motivation may be a factor requires that we can measure changes in the strength of this variable. We may be interested in how conflicting motivations interact or what effect strength of motivation has upon rate of learning, for example.

The methods we choose to apply for measuring motivation will depend both on the type of motivation and on the type of animal. We may try to measure the persistence of appetitive behaviour, or the frequency with which a consummatory act is performed. It is usually impossible to measure directly the internal state of an animal and we have to be content with measuring a response of some kind. This will be the result of the interaction between a stimulus and some mechanism which controls the performance of the response and whose properties will vary with the internal state. In fact we can keep the stimulus constant, and for the most part this indirect kind of measure is perfectly adequate to record changes in the internal state.

Feeding is certainly the best example to use when discussing various measures of motivation. All animals require food at regular intervals and it is easy to control their motivation by depriving them of food for varying lengths of time. The following list gives a few of the possible measures of the feeding tendency. They are conceived with a mammal such as the rat in mind, but some could be adapted for other animals. All of them could equally well be used to measure thirst.

1. AMOUNT OF FOOD EATEN.   This is basically a measure of the consummatory act. It is usually easier to weigh the amount of food eaten than to count the number of feeding movements, for example. Most animals, when presented with food *ad libitum* will eat their way through it at a fairly constant speed until satiated. One obvious disadvantage of this measure is that it reduces hunger as it proceeds. This might be a nuisance if one wanted to study how a measured level of hunger affected some other aspect of behaviour.

2. HOW BITTER FOOD CAN BE MADE BEFORE IT IS REFUSED. This measure attempts to block the consummatory act of eating and one tests how far the animal will persist in spite of this. Quinine is an intensely bitter substance to ourselves and other mammals appear to find it likewise. A rat is presented with a series of tiny food pellets or drops of condensed milk, each adulterated with quinine. The concentration is gradually raised and at a certain level the rat will test the food with its tongue but reject it as too bitter.

3. STRENGTH OF PULL TOWARDS FOOD. In this and the next measure, we might be said to block the animal's appetitive behaviour. A rat is fitted with a harness which can be engaged against a spring balance. If food is placed in full view at the end of a runway it is possible to measure how hard the rat pulls to reach it. It might also be possible to measure the speed of its run towards food in a similar apparatus.

4. LEVEL OF ELECTRIC SHOCK ACCEPTED. Again food is placed in full view, but to reach it the rat has to cross an electrified grid which it has previously learnt will shock its feet. By varying the level of shock it is easy to measure how much the rat will accept in order to reach the food.

5. RATE OF BAR-PRESSING FOR A FOOD REWARD. This too is a measure of appetitive behaviour, but it uses a behaviour pattern which the experi-

menter has purposely 'built in' to the food-seeking behaviour for the sake of convenience. The 'Skinner box' has already been described briefly (p. 25 and Fig. 2.1). It is a useful apparatus for quantifying a response which an animal has learnt. A rat is put into the box when hungry and is taught that it receives a small pellet of food when it presses a bar which protrudes into the box. (In Chapter 7 we shall discuss this sort of learning more fully.) When it has thoroughly learnt this, the apparatus is so arranged that rewards do not follow every press but come at irregular intervals averaging out at—say—one reward every 30 seconds. In psychological jargon this is called a 'variable interval reinforcement schedule', and it means that the rat never knows whether any particular bar press will give results. Somewhat surprisingly perhaps, they press the bar much more regularly with such a schedule than they do if every press gets its reward. Because the rate is so regular under these circumstances it is suitable as a measure of hunger; the rate can be tested after different lengths of food deprivation.

We have outlined a number of possible ways for measuring hunger— many more could be devised. Superficially they all appear to be measuring the same thing, but we can learn something of the nature of feeding motivation if we compare how these different measures change with length of food deprivation. There is no study which combines all the five measures we have discussed, but Miller[348] describes a number of experiments which show that three of them, amount eaten, quinine accepted and rate of bar pressing, do not all rise together.

Over the range from 0 to 54 hours of food deprivation the amount of quinine rats will accept steadily rises, so does their rate of bar pressing, but their food intake reaches a maximum after only 30 hours and actually declines slightly thereafter. Thus a rat goes on signalling its increasing hunger by accepting food that is more and more bitter although it eats less. Presumably its stomach cannot hold more than a certain amount and the rat stops eating when it feels full. However, certain rats will eat very large amounts of food yet show few other signs of 'hunger'. These are animals in which gross overeating—a condition called 'hyperphagia'—has been produced by making a small lesion in part of the brain which controls normal 'satiation'. (This result is discussed more fully later in this chapter.) Hyperphagic rats eat enormous quantities of food and become very fat but they are not 'hungry' in the same sense as food-deprived normal rats. They are greatly handicapped by their obesity in any physical tests, such as pulling towards food, but they can easily press the bar of a Skinner box. They will do this, but at a much lower rate than a normal rat deprived of food for the same time. Further, hyperphagic rats are very 'finicky' eaters. The least interference puts them off and, for example, they will not accept as much electric shock to reach food as will normal rats, nor will they accept as much quinine. Now it might be argued that rats with brain damage are

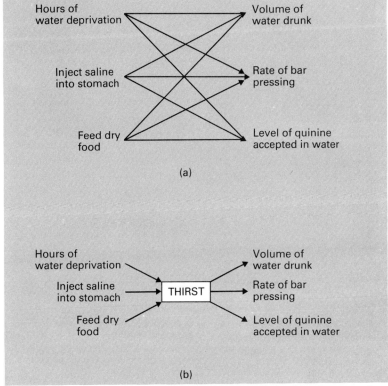

Fig. 4.6 Two possible ways of representing the relationships between three independent variables affecting drinking behaviour and three dependent variables which measure drinking. If the relationships depicted in (a) hold and each independent variable does have effects on each dependent one, then one might propose the model shown in (b) which postulates an intervening variable 'thirst'.

so abnormal that nothing can be deduced from their apparent anomalies. However it seems incontestable that what we commonly lump under the term 'hunger' may be a conglomerate of factors and brain damage elevates some whilst depressing others.

There is evidence of the same type for thirst. In Fig. 4.6a we set out a series of independent variables all of which, in commonsense terms, would be expected to affect the tendency to drink. In turn drinking should be measurable in terms of any of the dependent variables shown (there are, of course other dependent and independent variables we could choose). We might expect that each independent variable would correspondingly affect each dependent variable in the way shown. If they do so then we would

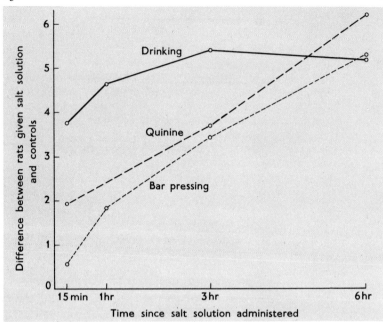

Fig. 4.7 How three different measures of thirst change in the period following placing 5 ml of strong salt solution directly into the stomach of a previously water-satiated rat. The units on the vertical axis are arbitrary; they are simply the difference between control and experimental rats on the various measures of thirst. (From Miller,[347] 1956, *Ann. N.Y. Acad. Sci.*, **65**, 318.)

have reasonable grounds for suggesting that a single intervening variable exists, which we would probably want to label 'thirst' as in Fig. 4.6b. We would then postulate that all the independent variables increase thirst and the motivation to drink, which thus leads to an increase in all the dependent measures of drinking.

In fact, very much in parallel with Miller's experiments on feeding mentioned above, the different measures of drinking do not correspond with each other very well. Figure 4.7 shows the results of an experiment by Choy, quoted by Miller[347]. He had rats with tubes implanted directly into their stomachs so that they could be given liquids without having to drink. Choy measured drinking in rats, previously satiated with water, after 5 ml of concentrated salt solution was put into their stomachs and he recorded the changes in our three measures of thirst over time. If we took bar pressing alone as a measure we would conclude that thirst is not increased for 15 minutes after giving salt. (There is no significant difference between experimental animals given salt and controls who were not after this time.) Thereafter, we would conclude, thirst goes on increasing for at least 6 hours. But this is only part of the truth; after 15 minutes the rats clearly are very thirsty in one sense because they begin to drink lots of water even though they will not press a bar any faster to get it. However, the amount they drink levels off 3 hours after the salt is given even though bar pressing continues to rise and so does their tolerance of quinine in the water. We need a combination of measures to get a reasonable picture of the effect of salt on thirst. As yet we do not know what lies behind the lack of correspondence between different measures, but it certainly means that to think of thirst or drinking motivation as a simple entity is not justified. This kind of data is an essential preliminary if we are to link physiology and behaviour and get at the real nature of motivation.

MEASURING OTHER TYPES OF MOTIVATION

Ethologists are often interested in measuring the strength of sexual, parental or aggressive motivation. There is plenty of evidence that, for example, the tendencies to attack and to escape may be aroused simultaneously and conflict with each other. Some measure of the strength of the tendencies is required even if we cannot equate such tendencies with fluctuations in the internal state of the body as we can for feeding and drinking. Indeed, the need for good behavioural measures is more urgent because we usually have no other suitable yardstick, such as time of deprivation, to apply. In the next section of this chapter we discuss some of these questions in relation to aggression in particular. We can probably learn little about the nature of the control system for aggression if we can express its activity only as a tendency to be aggressive. However, we can try to measure the tendency in a consistent manner and choose criteria which reflect the whole range of aggressive patterns the animal can show. This will certainly help in the study of conflict behaviour when, as discussed in Chapter 5, the aggressive tendency may be interacting with the escape tendency.

Sometimes it is possible to use measures very similar to those described for feeding and drinking. Female rats will cross an electrified grid to reach a male. The strength of shock needed to stop them shows cyclical changes corresponding to their oestrus cycle, it is highest during oestrus or 'heat'. This, then, is a measure of the 'strength' of their sexual appetitive behaviour. Similarly a mother rat's parental tendency might be measured by making her pull against a spring balance to reach her litter.

Specialized courtship behaviour is a common preliminary to actual mating in some animals. In one sense courtship might be regarded as sexual appetitive behaviour and because it may be repeated many times before mating, it has been used as a measure of the sexual tendency. Mating itself—the consummatory act—is often less useful because a single mating may be followed by a long period of quiescence. This precludes using 'frequency of mating' as a measure unless we can make observations over a very long period. A group of workers, mostly in the Netherlands, e.g. van Iersel,[246] Sevenster[432] and Sevenster-Bol[433] have made an extensive study of stickleback reproductive behaviour. They have developed a series of measures which have proved useful for analysing the interactions between sexual, aggressive and parental tendencies in the male. To measure sex and aggression they present to a male on his territory, a test fish (confined in a glass tube) which is either a rival male in full nuptial colours or a receptive female. They count the number of 'zig-zag' courtship movements (see Tinbergen[460] for a full description) or bites directed towards the test fish within a period of 1 minute.

With sticklebacks, for the reasons mentioned above, it is impractical to use the consummatory act of sexual tendency—fertilization—as a measure. Instead it is found that the frequency of zig-zags performed during a 1 minute test with a confined female is positively correlated with a male's tendency to perform later parts of the sexual behaviour sequence. These normally end with fertilization if the female is receptive. The frequency of bites used in tests of the aggressive tendency may be considered to measure the consummatory act directly, as far as this term has any meaning for aggressive behaviour.

Simple frequency measures of this type take no account of the 'intensity' of performance—equivalent perhaps to the effort the rat expends pulling against the spring balance to reach food. Sevenster[432] admits that both zig-zags and bites do vary in their intensity but it is very difficult to measure this quality. The lack of an 'intensity' measure is unlikely to upset the validity of one using frequency alone, because there is probably a positive correlation between intensity and frequency. It is unlikely that some highly aggressive male stickleback would content itself with a few, but very powerful bites to a rival!

Sometimes frequency is not a suitable measure and, for example, the parental tendency of the stickleback is usually measured as a duration of performance. This is dictated by the nature of the behaviour. During 'fanning' the male positions himself in front of the nest entrance and hangs there, beating vigorously forwards with his pectoral fins and backwards with his tail. The net result is that the fish stays still and a current of water is propelled through the nest and over the developing eggs. This keeps them well oxygenated and van Iersel[246] has shown that the presence of CO_2 in the water around the nest is a powerful stimulus to fanning. It would be possible, though difficult, to count the number of fin beats and thus record fanning as a frequency, but since the rate is relatively constant

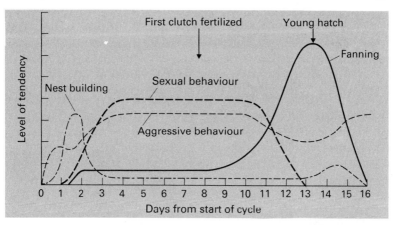

Fig 4.8 Long-term fluctuations in the tendencies to perform nest building, sexual behaviour, aggressive behaviour and nest ventilation ('fanning') during the reproductive cycle of the male stickleback. (From Sevenster,[432] 1961, *Behaviour Suppl.*, 9, 1.)

it is far easier to time the length of fanning bouts, add them together and express the parental tendency as total time spent fanning per standard test period.

Using the measures just described, and others like them, it has been possible to record long-term fluctuations in the motivation of the stickleback male during his reproductive cycle (see Fig. 4.8). The cycle starts with a phase of nest building which then declines to a low level whilst the aggressive and sexual tendencies rise rapidly and both remain high for a week or more. After one or more fertilizations, sex declines and the parental tendency rises with the male spending long periods fanning at the nest.

Apart from these long-term changes there may be short-term, minute-to-minute fluctuations in the sexual and aggressive tendencies. Averaged over a period of hours or days, both tendencies are continuously high during that phase of the reproductive cycle when the male is courting females but also defending his territory against other males. However, at any given moment during this period, spot checks of sex and aggressive tendencies show that there is a negative correlation between them. A male who will perform a lot of zig-zags to a female will not direct many bites to a male and *vice-versa*. This may be because the levels of both tendencies undergo spontaneous fluctuations, with one rising as the other falls. More probably the arousal and performance of one activity which inevitably accompanies the measurement of its tendency, has an inhibitory effect upon the other.

THE NATURE OF AGGRESSION

Aggression presents many problems of description and measurement and it is important to examine some of them here. The aggressive behaviour of animals is currently attracting a great deal of attention from psychologists and psychiatrists as well as ethologists. This is because questions relating to the nature of human aggressiveness are of vital importance in modern society. How far has aggressiveness an inherited basis in man, descended to us from our ape-like ancestors? Is the expression of aggression inevitable or can it be reduced or even eliminated by certain types of upbringing? Discussion of such questions in relation to man would carry us far beyond the scope of this book, but the type of observations on animals which we have been discussing have a direct bearing upon them. In particular there is a necessity to develop appropriate and objective measures of aggressive behaviour.

Recently such widely reviewed books as Ardrey's *Territorial Imperative*[17] and Morris's *The Naked Ape* and *The Human Zoo* have used ethological observations to justify a view of human aggressiveness which relates it very closely to man's biological past. Basically they all stem from Lorenz's view, developed in great detail in his book *On Aggression*,[309] that aggression in animals and men is the result of an inherited, spontaneous tendency whose properties are much the same as the biogenic drives to eat and to drink. Set against this is the view that, although aggression may have an inherited basis, there is nothing inevitable about it and its expression depends as much upon experience and external factors as the internal state of the animal. Amongst the chief exponents of this view are Barnett and Scott and essays by them and others reviewing Lorenz's book have been conveniently collected together by Montagu.[353] There is also an extremely balanced and clear approach to the whole problem of comparing animal and human behaviour in Reynolds[393] while Johnson[259] provides a useful survey of the whole field of aggression.

What sort of evidence from animals has bearing on the problem? The question of how to define aggressive behaviour rises at the outset. Animals usually have recognizable and often stereotyped behaviour patterns for attack, but they may be used in several rather different contexts. In some cases there is a clear biological function for aggression as when animals fight to win a territory, to defend their young or to obtain food. Some species have a social organization based upon a stable hierarchy of dominance (see Chapter 8) and they may have to fight to maintain their status within such a group. Other contexts in which we observe aggression are less easy to explain. Pain arouses aggression; rats given small electric shocks attack a cage mate whom previously they ignored. Frustration of various types has a similar effect and one school of psychologists have suggested that all aggression is the result of frustration (see Miller[345]). Clearly frustration of various kinds is one cause of aggressiveness in animals. A rat in a Skinner box will attack another rat tethered nearby if the bar he has learnt to press ceases to yield him the expected food rewards. In man too we are familiar with the effects of frustration—bad tempers abound in traffic jams. Nevertheless, it is extremely difficult to explain all animal aggression in terms of frustration and a more direct, biological origin seems certain.

In predatory animals the patterns of intra-specific fighting may closely resemble those used in catching and killing prey. In spite of this, it is not usual to consider inter-specific predatory behaviour as aggressive and one assumes it is part of feeding behaviour. It seems best to consider aggressive behaviour patterns in the same way as one regards patterns of locomotion, as 'available' at the motor level of organization to more than one motivational system. If we exclude predation from our definition then the common link for the performance of aggressive patterns would appear to be a functional one; aggressive behaviour serves to displace another individual by causing injury or at least threatening to do so. However, we should not assume that the motivational systems are completely separate. Huntingford[241] has shown that in sticklebacks there must be some common causal factors between the aggression shown to rival males and the aggression—or perhaps we should call it defensive behaviour—shown towards predators such as small pike. Fish which behaved in a 'bolder' fashion towards pike were those which showed the highest levels of fighting in territorial disputes.

It is not uncommon to find that closely related species apparently differ in their overall level of aggressiveness, which affects many aspects of their behaviour. There are examples from animals as diverse as sticklebacks and primates (see Huntingford[242] for a review). Clearly natural selection can set a species aggression at the level which is best adapted and this may have to be a compromise between the needs of different behavioural systems. We shall return to this question in the chapters on conflict and evolution.

Even with these complications it is possible to arrive at a reasonable definition of aggression in animals which relates to its displacing other individuals. It is much more difficult to arrive at a satisfactory definition of human aggression, because our behaviour takes such variable forms. In Carthy and Ebling's book[87] various authors propose that a whole range of behaviour from nail-biting through verbal insults to suicide are all manifestations of aggression. The aggressive behaviour of a street gang may involve actual violence which is related to their concept of 'territory' and therefore, at the functional level of analysis, is comparable to some animal aggression. Such behaviour will also involve the participants in the same kind of physiological arousal that we can measure in animals. Not all behaviour which is commonly called aggressive shares these features. In modern warfare an individual's act of pushing a button may lead to the destruction of other individuals at a great distance and of whom he has no direct knowledge. In a biological sense the button-pusher is not aggressively aroused and such behaviour defies any simple, biologically-based definition. Lorenz[309] and Tinbergen[467] have drawn attention to the dreadful dilemma of modern man whose technology permits him to burn alive people whom he never sees, and thus precludes any human contact which might inhibit his taking such an action.

From the question of defining aggression, we may return to examine some of the evidence which bears on the original problem of how far aggression can be compared with a biogenic drive such as feeding. Following Hinde[217] we can suggest several inter-related questions here, discussion of which helps to characterize aggressive behaviour more clearly.

Has aggression an inherited basis?

There is now a great deal of circumstantial evidence that, whatever their previous experience, many animals have a tendency to respond aggressively when first placed in certain situations or given certain stimuli. Often the situations and stimuli that evoke aggression relate to the setting up of a territory at the beginning of the breeding season and we shall be discussing territorial behaviour in more detail in the next chapter. Here we should note that aggressive responses are far commoner amongst males (and as mentioned earlier the male hormone testosterone tends to increase aggressiveness) and that the stimuli to which they respond are often those provided by a rival male. Thus Cullen[113] has shown that sticklebacks reared in complete isolation from the egg, set up territories and attacked rival males in the typical manner. Isolation experiments have certain limitations, as we discussed in Chapter 2, but experiments of this type and the role of aggression in the organization of territorial behaviour in such a wide variety of animals, certainly suggest that there is an important genetic component in the development of aggressive behaviour.

We know that selective breeding can change levels of aggression. Game-fowl and Siamese fighting fish have both been specially bred for abnormally high levels of aggression. Lagerspetz[285] has successfully bred for high and low aggression in mice, and the two selected lines differ greatly in their readiness to attack intruders and the intensity of fighting. These examples show that animal populations normally carry many genes which affect their levels of aggression and thus selective breeding can change them in either direction. As we discuss further in Chapter 6, selection of this type usually affects behaviour only in a quantitative fashion. It cannot tell us much about the inheritance of the basic patterns of aggressive behaviour themselves.

However, even if we accept that the potential to perform aggressive behaviour has an inherited basis, this is far from establishing aggression as a biogenic drive. On page 112 we mentioned some of the difficulties in identifying either regular fluctuations in aggressiveness or the phases of appetitive behaviour, consummatory act and quiescence, such as characterize feeding and drinking. This leads to a second question.

Is there evidence for aggressive appetitive behaviour?

In other words, do animals actively move about looking for fights? We are all familiar with the sight of an aroused dog or cat pursuing another which it has just put to flight and clearly making no attempt to break off the encounter. But will a hitherto restful cat get up and move off to seek an opponent when none is visible? Birds are sometimes said to be 'patrolling their territories' and they will attack any intruder they come across, but it is impossible to be certain that it was aggressive motivation that initiated their patrolling. We would have to demonstrate a lowered threshold for aggressive responses just as the bird set off and before it had seen a rival and such measurements are difficult to make.

However, approaching the question from a slightly different angle, there can be no doubt that under some conditions aggressive behaviour is reinforcing, i.e. an aroused animal will perform behaviour giving it the opportunity to respond aggressively. Perhaps the most elegant examples of this come from the work of Thompson with game fowl[453] (see p. 164) and Siamese fighting fish[452] (both, significantly, animals which have been selectively bred for their aggressiveness). Fighting fish males will show the full aggressive display with raised gill covers and extended fins towards a mirror image or a model, and they will also attack and bite such stimuli when first shown. Thompson kept males alone in tanks where it was possible to make the male's image appear rapidly simply by switching off an outside lamp so that one wall of the tank functioned as a mirror. He then conditioned the fish in a manner exactly analogous to the rats in the Skinner box described on page 25. The latter had to press a bar to get a food pellet and soon learnt to do so. Thompson's fish had a ring suspended in the water and if the fish swam through it the lamp was automatically switched off for a few seconds. An image then appeared on the glass, to which the fish displayed aggressively. Under these conditions, the males learned to swim through the ring and repeatedly did so, some hundreds of times each day. In other words, given the opportunity to control its own situation the males chose one where they would be behaving aggressively for much of the time. In a similar fashion game fowl cocks would learn a response which gave them the opportunity to perform their aggressive display. It might be possible to argue that the presence of the ring in the tank became an aggressive stimulus and kept the fish aroused. Thompson has shown that artificial objects associated with the appearance of a rival male do themselves come to elicit attack. Even so, the conclusion that aggressive behaviour is 'rewarding' to the fish once aroused seems inescapable and does, in this respect, justify a comparison between the tendency to fight and that to feed or to drink.

Is there a phase of quiescence following the performance of aggressive behaviour?

The stereotyped patterns of aggressive displays and attacks would seem to be the equivalent of the consummatory acts directed towards the goal object—i.e. the rival. However it is difficult to detect any regular phase of quiescence following their performance. With other types of behaviour, even feeding and drinking, it is common to find that there is a short period of arousal and warm-up after performance begins (see p. 8). Mice, for example, feed in short bouts interrupted by pauses and the length of these bouts increases over the first few minutes before the mouse becomes full.[499] Thus the intensity of the behaviour may transitorily increase before feed-back signalling 'goal-achieved' begins to take effect and leads to quiescence. With aggressive behaviour such arousal is intense and once attack begins the animal's threshold for further attacks often falls considerably.

Sevenster[432] and Wilz[508] have measured aggression in male three-spined sticklebacks using the method outlined on p. 132. One significant finding is that a male's tendency to bite a rival confined in a glass tube is *higher* at the end of a 10-minute test than at the beginning. We would certainly not expect this were an animal given ad lib food for 10 minutes, and in this respect the striking arousal of aggressiveness once attack has begun, is in strong contrast to biogenic drives. Wilz has good evidence that for several minutes following an aggression test, a male stickleback is so aggressively aroused that he is unable to respond sexually to a female. In fact, if a female is presented when the aggressive stimulus is removed, males perform several activities whose function seems to be connected with enabling aggression to be reduced so that the sexual motivational system can gain control. Thus a quiescent stage following the performance of aggressive behaviour and driving off an opponent is not at all evident. In the continued presence of an aggressive stimulus attacks eventually cease but a change of stimulus will often start them again, just as a change of stimulus immediately restores a bull which is apparently exhausted sexually. It is probably most appropriate to regard this waning of response as habituation (see p. 233) rather than as quiescence resulting from a fall in motivation. Thus the data from animals indicates that unlike the biogenic drives to feed and to drink, the tendency to attack often *increases* as a result of the performance of aggressive behaviour. This conclusion is of considerable theoretical and practical importance in the controversy about the nature of aggression because of Lorenz's view of the ways in which both human and animal aggression can be controlled. Discussion of this can be directed to a fourth question.

Is aggression inevitable?

The reason for phrasing the question in this way relates to Lorenz's application of his psycho-hydraulic model to aggression. It will be recalled that with this model the only way to reduce drive was to perform behaviour and in the absence of an 'outlet', drive accumulated and became stronger.

If human beings have an inherited aggressive tendency which operates in this way then clearly we shall have to accept that it will be impossible to prevent all manifestations of aggression. The best policy would seem to be one of encouraging the sublimation or redirection of aggressive tendencies into outlets less harmful than those of physical conflict. This view is, in essence, that held by Lorenz and Ardrey.

There are a number of experiments which are often quoted as showing the inevitability with which aggression expresses itself, but they are all open to other interpretations. Animals reared in total isolation are sometimes highly aggressive. This has been clearly shown in mice and in jungle fowl, where in the latter species, Kruijt[278] found that after months of isolation the birds would attack feathers and have prolonged, circling fights with their own tails. However the behaviour of isolates is profoundly altered in a number of ways and one is not justified in interpreting these observations as the sole result of an accumulating aggressive drive. Thus isolated rodents are generally highly excitable and are known to undergo hormonal changes—probably resulting from the stress of their situation—such that the gonads of males may be stimulated to secrete more testosterone. This, by itself, might cause an increasing tendency to attack.

Lorenz[309] described observations on a highly aggressive cichlid fish, *Etroplus maculatus*. Fish breeders have found by trial and error that in order to get a pair of *Etroplus* to breed successfully, it is necessary to have one or two other non-breeding males in the tank who serve as 'whipping boys'. The breeding male attacks them from time to time, but little aggression is seen within the pair. If a pair of *Etroplus* are kept alone, they rarely breed successfully because the male constantly attacks the female. These observations have been confirmed under well-controlled conditions by Rasa[391] and Fig. 4.9 illustrates her results. There is a huge increase in attacks on the female if the male has no other fish to attack. Lorenz[309] and

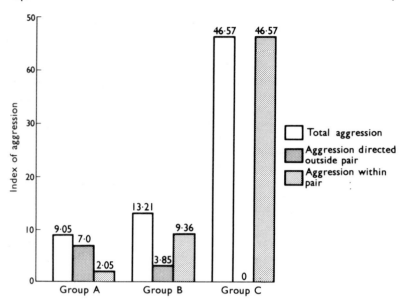

Fig. 4.9 Rasa's data on aggressive behaviour in the cichlid fish, *Etroplus macula-tus*. The 'index of aggression' represents the average frequency of chasing attacks made by both fish of a pair within five minute periods. In group A pairs were kept in communal tanks with other fish ; in group B pairs were kept separately but could see other fish through a glass screen ; in group C pairs were completely isolated. Aggression is classified into two categories, that directed towards other fish and that between the members of a pair—the only category in group C. (From Rasa,[391] 1969. *Z. Tierpsychol.,* **26**, 846.)

Eibl-Eibesfeldt[143] interpret this result in terms of accumulating aggressiveness in the male which must find an outlet. However the lowering of the attack threshold does not necessarily imply accumulation. As we shall be discussing more fully in the next chapter, in territorial animals the presence of a female commonly arouses aggression in males in addition to sexual responses. Consequently the male *Etroplus* are likely to be aroused under all Rasa's experimental conditions. Their aggression is inhibited to some extent because the female also arouses sexual tendencies and where other fish are present, most of the attacks are 'redirected' towards them. Redirection is familiar in the aggressive behaviour of animals and men when a stimulus which arouses aggression is also, for some reason, not available for attack. The boss reprimands his clerk, who in turn cuffs the office boy.

From Rasa's experiments we may note that the actual incidence of attacks rose enormously in the isolated pair condition. She suggests that this is because the male *Etroplus*, like the stickleback, is most aggressive close to his nest. This is where the female tends to remain even when attacked and consequently we may expect the male to be maximally aroused. Further, in the normal condition the distant attacks he makes upon other fish will lead him away from the nest. Then, because the other fish retreat out of sight as far as they can, the *Etroplus* male's aggressiveness may have more chance to subside before the next stimulus appears. If we are to examine critically the concepts of accumulation and inevitability we need much more information on the behaviour of fish in which arousal is kept to an absolute minimum for varying periods of time.

Several studies, with various species of fish, have tried to do this. Clayton and Hinde[100] watched the recovery of fighting responses in Siamese fighting fish who had had prolonged exposure to their mirror image and therefore ceased to respond to it. Recovery took many days and, although one could look upon this as the gradual accumulation of aggressive motivation, it can more easily be regarded as the waning of habituation since we know how stimulus-dependent aggressive behaviour is. Similar measures have been made by Wilhelmi[501] who studied the recovery of fighting in swordtails allowed to fight rivals until one gave up. Again isolation from one up to eight weeks led to the gradual recovery of responsiveness. Wilhelmi suggests that either drive accumulation or habituation could be the cause of this change.

Somewhat in contrast, Heiligenberg's[208] studies on fighting in a cichlid, *Pelmatochromis*, showed that the tendency to fight *decreased* over a few days in the absence of any arousal. His fish had not been fought to exhaustion to begin with, but his results lend no support to the idea that aggression is always accumulating in the absence of an opportunity to fight.

Lorenz's concepts have also been criticized by several of the mammalian workers because he has tended to underestimate developmental factors affecting aggressive motivation. Scott[426] gives a very clear survey of numerous experiments with rodents which show how levels of aggressiveness can be changed in a most dramatic fashion by regulating early experience. It is relatively easy to train one mouse always to attack a strange animal, whilst another of the same strain can be trained to remain completely placid. In fact the changes in aggression which result from such differences in experience resemble quite closely those which Lagerspetz produced in her mice by selective breeding. In view of such clear effects of upbringing and considering how equivocal the animal data is in other respects, there seems no reason why we should just accept human aggressiveness as being inevitable. We certainly have to accept that man has a potential—probably inherited—for aggressiveness, but the long period of childhood development and the powerful influences that both parents and the rest of society can bring to bear on the individual offer us a solution.

THE PHYSIOLOGICAL BASIS OF MOTIVATION

Continued analysis of this behavioural type is essential, but it is also important to examine motivation at the physiological level and try to link up behaviour with events in the nervous system. Some of the most promising bridges between neurophysiology and behaviour have developed from the work of physiological psychologists—most of them Americans—on motivational problems. Grossman[182] provides a good introduction to the whole field of physiological psychology. We must now turn to consider some of this work and begin with some description of one area of the vertebrate brain which has proved to be of major importance in the control of motivation—the hypothalamus.

The hypothalamus

This relatively minute volume of brain tissue—in the human brain it is smaller than the last joint of the little finger—is of primary importance in a whole host of reactions. There is an excellent general account of its physiology in Walsh[482] who says of the hypothalamus, '. . . this small centre plays a dominant role in determining the use that is made of the resources of the body. . . . It is difficult, indeed, to think of any function of the body that is not dependent, directly or indirectly, upon the hypothalamus.'

A little neuro-anatomy is needed at this point; the reader is referred to a clear, concise account in Romer.[397]

The brain of all vertebrates is constructed on the same basic plan, and at an early stage in embryology consists of three swellings at the anterior end of the spinal cord. These are called the prosencephalon, mesencephalon and rhombencephalon or more simply the fore-, mid- and hind-brain. Primitively these swellings arose to cope with the increased amount of sensory information which flowed into the central nervous system from the sense organs of the head. The fore-brain originally dealt with olfaction, the mid-brain vision and the hind-brain balance and hearing. In most living vertebrates their original functions have become greatly extended and complicated, but still the appropriate sensory data are led first to these regions, even if subsequently they are passed on elsewhere.

The fore-brain is easily sub-divided into two portions, as shown in Fig. 4.10. The anterior portion has arising from its roof the cerebral hemispheres, which primitively were olfactory areas but have now come to dominate the whole nervous system in mammals. The posterior portion of the fore-brain —called the diencephalon—has on its dorsal surface the pineal organ. This was once associated with a light receptor or pineal eye which can still be seen in some living reptiles. The side walls of the diencephalon are thick and form the thalamus, an important 'staging place' in the brain where fibre tracts link up with one another in numerous 'nuclei' or clusters of

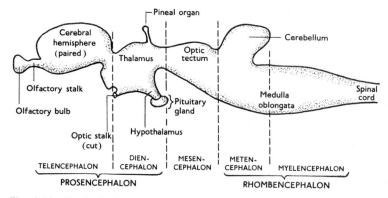

Fig. 4.10 The basic divisions of the vertebrate brain. The brains of all vertebrates pass through a stage rather like this during development, but in mammals and birds in particular, the adult brain is dominated by the enormous growth of the cerebral hemispheres and cerebellum. These come to overlie all the rest and obscure the original lay out ; see also Fig. 9.1. (Modified from Romer,[397] 1962, *The Vertebrate Body*, 3rd edn. W. B. Saunders, Philadelphia and London.)

neuron cell bodies. On the floor of the diencephalon, below the thalamus as its name implies, is the hypothalamus.

There are nuclei in the hypothalamus, but they are not so well defined as in the thalamus above. However, there are several well-marked fibre tracts entering and leaving, and these put the hypothalamus into connection with the cerebral hemispheres and also with more posterior parts of the brain. From the behavioural point of view, one of the most significant structural features of the hypothalamus is its intimate connection with the pituitary gland. This endocrine gland, which controls the whole hormonal system of the body (see page 153), develops from the fusion of a downgrowth of the embryonic hypothalamus with an upgrowth from the roof of the embryonic mouth cavity. The pituitary stalk, which joins it to the hypothalamus, contains both nerves and blood vessels. The hypothalamus itself has a very rich blood supply and some of its cells are even penetrated by capillaries.

From its connections with other parts of the brain, its rich blood supply and its links with the pituitary gland, the hypothalamus is well adapted both to measure changes in the metabolism of the body and to set in motion activities which will rectify them. It is, in other words, well suited to serve as part of a homeostatic control system and there is plenty of physiological evidence that it does so. For example, the control of body temperature is one of the most delicate homeostatic systems in a mammal or a bird. There are areas of the hypothalamus which are highly sensitive to changes in the temperature of the blood. If these areas are heated artificially by implanted wires, the animal starts sweating and panting. Sweating is controlled peripherally by the autonomic nervous system and the hypothalamus can initiate its activity. The reverse effect is produced when the temperature-sensitive areas are cooled; now the animal shivers—another autonomic response (see Walsh[482]).

All this may appear to be pure physiology, having little to do with the study of behaviour. But homeostasis is one of those topics where the boundaries between traditional fields of study break down. It is not helpful to distinguish between physiology and behaviour in the matter of temperature control. If a rat is briefly cooled in the manner described, shivering is started to generate some heat. We might classify this as a reflex activity and consign it to the realms of physiology, as described in Chapter 1. However, if the rat is cooled for a longer period shivering alone is inadequate and, if given the material, the rat begins to build a nest or to enlarge the one it already has, in order to insulate itself. The reflex response is now supported by a complex behavioural one. Both are initiated by the hypothalamus and both form part of the rat's homeostatic system, although nest building will involve a more elaborate neural mechanism than does shivering.

The hypothalamus and motivation

It is from studies of the role of the hypothalamus that some of the most important links between brain and behaviour have developed. Modern physiological techniques allow parts of the brain to be explored with electrodes, controlled injection of chemicals or by the destruction of very small selected areas.

As a typical example of the way such studies reveal the underlying physiology of motivation we may consider the role of the hypothalamus in thirst. There are cells in its lateral areas which respond to increased concentration of the circulating body fluids. They can set in action two compensating systems. The first, acting via the links with the pituitary gland, causes the secretion of antidiuretic hormone (ADH) which increases the resorption of water by the kidneys. The second system causes the animal to seek out and drink water. If the link between the hypothalamus and the posterior lobe of the pituitary gland is damaged, ADH may never be secreted. In such a situation the kidneys continue to excrete copious quantities of urine and to compensate the animal drinks large quantities of water—a condition known as *diabetes insipidus.*

Normally the amount of water taken in is exactly adjusted to the animal's needs, but if the lateral hypothalamic detector area is artificially stimulated either electrically or by injecting hypertonic saline, drinking is greatly increased. Andersson[9,10] prepared a number of goats with fine hollow needles penetrating into the hypothalamus. He allowed them to drink as much water as they would take and then injected concentrated salt solution. There was little effect unless the needle tip was in the lateral hypothalamus. In this region the salt caused the goats to drink frantically within a minute or two of injection. These animals, previously satiated with water, would drink salt and bitter solutions which they would not normally touch even if extremely thirsty. This condition is not comparable to *diabetes insipidus*, because the goats are not compensating for water lost through the kidneys. They are drinking water which is surplus to their physiological needs.

Andersson[11] could produce the same effect if he stimulated the lateral hypothalamus electrically and such drinking behaviour is often described as 'stimulus-bound', because it continues only whilst the stimulus (saline or electrical) is actually being applied. Stimulus-bound drinking has also been produced in the rat and the importance of the lateral hypothalamus for drinking is further emphasized by the behaviour of rats with damage to this part of the brain [450] Such animals may stop drinking altogether. They will not drink even though in the last stages of dehydration and will eventually die in the presence of water, unless this is given artificially through a tube into the stomach. It is fascinating that a rat with a lateral hypothalamic lesion is not just disinterested in water, it becomes actively averse to it. If water is placed in its mouth it will not swallow but allows the water to run out, making tongue and lip movements identical to those made by a normal rat towards an intensely bitter liquid.

These effects of stimulation and ablation might lead us to conclude that the lateral hypothalamus is responsible for the state we observe behaviourally as a tendency to drink. How far is this conclusion justified? Before discussing this question we can add comparable results which also implicate the hypothalamus in the control of feeding.

Mammals and birds normally keep their weight very constant and adjust the amount they eat accordingly. If rats are given a super-rich diet they eat less; if their food is mixed with non-nutritive cellulose they eat more. We have already mentioned that rats with damage to central areas of their hypothalamus (the ventromedial nucleus, in fact) lose this sensitive control of their eating. Fig. 4.11 shows the food intake of a rat whose ventromedial nucleus has been ablated compared with that of a sham-operated animal. After a few days of post-operative depression of food intake, the brain-damaged rat begins to eat huge amounts of food—at least four times the normal amount. This so called 'dynamic phase' of hyperphagia lasts for 3 weeks or so. Beyond this, food intake slowly declines and eventually settles down with the animal eating about double the normal amount. Needless to say hyperphagic rats become grotesquely fat and are very inactive.

Electrical stimulation of the ventromedial nucleus of normal rats depresses their feeding, so that this region of the hypothalamus has often been called a 'satiety centre'; it measures when the animal has eaten enough

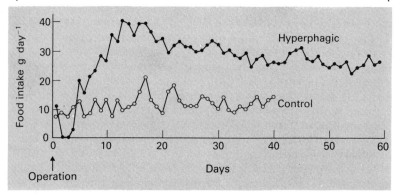

Fig. 4.11 The daily intake of normal rats and those with bi-lateral lesions in the ventromedial nucleus of the hypothalamus. Further explanation in text. (From Teitelbaum, P., 1955, *J. Comp. Physiol. Psychol.*, **48**, 156–63.)

and inhibits further eating. (The term 'centre' here means a group of nerve cells with a common organizing function; see p. 148 for further discussion of centres.) We may note that hyperphagic rats do not lose *all* control of their food intake. Fig. 4.11 shows that eventually they are eating considerably less than they did during the early dynamic phase. Satiation is not abolished with the ablation of the ventromedial nucleus, but its threshold is greatly raised.

A centre complementary in function to the ventromedial nucleus and which promotes feeding is located in the lateral hypothalamus closely associated with the drinking area. In rats, experiments similar to those described for thirst have shown that this 'feeding centre' does initiate the appetitive behaviour of eating; stimulus-bound feeding can be elicited by implanted electrodes and damage to the area can lead to starvation in the presence of food. The feeding and drinking centres are closely linked anatomically and they interact with each other in a complex fashion. For details of the behaviour associated with this interaction the reader is referred to the admirably clear account by Teitelbaum and Epstein.[450]

How far are we observing the operation of the 'normal' drinking and feeding systems during stimulus-bound behaviour—remembering that this term does not refer to 'normal' stimuli but only to artificial stimuli applied chemically or electrically within the brain itself? It would be important to see whether an animal shows normal appetitive behaviour when being stimulated, because it might be argued that their feeding or drinking was merely a reflex response to the stimulation of neural pathways controlling the motor patterns concerned. It is possible to get fairly well co-ordinated lip and tongue movements by stimulating other areas of the brain—parts of the motor cortex of the cerebral hemispheres, for example. Some of Andersson's observations appear to rule out this explanation because, when stimulated, his goats showed all the signs of normal thirst. They walked over to the corner of their pen and searched around for the water bowl, i.e. they showed normal appetitive behaviour and not just 'forced drinking'. Again with feeding in rats, Coons et al.[103] have shown that stimulus bound feeders will learn a new response (bar pressing) in order to acquire food, and that this response is transferred and used when subsequently the same animals are made normally hungry.

Such results are very convincing, but some doubts still remain and have been emphasized recently by Valenstein[475] and others. There are a number of experiments which have shown that stimulus-bound feeders and drinkers are much more easily dissuaded from their goal than normally motivated animals. At least this is the case in rats, where slight adulteration of food or water with quinine is usually enough to stop them responding. (We have already mentioned how easily hyperphagic rats are dissuaded by quinine.) Further, the rate at which rats drink by lapping from a tube is normally very constant. Normal water deprivation simply leads to lengthening their bouts of drinking but does not affect their lapping speed. However White et al.[496] report that stimulus-bound drinkers do change their rate of lapping with the intensity of electrical stimulation, which certainly suggests that the motor-organization centres for drinking are being affected.

We shall need more facts to be able to resolve this question and, even from the evidence we have at present, we must not expect the same details to apply to all vertebrates or even to all mammals. Nevertheless there can be little doubt that the basic function of the hypothalamus as a detector of physiological imbalance remains constant. Its cells will meter the temperature, food, water and hormonal concentrations of the bloodstream. As a result of any particular imbalance, the detector sites will initiate both physiological and behavioural action, as we have outlined earlier in this section.

The behavioural problems then become more obscure, because whilst we may accept that the hypothalamus is essential for the initiation of motivational states, we cannot claim that it alone is essential for their control. Clearly this control involves many other areas of the brain also. Grossman[183] reviews the extensive evidence showing that other parts of the forebrain are involved in feeding, drinking, sexual and aggressive behaviour. Many experiments have involved making lesions and observing their behavioural effects. It has been found that damage to most areas of the cerebral hemispheres does not affect an animal's motivation in a specific way, but there are some significant exceptions. Many of these concern lesions in the extreme frontal parts of the hemispheres and also in a complex series of fibre tracts called collectively the rhinencephalon or limbic system. These are situated near the base of the cerebral hemispheres and have links with the hypothalamus. Parts of the limbic system used primitively to be connected with the sense of smell, but in higher mammals it is clearly involved in more general aspects of their behaviour and appears to modify the activity of hypothalamic centres.

Lesions in the frontal lobes of the hemispheres and in the limbic system have been found to affect the sexual and feeding tendencies of dogs, cats, and other mammals and sometimes in a highly dramatic manner. For instance, dogs with damage in parts of the frontal lobe show ravenous eating, devour inedible materials and become grossly obese. Cats with lesions in the region of the amygdala (a part of the limbic system) became hypersexual, mounting and attempting to copulate with a wide variety of animals and inanimate objects (see Grossman[182] for further details).

There is some evidence that a loop of neural tracts in the forebrain is involved in the control of feeding and drinking. Tracts from which such behaviour can be elicited on stimulation leave the hypothalamus, proceed through various limbic structures and can then be traced back to the hypothalamus again. Of course such a circuit is not closed—it will have outside connections along its length—but perhaps continuing activity within such a loop is one basis for a motivational state.

Summarizing the evidence on control centres

In conclusion we may recapitulate the evidence on the neurophysiological basis of motivation and consider, in particular, whether specific motivational states can be initiated by the activity of discrete centres in the hypothalamus.

The concept of a 'centre' is rather imprecise but it is usually taken to mean a relatively small group of nerve cells with a common function. Various techniques have been used to explore the organization of the brain and, as we have seen, one of the commonest techniques used by physiological psychologists is to make lesions and study their effects on behaviour. There are dangers in trying to locate centres by making lesions because the brain is an amazingly complex structure made up of minute units. Burning holes in it or cutting pieces out is inevitably a rather crude technique. In addition to removing many cell bodies, nearly always some fibre tracts are damaged which lead from or through the area of the lesion. Other areas, supplied by such tracts but not necessarily close to the site of the lesion, may be affected by it and this may make interpretation of the results difficult. For instance, Reynolds[392] has suggested that the ventromedial nucleus of the hypothalamus is not a 'satiety centre'. He considers that the lesions which are made to ablate it result in constant 'irritation' of tracts

Table 4.1   Possible techniques to be employed in exploring a postulated 'centre' controlling a particular type of specific behaviour.

Technique	Expected Result
1. Ablate area of brain	Specific behaviour not shown even when conditions are optimum for it.
2. Stimulate area {electrically chemically	Specific behaviour shown even when completely inappropriate.
3. Depress area chemically	As for (1)
4. Record normal electrical activity from area in conscious, freely-moving animal.	Activity high when specific behaviour is being performed. Activity low when not being performed.

linking this nucleus to the lateral hypothalamic 'feeding centre', and it is stimulation of this latter area which causes hyperphagia.

Gregory[178] uses a vivid analogy to emphasize the problems of interpreting the results of lesions. Discussing how far comparing the brain with various types of machine is helpful for understanding its operation, he says, 'Thus the removal of any of several widely spaced resistors may cause a radio set to emit howls, but it does not follow that howls are immediately associated with these resistors, or indeed that the causal relation is anything but the most indirect. In particular, we should not say that the function of the resistors in the normal circuit is to inhibit howling. Neurophysiologists, when faced with a comparable situation, have postulated "suppressor regions".'

The evidence gained from lesions needs support from other types of experiment if we are to identify control centres with any confidence. Suppose we are trying to demonstrate that a certain area of the brain is responsible for starting one type of specific behaviour, related to a particular goal. Ideally we might hope to use the techniques set out in Table 4.1 and record the equivalent results. These are a somewhat hypothetical series of experiments and the postulated results are over-simplified, but there is information comparable to this for the lateral hypothalamic 'feeding centre' and for the corresponding ventromedial 'satiety centre'*

If we can obtain such results we can identify *one* aspect of a brain region's function with fair confidence, but we must never assume that this is all it does. For instance, lesions in the ventromedial area certainly affect other aspects of a rat's behaviour. It becomes less fearful, shows increased loco-

---

* It is interesting to note that these centres thus appear to have properties analogous to the 'link' and 'analyser' respectively of Deutsch's behaviour model (see p. 123 and Fig. 4.5).

motion in novel situations and it becomes more responsive to painful stimuli (Grossman[184]). Some of these effects may well contribute to the 'hyperphagia syndrome' upon which we have been concentrating and which, as we have seen, is not easy to understand simply in terms of increased feeding motivation. In other words, there is no substitute for a really thorough investigation of all the behavioural effects of brain lesions and brain stimulation. Physiology and ethology can help one another here.

The evidence for centres controlling other behaviour systems is not so complete as that for feeding. We can locate areas in the anterior hypothalamus which pick up sex hormones from the blood and control the onset of sexual behaviour, but aggression or fear centres are much more problematical. Attack and 'rage' responses in cats, for example, can be obtained by stimulating many different areas of the hypothalamus and limbic system. It is necessary to be particularly careful with interpretation here because all sorts of painful stimuli can provoke rage and the electrodes may sometimes be stimulating pain pathways. Nevertheless, whilst individual elements of the cat's 'rage' response such as spitting, hair erection and arching of the back can be elicited from other areas of the brain, the hypothalamus must be intact if these are to be integrated into the full pattern.

Recently Valenstein[475] and his collaborators have questioned the specificity of these proposed centres. In part, their approach takes us back to the discussion of general versus specific drives with which this chapter opened. They have found that *different* behaviour—feeding or drinking—can be elicited from the same electrode in the lateral hypothalamus, depending on the experience of the animal in the situation. They suggest that stimulation arouses the animal in a general fashion and facilitates any responses which—to use their term—are 'prepotent'. This prepotency will be affected by the internal state of the animal and the external stimuli. The exact experimental procedures and the evidence they provide are too complex to discuss here. A very good picture of them is gained by reading the dialogue between Wise[509-10] and Valenstein et al.[473-4] in *Science*. Again the need for good behavioural studies and a full knowledge of how a rat responds in the different experimental situations must be emphasized.

Wise argues the case for specific drive centres, although he does not deny that the feeding and drinking centres are very close and that both are liable to be affected by diffusing current from the same electrode. Such diffusion is always a potential hazard of brain stimulation work and for this reason Miller[349], Grossman[181] and others have advocated the use of chemical stimulation. The central nervous system is highly diverse biochemically and this could make chemical stimulation much more selective than making lesions or stimulating electrically. Both electric charge and chemicals will diffuse out from their points of origin and affect several adjacent 'circuits' but, whereas all neurons respond to electrical stimulation, each circuit may respond only to a few specific chemicals. This means that successive injections of a series of different chemicals may affect 'circuits' one at a time and such results are more easily interpreted. Using such a technique Grossman succeeded in selectively increasing either feeding or drinking by specific chemicals injected through the same needle into the lateral hypothalamus. Since these effects were so distinct, it seems unlikely that the lateral hypothalamus is as undifferentiated as Valenstein implies.

Whatever view is taken in this controversy, most workers would agree that it is impossible to pin down drive centres to a small, unique group of neurons. Indeed it is probably misleading to think of centres in these terms, because the brain probably functions using larger and more diffuse groupings of its units. The feeding and drinking centres are probably the nearest approach to the ideal, but even here exact localization is not possible.

The elegant work of Teitelbaum and Epstein[450] shows that there is considerable plasticity in the cells of the lateral hypothalamus. Rats which have stopped eating altogether following a lesion in this area can be coaxed back into normal feeding again by starting with an attractive liquid diet. They regain first the desire to eat and later the ability to regulate their food intake. This must mean that neurons outside the original feeding centre have come to take over its function. The ability to compensate for the effects of damage is widespread in the brain; it rules out too rigid a localization of function. This conclusion is reinforced by the awkward fact, which we have already mentioned in respect to Valenstein's work, that stimulation at the same site on different occasions may not always produce the same results. Von Holst and von Saint-Paul[230] describe numerous examples of this from their work on brain stimulation in chickens. Often they could correlate the changed effects with other spontaneous changes in the birds' responsiveness to particular stimuli. The activity of one control system will affect that of others and may change the threshold at the site of electrical stimulation, thus determining into what channels stimulation spreads.

We must not expect that artificial stimulation—nor the natural external stimulus situation—will arouse only one, isolated motivational system. In the next chapter we shall consider in more detail what happens behaviourally when two different systems are aroused simultaneously.

HORMONES AND MOTIVATION

We have just been discussing a number of ways in which an animal's behaviour is linked to the metabolic state of its body. Food and water deprivation result in behaviour which is designed to restore such deficits and we noted that together with behavioural responses there may also be hormonal ones; thirst leads to the release of antidiuretic hormone from the posterior pituitary. We can now consider the interaction between hormones and behaviour in more detail. This topic deserves closer attention because in many ways the endocrine organs, which produce the hormones, and the nervous system share the common functions of communication and co-ordination both within the animal and between it and the outside world. The hormones form a chemical message system which is probably as old as the nervous system. Indeed the one may have developed from the other in part. Throughout the animal kingdom we find neurosecretory cells in the nervous system. These are modified neurons which can pass special chemicals down their axons and into the bloodstream. Often these cells are clustered together to form glands, such as the corpus cardiacum of insects, which have close connections with both nervous system and bloodstream. The vertebrate pituitary gland develops from the fusion of neural and epithelial tissue and remains closely connected to the hypothalamus. The pituitary regulates by its various secretions all the other endocrine glands and is in turn regulated by the nervous system. This system of control enables environmental changes picked up by the nervous system to be matched by an appropriate hormonal response. The most familiar example is the control of breeding seasons in mammals or birds. Changing day length, perceived by the eyes, results in changed activity in the hypothalamus which stimulates the pituitary. This in turn secretes hormones which start the various growth changes in the body associated with the onset of breeding condition.

Throughout all their interactions the functions of the two communication systems—endocrine and nervous—remain essentially complementary. The nervous system can only pass information by trains of nerve impulses. Its state can change very rapidly, but it is clearly less suited to transmit a steady unchanging message for a long period; it operates on a time scale from milliseconds to minutes. The endocrine system cannot respond so rapidly, but its cells can maintain a prolonged steady secretion into the bloodstream lasting for months if necessary. Moreover hormones can reach every cell in the body via the bloodstream, whereas the nervous system generally controls only the muscles.

Circulating hormones are commonly regarded as prime motivating factors in animal behaviour and ethologists have often assumed that they act directly on brain control centres to increase drive. Certainly there are dramatic examples of hormones apparently 'forcing' behaviour from an animal even in the most inappropriate circumstances. For instance, Blum and Fiedler[58] describe how injections of the pituitary hormone prolactin (see below) will cause isolated males of the fish *Crenilabrus ocellatus* to perform the parental fanning movement. This is similar in form and function to the stickleback's fanning described on p. 132. Injected *Crenilabrus* males will fan in a bare tank devoid of all the normal external stimuli for fanning, CO_2 in the water, fertilized eggs, nest, nest site etc., and the amount of time they spend fanning is dependent on the dose of prolactin. Examples of this type are convincing evidence for the central motivating role of hormones but they can also affect behaviour in other ways.

Before discussing examples of hormone action it is necessary to give a brief outline of those aspects of the vertebrate endocrine system which are most important for behaviour. Fuller accounts are given by Gorbman and Bern[170] and by Austin and Short.[18]

The pituitary gland

The pituitary gland secretes several hormones which affect the output of other endocrine organs and in this way the pituitary effectively controls the whole endocrine system. The hormones we shall be most concerned with are the **gonadotrophins** which act upon the gonads and promote both the growth of germ cells and the tissues of the gonads which secrete the sex hormones. There are two main gonadotrophins—**follicle stimulating hormone** (FSH) and **luteinizing hormone** (LH)—both were named after their action on the female gonads or ovaries but they are secreted by, and have similar functions in males. In females both are necessary for the growth of eggs and for their release into the oviduct ready for fertilization.

A third pituitary hormone important for behaviour is **prolactin**, also known as **lactogenic hormone** or **luteotrophic hormone** (LTH), which has a variety of physiological effects. We know that it is secreted by various classes of vertebrates but it often has completely different functions and 'target organs' (those parts of the body whose growth or functioning is affected by the hormone). Interestingly, most of its effects are on 'parental behaviour' in the broadest sense. As just mentioned, prolactin stimulates fanning of the eggs by male sticklebacks and some other fish; it promotes broodiness in chickens (although not all birds), the secretion of crop milk in pigeons and the growth of the mammary glands and milk secretion in mammals.

The name luteotrophic hormone refers to another important function of prolactin in mammals, for it is required for the maintenance of corpora lutea in the ovary and stimulates then to produce progesterone (see below).

The gonads—ovary and testis

It has been known for centuries that castration has profound effects on the behaviour and body form of vertebrates. This is in contrast to many invertebrate groups in which castration has little or no outward effect. Only in the vertebrates are the gonads important endocrine organs where, under stimulation from pituitary FSH and LH, they produce the sex hormones from special secretory cells. The female hormones are collectively called *oestrogens* and the male hormones, **androgens**. All these hormones are steroids which are closely related in chemical structure, and although different vertebrate groups secrete slightly different steroids, those from one group are usually quite effective in another. The commonest androgen secreted by mammals is called *testosterone*.

The sex hormones are responsible for the development of the secondary sexual characters and for the growth of the reproductive system in preparation for the shedding of eggs and sperm. Usually there are permanent differences between the body form of male and female, which are maintained throughout adult life. These are often augmented by the seasonal growth of secondary sexual characters under the influence of increased sex hormones. Stags can be distinguished from hinds all the year round but in addition they show seasonal growth of antlers.

Finally we must mention another steroid hormone produced by the ovary. After a mammalian egg has been shed, its empty follicle enlarges and forms a prominent yellowish structure on the ovary surface—the corpus luteum. This begins to secrete **progesterone** under whose influence the lining of the uterus is prepared for receiving the egg after fertilization and development into a blastocyst. Progesterone also inhibits the contraction of the uterine muscles, which must be avoided if pregnancy is to continue. It is justly called 'the hormone of pregnancy', but it is not an exclusively mammalian hormone. Structures like the corpora lutea form when eggs are shed in fish, amphibians and reptiles but we know little about the presence or action of progesterone in these classes. Birds have progesterone which is almost certainly secreted by the ovary although their corpora lutea are not conspicuous: male birds of several species are also known to produce progesterone, probably in the testis.

THE BEHAVIOURAL EFFECTS OF HORMONES

As mentioned earlier, behaviour workers have usually concentrated on the central action of hormones. They are, however, extremely potent chemicals which often produce rapid growth changes in their various target organs. Such changes may affect behaviour indirectly in ways that may be overlooked.

For example, in female mammals oestrogen causes growth of the uterus, changes in the epithelium of the vagina and growth of the mammary ducts. In birds oestrogen, sometimes acting together with progesterone, causes enlargement of the oviduct and reproductive tract. In some species it also causes the shedding of ventral feathers and increased vascularization of the ventral skin to form a brood patch. The mammary glands of a pregnant mammal and the crop of pigeons during the latter stages of incubation grow and become engorged under the influence of prolactin.

All these target organs have a nervous supply, and information on their growth may feed back to the nervous system and modify behaviour. Lehrman[287] has shown that the engorgement of their crops with 'milk' is one of the factors that predisposes doves to feed their squabs. When he anaesthetized the nerves of the crop wall the doves' tendency to feed was reduced, presumably because they could no longer perceive crop distension.

Again, bodily changes produced by hormones may render an animal more responsive to certain types of stimuli. Komisaruk, Adler and Hutchison[273] have shown that, under the influence of the oestrogen which increases in her body up to the time of oestrus, the sensory field of the perineal nerve supplying the genital region of a female rat extends markedly. This means that at the peak of oestrus she is more easily stimulated by the probing thrusts of the male's penis and can orientate her body so as to help intromission. Correspondingly, Beach and Levinson[43] have shown that testosterone causes a thinning of the skin covering the glans penis in male rats. This means that the tactile sense organs are more exposed and, under the influence of the hormone, the males become more sensitive to the movements of the female and during intromission. We have just mentioned how the brood patch of some birds develops with oestrogen secretion. Hinde and Steel[222] find that a female canary is more responsive to stimuli from the nest cup after its brood patch has developed, and this affects its behaviour during nest building and incubation.

Peripheral changes must not be ignored, but from the behavioural viewpoint the most important target organs for hormones are regions of the central nervous system itself. With improving techniques the study of the central action of hormones has advanced very rapidly over the past few years. It is possible to implant hollow needles into the brains of animals as small as a rat or a pigeon without interfering with their freedom to behave. Brain atlases of several species are now available from which, using stereotaxic instruments, the tip of a needle can be placed within a fraction of a millimetre of the desired spot. The exact position can be checked later by sectioning the brain after behavioural tests are finished. Two basic techniques have been used. Firstly hormones in solution have been injected into neural tissue or into the ventricles or cavities in the brain; one of these —the third ventricle—is adjacent to the hypothalamus. Secondly, and more commonly, the needle before insertion has its tip coated with a waxy or crystalline form of the hormone (see Fig. 4.12). Refinements of this second technique have made it possible to control very precisely what

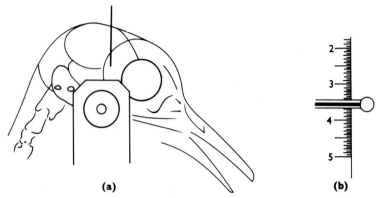

Fig. 4.12 (a) Lateral X-ray picture of the head of a dove showing the skull, held fast in the clamps of a stereotaxic instrument, and with a fine tube carrying a hormone implant penetrating down past the cerebral hemispheres into the anterior hypothalamus; (b) shows a hormone implant fused on to the tip of fine steel tubing. The large units on the scale are millimetres. (Drawn from photographs in Hutchison,[243] 1970. *J. Reprod. Fert., Suppl.* **11, 15.**)

surface area of hormone on the needle tip is in contact with the nervous tissue, and thus to control the rate at which it diffuses away from the tip.

In a now classic study of this type, Harris and Michael[204] investigated the role of oestrogen in the sexual behaviour of female cats. When on heat or in *oestrus*, as this receptive stage is called, female cats show a very characteristic posture which resembles in some respects that of the receptive female rat (Fig. 2.2). The rump is elevated, the tail deflected to one side and the hind legs making treading movements. A cat will assume this posture as soon as a male approaches and will submit to being mounted. This is in sharp contrast to an unreceptive female—in *anoestrus*—who will lash out viciously if a male gets too close. Female cats normally show three oestrus cycles a year and each period of behavioural oestrus is preceded by slow changes in the reproductive system. The uterus grows and its walls thicken, the lining of the vagina thickens and begins to slough off cells. Only when these bodily changes are about complete, does the cat begin to show the behaviour typical of oestrus. It is possible to inject repeated small doses of oestrogen into the bloodstream of castrated females over a long period and produce complete growth of the reproductive tract without changing behaviour.[344] This immediately suggests that in cats the initiation of oestrus behaviour does *not* depend on nervous feed-back from the enlarged reproductive system; i.e. the peripheral effects of oestrogen are not involved in the behavioural changes.

Harris and Michael have produced clear confirmation of this by hormone implants. If a needle whose tip is coated with a slowly dissolving oestrogen is placed in certain parts of the hypothalamus, castrated cats often show strong oestrus behaviour. Most importantly they show full oestrus behaviour whilst their reproductive system remains completely undeveloped. The needle inevitably tears through a considerable volume of brain tissue when it is inserted and clearly it is essential to have a series of control experiments before it can be claimed that the observed effect is due solely to oestrogen acting on the hypothalamus. Harris and Michael provide these. In one set of experiments they got oestrus behaviour in 13 out of 17 castrated cats with implants in the hypothalamus, but no case of oestrus from 18 implants in other parts of the brain. Further, oestrus behaviour was shown only when the needle was coated with oestrogen; other substances had no effect.

Towards the posterior part of the hypothalamus, oestrus can be evoked from a fairly wide area, but further forward only those implants which touch on a tiny area just dorsal to the optic chiasma (see Fig. 4.10) are effective. This looks like an example of the phenomenon which we discussed earlier where the same effects can be produced by stimulation at a number of different sites because they all impinge on the same converging or diverging fibre tracts. Certainly the initiation of sexual behaviour is relatively localized but it probably involves more than one group of neurons. These neurons appear to have a special affinity for oestrogen which they pick up from the bloodstream. Michael[343] found that when he injected radioactive oestrogens, whose exact position can be picked up later in sections of the brain, they were concentrated in just these same areas of the hypothalamus.

One of the problems posed by Harris and Michael's experiments concerns the delay between implanting oestrogens and the appearance of behavioural oestrus. This varied from 4 to 106 days, but once begun oestrus behaviour was just as persistent in cats with a long latency, often lasting for 6 or 8 weeks. Presumably the latency has something to do with rate of diffusion of the hormone away from the needle tip and into responsive neurons.

Recent studies with other animals have confirmed the general conclusions derived from the cat experiments. For example the work of Hutchison[243] and Komisaruk[272] with implants in doves has revealed regions in the brain which are especially sensitive to hormones. The implantation of testosterone in the anterior hypothalamus and the pre-optic nucleus (so called because it lies just anterior to the point where the optic chiasma enters), causes castrated male doves to show courtship behaviour and aggressiveness. Progesterone implanted in the same places suppresses these behaviour patterns, and causes an increased tendency to incubate— a direct demonstration of the antagonism between progesterone and testosterone which we mentioned earlier.

There is also neat confirmation of the role of the hypothalamus in initiating not just courtship and copulation, but the whole pattern of reproductive behaviour. Thiessen and Yahr[451] have worked with the Mongolian gerbil in which males mark their territories with an oily secretion from a gland on their ventral surface. Castration leads to regression of the gland and to the disappearance of marking behaviour. Testosterone implants—again in the pre-optic area already mentioned—or minute injections into the lateral ventricles cause castrated male gerbils to go through the full patterns of marking. Their ventral glands, however, remain completely undeveloped because not enough testosterone gets out into the blood stream. Thus there can be no doubt about the central control of marking behaviour.

Many problems remain concerning the central action of hormones on behaviour. For example there are anomalies concerning the latency of action. As mentioned, oestrogen implants in cats often take many days to produce their effects but oestrogen injections into the wing muscles of doves produce their maximum effect upon behaviour within 30 minutes, (Vowles and Harwood[480]). Before we can interpret such diverse results we shall need to know much more about the nature of the central nervous target organs and how the hormones get to them.

On p. 152 we described experiments which linked the amount of time a male fish spent fanning with the amount of prolactin he had been given. Some of Hutchison's[244] work with implants of testosterone in doves provides evidence that as the concentration of hormone rises in the hypothalamus, so a male's behaviour changes, not just quantitatively but qualitatively also. The earliest part of the sexual behaviour sequence—the nest soliciting display—comes in at the lowest concentrations to be replaced by the higher intensity elements—bowing and chasing—when larger implants are used. This suggests that sometimes there is a quantitative relationship between a hormone level and motivation, defined in behavioural terms. Work with the behaviour of female doves confirms this. Cheng[91-2] found that when a female's ovaries are removed depriving her of oestrogen, her different sexual behaviour patterns disappear in a fairly regular order. They reappear in the reverse order as oestrogen is given in slowly increasing doses by injection.

Elaborate sequences of behaviour usually involve more than different levels of one hormone. This is because there are always interactions and just as hormones affect behaviour so, in turn, behaviour affects the level and type of hormone secretion. A relatively simple case is seen in those female mammals—cats and rabbits for example—which will not ovulate unless mated. The stimulus of copulation is needed to trigger off a pulse of·LH from their pituitary gland and cause ovulation.

Some of the best examples of interactions between hormones and behaviour again come from doves and pigeons. It has been known for more than 50 years that female pigeons do not normally lay eggs if kept alone or with other females, but begin to lay soon after a male is introduced. This must mean that pituitary secretion is influenced by the male, and simple tests show that the sight of a male is enough to cause ovulation, provided he is able to court the female. Thus a male who performs the typical courtship bows, even though separated behind a glass screen, is much more effective than a castrated, and therefore inactive, male in the same cage.

Lehrman and his co-workers have investigated the behavioural and endocrine basis of the reproductive cycle in the ring dove. When males and females are kept separately in large stock cages, they show few signs of reproductive activity. However, if a pair is put into a cage together courtship begins within a few minutes and this marks the beginning of a reproductive cycle. The birds court and mate, build a nest, lay eggs, take turns to incubate them, feed the squabs when they hatch until they leave the nest and can feed themselves, and then begin courtship again and the cycle repeats. This elaborate sequence occupies about 6 weeks and at each stage the bird's behaviour and physiological condition are both adapted to current needs and preparing for the next stage. The consideration we have given earlier on the time scale for neural and endocrine control certainly suggests that there must be underlying hormonal changes, and Lehrman's work reveals how these interact with external stimuli and the behaviour of the doves. We must condense the results of a whole series of elegant experiments; Lehrman[288-9] gives fuller accounts with references.

A male dove is usually ready to court the moment he is put with a female, which probably means that his testes are active and secreting testosterone. Visual stimuli from his courtship activate hypothalamic centres which control pituitary secretion in the female and induce the release of FSH and LH. These stimulate the growth of her ovary which in turn secretes oestrogen under whose influence her reproductive tract begins to grow. After a day or so both birds select a nest site and begin to build. The start of nest building may be related to increased secretion of sex hormones, but this point is still uncertain. During nest building courtship continues and the birds copulate. When the nest is complete the female becomes increasingly attached to it and she soon lays her first egg. This follows the release of LH from her pituitary, and for a few days previous to this her ovary has begun to secrete progesterone which induces incubation behaviour. The male, too, secretes progesterone which acts antagonistically to testosterone so that his courtship and aggression die away and he becomes willing to incubate eggs. The birds take turns to sit on the eggs and after a few days' incubation the first signs of growth in their crops can be seen which indicates that their pituitaries have begun to secrete prolactin.

The act of incubation itself probably brings this about by feed-back to the hypothalamus which 'switches on' the pituitary. Prolactin not only leads to crop growth but also inhibits the secretion of FSH and LH which, in turn, shuts off the secretion of sex hormones and sexual behaviour between the pair dies down. After 14 days of incubation the eggs hatch and the squabs are fed on crop milk. This is pure at first but after a few days an increasing proportion of solid food appears in the milk which matches the growth of the squabs. After 10 to 12 days the young birds leave the nest, by which time parental feeding and prolactin secretion have begun to decline. As prolactin levels in the blood fall so FSH and LH can be secreted and the male soon starts to court the female again.

Some of the experiments involved in working out this sequence are quite dramatic. Thus the secretion of prolactin normally follows when the doves have been sitting on eggs for a few days, but a male dove's crop develops even if he does not actually incubate so long as he can see his mate incubating! Here then a specific visual stimulus leads to the secretion of a specific hormone, but it will do so only if the male is in a particular physiological state. He must previously have participated in nest building, because if he is separated from his mate earlier in the cycle his crop does not develop, even if he can see her incubating.

The general organization of the dove breeding cycle is typical of all birds, although the details of hormone action at the different stages vary. Thus in the dove progesterone secretion is certainly important for the start of incubation and prolactin is almost ineffective. In chickens on the other hand, it is clear that prolactin alone can initiate incubation behaviour and produce a 'broody' hen.

Clearly the relationship between hormones and motivation has been subject to change during evolution. The external stimuli which operate as cues in the reproductive cycle also vary from species to species. They are adapted to the particular circumstances which operate during the normal breeding of the species. Natural selection can provide some remarkable variations on the basic theme. Female cuckoos do not build nests but respond to the sight of their host's nesting behaviour. A cuckoo's ovulation is so timed that it can lay an egg in each of several nests just as the host bird completes its own clutch (Chance[89]).

5

Conflict Behaviour

In this chapter we shall be considering situations in which the smooth course of behaviour is interrupted in various ways. Firstly, an animal may be trying to achieve some result but its attempts are thwarted. A hungry rat can see food at the end of a runway, but finds its path blocked by a glass plate. A bird trapped in a room makes futile attempts to escape through the window panes. Secondly, we have situations where two mutually incompatible tendencies are aroused simultaneously. Suppose the hungry rat in the runway can see food and its path is not blocked, but it has previously received an electric shock at the place where the food is. The food stimulates it to approach, but memory of the shock keeps it back.

We might label the first examples as 'thwarting' where only a single tendency is aroused, and the latter 'conflict' where two tendencies oppose one another. In practice there is a good deal of overlap, and it has been suggested that 'pure' thwarting does not exist since frustration may lead to avoidance of the frustrating situation and thus a kind of approach/avoidance conflict may be set up. At the moment we have insufficient evidence to decide this, but in any case thwarting and conflict can be considered together because both have similar behavioural results and both tend to produce the same physiological changes in the body.

Conflict is an important topic because, contrary to popular belief, it is not only man who is exposed to frustration and torn by opposing aims. Things do not always run smoothly in the animal world and, as we shall see, natural selection has produced a number of ways for dealing with conflict situations. In general, conflicts in nature are of rather short duration, but it is easy to produce a chronic conflict situation in a laboratory animal. Such situations have been extensively studied by psychologists because they may throw light on human neuroses, some of which may originate from just such a chronic conflict.

Here we shall concentrate on 'natural' conflicts in their biological setting, dealing mainly with the work of ethologists on vertebrates, but in conclusion we shall try briefly to relate this to some of the work on animal neuroses.

'Stress'

At this point it is necessary to consider the bodily changes which are likely to occur in conflict situations. We can lump them under the term 'stress', because the body's response to a wide range of 'stressors' (thwarted escape, overcrowding, extreme cold and burns, for example) is very similar. Barnett[29] describes the physiological changes involved and points out that most of them are attempts to restore the delicate balance of the body's metabolism when it has been upset.

In moderate stress we can detect increased activity in the autonomic nervous system (see Romer[397] for details of its structure) which supplies the viscera and smooth muscle. It also supplies the adrenal glands, endocrine organs close to the kidneys which have a double structure, an internal medulla (supplied by autonomic nerve fibres) and an external cortex. Stimulation of the adrenal medulla via its autonomic nerve supply causes it to release the hormone **adrenalin** into the bloodstream. This causes changes in numerous parts of the body; the sweat glands of the skin begin to secrete, hair becomes erected, the heart beats faster, breathing becomes more rapid and deeper, and blood gets diverted to the muscles from the alimentary canal. We have already mentioned these changes in the last chapter, because they also accompany the strong arousal of tendencies such as attack, escape and sex; they prepare the body for violent action of any type required. In brief conflicts there will perhaps be a rapid flush of adrenalin through the animal which then subsides, but if the stressful situation persists, then a further reaction begins. This involves the adrenal cortex which is stimulated to release its hormones, not directly by nerves as is the medulla, but by another hormone, **adreno-cortico-trophic-hormone** (ACTH) produced by the pituitary gland. Here, as with the release of adrenalin, the·nervous system initiates the response. Stress activates cells in the hypothalamus which itself then stimulates the pituitary to release ACTH. It is not yet fully clear how the adrenal cortex hormones help an animal to adapt to stress. Some of them are concerned with glucose metabolism and may serve to mobilize the body's long-term food reserves. There is now some behavioural evidence that ACTH itself helps to reduce 'anxiety'. Whatever their action, the release of adrenocortical hormones is most dramatic. The cortical cells become drained of their contents and, if stress persists, the adrenals enlarge, sometimes by 25%, and all this growth is found to be in the cortex. Animals under chronic stress become really ill and may die. Barnett[29] has shown how a wild rat, unable to escape from the territory of a dominant, resident male, may die after a few hours of intermittent attacks, even though it has no wounds. Persistent stress due to overcrowding has been suggested as one reason why natural populations may show a rapid decline after a period of unusually high numbers (Chitty[93]; Christian[94]). Death associated with stress is certainly a normal feature of the life cycle of some short-lived animals. A most remarkable example is the small carnivorous marsupial *Antechinus stuartii* where males live for just under a year and they all die over a three-week period following mating. During the mating season there is a great deal of aggressive interaction between males, and it is known that their metabolism of adrenal cortex hormones is altered by this behaviour. If *Antechinus* males are kept alone in cages, they live well beyond the normal life span (Lee, Bradley and Braithwaite[285a]).

The diverse physiological changes we have been discussing can be regarded as successive stages on a scale of 'arousal'. In Chapter 3 we mentioned how any incoming stimulus leads to the activating of the brain's higher centres via the reticular formation; at this stage an animal will begin to 'pay attention'. If the stimuli are strong and persistent the hypothalamus initiates activity in the autonomic nervous system; adrenalin begins to circulate and the animal becomes highly aroused and perhaps shows 'emotion'. The final stages on this scale are those just described, with intense and persistent arousal leading to the release of adrenocortical hormones and the animal showing signs of severe stress (see Hokanson[228] for further details).

We shall find that a knowledge of the underlying physiological changes helps in the interpretation of some behaviour which appears in conflict situations.

Territorial conflicts

Perhaps the best way to approach the study of conflict as it occurs naturally is by way of territorial behaviour which, as we shall see, almost inevitably leads to conflict where one territory meets another. The best definition of territory is 'any defended area'. Commonly, at the beginning of their breeding season, males establish and defend an area against other males of the same species. Such behaviour is widespread among the vertebrates. Many fish and lizards are territorial, so are the great majority of birds and many mammals, especially carnivores. Perhaps territoriality is rarest amongst amphibians but even in this class there are examples. Territory crops up sporadically among invertebrates, in such insects as crickets and dragonflies and in the fiddler-crabs, but it is relatively rare.

We are still far from a complete understanding of the ultimate functions of territory. Something whose establishment and defence takes up so much of an animal's time must have a powerful selective value. A host of possible functions have been discussed, the majority of which suggest that this spacing out of the population is related to the available supply of food. The exact nature of this relationship is still a matter of dispute and the various arguments can be followed in Lack,[284] Wynne-Edwards[514] and also in Volume 98 of the ornithological journal, *Ibis*, which contains a series of articles on the role of territory in various bird species. There is some further discussion of this problem in relation to social organization in Chapter 8.

Certainly the relationship between territory and food cannot always be the same because the form which territories can take varies so greatly. Some birds of prey defend a large area a mile or more square from which they get all their food. The herring-gull's territory is a couple of square yards around its nest in the gull colony. In the bird of prey, food and territory are directly related but the gull gets no food from its territory. However, its territorial behaviour will limit the size of a gull colony and thus control the number of herring-gulls feeding on the nearby sea-coast.

Apart from food supply, there are other selective advantages to holding a territory which in some species may have played an important part during evolution. Among them are freedom from disturbance during pair formation and the avoidance of predation and disease.

Here we must concentrate on the behaviour associated with holding a territory. The first essential for a territorial animal is that it must be aggressive towards others of its kind. As we discussed in the previous chapter, aggression is seen in other situations as when animals fight over food or establish a 'peck order' within a group, but it is most conspicuous at the start of the breeding season when territories are being established.

Aggression alone is not sufficient. If animals were simply aggressive they would spend far too much time and energy in fighting and damage themselves. Further they might behave too aggressively towards their mate or their offspring as well as to a rival. Snow[442] records the case of one male blackbird who was extremely aggressive early in the breeding season and won a large territory. He attracted a female, but instead of the rapid waning of aggression between members of a pair which normally occurs, this male continued to attack his mate for over a month. His attacks were so persistent that she failed for several days in her attempts to start a nest.

Clearly this is aggression carried to a maladaptive level and we can observe an even more extreme case in one breed of the domestic chicken. Cock-fighting is still a popular sport in many parts of the world where deliberate selection for fighting abilities has been practised for centuries. Game-fowl have extremely powerful leg muscles, long spurs and a greatly reduced comb—a vulnerable spot for damage in normal breeds. They have the temperament to match their physique and are fantastically aggressive. A normal rooster on his home territory will attack and drive off any new-comer who is placed there—the latter normally beats a retreat with no delay. However a game-cock placed on the territory of a dominant cockerel of another breed will instantly attack the territory holder and drive him off. The inevitable result of this abnormal aggressiveness is that game-fowl are difficult to breed, because females are aggressive too and much time is wasted in fighting.

Game-fowl would have little hope of surviving in the wild, where natural selection has prevented aggression from reaching too high a level and has favoured the evolution of a compensating tendency—that of escape. The majority of animals show escape responses to predators, but here we are concerned with escape evoked by a territorial rival.

After a few days of strife in early spring a population of male birds will begin to settle down, each with a territory in which he is dominant. Males which appear later and try to establish themselves will have to fight hard

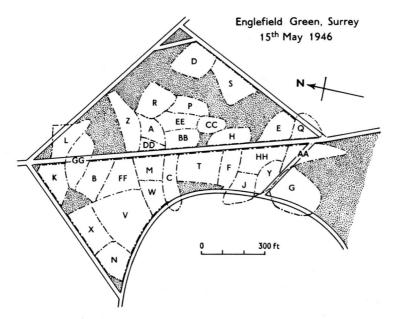

Fig. 5.1 A territory map of a population of willow-warblers in birch woodland. The size of territories varies from under 1,000 square yards to more than 5,000 square yards. Shaded areas were unoccupied. (After May,[338] 1949, *Ibis*, **91**, 24.)

and may win only a small area. Territories have been likened to elastic discs —the more they are compressed, the more they resist further compression. Each male is most aggressive near the centre of his territory. As he moves away from the centre his attacks upon a trespassing neighbour are less violent, and there comes a point at which he is equally likely to attack or escape when his neighbour approaches. This point we can call the boundary of his territory and careful observations on a population of males enable one to construct a map with boundaries drawn through the points at which neighbouring birds show this balance between attack and escape. Fig. 5.1 shows such a map for a population of willow-warblers; similar maps have been made for a group of sticklebacks in a large aquarium.

In summary, the establishment of a territory involves both attack and escape behaviour, and if an animal meets its neighbour at the boundary between their territories, both attack and escape tendencies are probably aroused and form the basis of a conflict situation.

In such a situation an animal may do any one of a number of things. (The term 'agonistic behaviour' is often used to cover all the different types of response seen in fighting and territorial behaviour generally.) We may observe actual attack or escape or their intention movements; the animal may alternate between the two or show some kind of compromise movement or posture.

Some types of agonistic behaviour have become modified by evolution into displays which probably serve both to intimidate a rival and to reduce actual fighting to a minimum. One of these displays—'threat'—we shall discuss in some detail, but first we may consider fighting itself.

'Pure' attack and escape

During fighting we can observe attack and escape behaviour, each operating to the virtual exclusion of the other. Prolonged fighting is extremely rare under natural conditions for the weaker of two combatants usually breaks away and escapes before any serious damage is done. Clearly selection favours individuals who quickly recognize defeat if it is inevitable. 'He who fights and runs away, lives to fight another day.' Sometimes fighting behaviour has evolved into an almost formalized contest in which two rivals can assess each other's strength without damage. Lorenz[309] describes the mouth-fighting of various cichlid fishes. In *Tilapia natalensis*, rivals approach each other displaying wide open mouths, seize their opponent's jaw and then pull against each other (Fig. 5.2). In the related *T. mossambiqua* the rivals oppose their opened mouths and push. In each case the weaker fish soon breaks away and accepts defeat. In an analogous fashion some male ungulates, e.g. deer and goats, engage their antlers or horns to push against each other. Commonly these appendages grow larger with

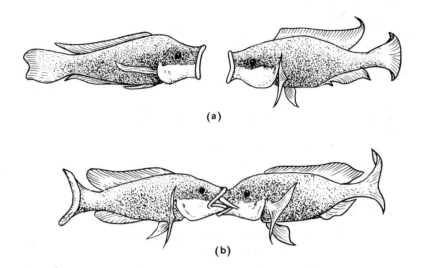

Fig. 5.2 Boundary fighting in *Tilapia natalensis*. The rivals approach with open mouths displayed (a), and then pull with jaws interlocked (b). (From Baerends and Baerends-van Roon,[22] 1950, *Behaviour Suppl.*, **1**. 1)

each successive year of a male's life. Geist[166] has shown that sometimes the size of horns alone comes to have symbolic value in disputes between males and that a male will not attempt to contest with another whose horns are much larger than his own.

'Pendulum' fighting is sometimes seen at the boundary between two territories, with attack and escape rapidly alternating. Such fights can be observed in cichlid fish who dash back and forth across an invisible line in the tank they share. They are also familiar in birds; one male flies unhesitatingly at an intruder who flees, but the momentum of the chase carries both birds over the boundary between their territories and promptly their behaviour changes with pursuer becoming pursued. The rapidity with which attack can succeed escape in pendulum fights demonstrates how closely both tendencies are dependent on the external situation.

Quite commonly a victorious animal whose rival has been driven off will remain highly aggressive for a time. In such a state he may show the redirected aggression which was discussed in the previous chapter and attack another neighbour or even his mate who would not normally become involved.

THREAT DISPLAYS

'Compromise' behaviour of various kinds may be seen when the tendencies to attack and to escape, or, more generally, to approach and to avoid, are both aroused simultaneously and to about equal extents. An animal may remain rather still in one place, perhaps making abbreviated intention movements to advance or retreat. Sometimes, instead of facing one way or the other it will turn its body sideways, though the head may still move from side to side. Behaviour of this type can be seen if one walks slowly towards a bird when it is feeding young in the nest. A similar conflict is shown by an octopus which has been given electric shocks on a number of successive trials when it attacked crabs at one end of its tank. It approaches the next crab in a hesitant, sidling fashion; half its tentacles remain attached to the shelter in which it lives, the other tentacles reach out towards the crab, but they approach it laterally rather than directly. The octopus expresses most vividly a compromise between advance and retreat in the actual way it holds its body.

Many territorial animals have special postures which are most commonly seen during boundary disputes. For example, two neighbouring lesser black-backed gulls may approach one another but do not fight. Each adopts the rather 'forced'-looking posture shown in Fig. 5.4a, and may stand facing its rival or turn sideways and walk parallel to him. In a similar situation two male cichlid fish, *Cichlasoma meeki*, slow down as they approach one another and remain facing with their gill covers fully raised (Fig. 5.3a). Two hostile male rats turn sideways when they meet and twist around each other with arched backs and partially raised fur (Fig. 5.3b).

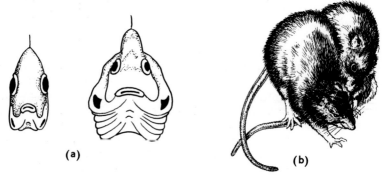

Fig. 5.3 Some threat postures. (a) Frontal threat display of *Cichlasoma meeki* (right) compared with the normal front view (left). (From Baerends and Baerends-van-Roon,[22] 1950, *Behaviour Suppl.*, 1, 1.) (b) Lateral threat display between male rats. (From Barnett,[28] 1963, *A Study in Behaviour*. Methuen, London.)

Ethologists call behaviour of this type 'threat', and boundary disputes often go no further than mutual threatening followed by mutual withdrawal. Threat forms an admirable substitute for actual fighting and one writer has neatly summed up its role in animal disputes. It is not, 'speak softly but carry a big stick', but rather, 'shout your head off but run away if the other fellow isn't impressed'!

In classical ethological theory threat is considered to result from dual motivation—a conflict between simultaneously aroused tendencies to attack and to escape when neither can find separate expression. Three main lines of evidence are presented to support this conclusion.

1. Threat is most commonly observed at the boundary between territories where, as we have seen, there is good reason to believe both tendencies are aroused simultaneously. This might be called evidence from situation or context, and we can couple with this the observation that threat behaviour is most commonly preceded and followed by 'pure' attack or escape. This again, is what would be expected if these two tendencies were involved in its performance. Presumably a shift in the threshold of one or the other can lead to the often rapid transitions in behaviour which are observed.

2. Linked with this conclusion, there is some evidence on the causation of threat which comes from independent manipulation of the levels of attack and escape tendencies. Blurton-Jones[59] had a group of completely tame Canada geese which ignored him when he was dressed in familiar old clothes. If he wore a white coat the geese attacked him uninhibitedly, whilst if he appeared carrying a broom (used to drive the geese into their house for the night) they would flee. The familiar threat postures of the goose (lowered head on outstretched neck, hissing etc.) only appeared when Blurton-Jones combined wearing a white coat with carrying a broom!

3. Finally there is also evidence gained from a close examination of the form of threat postures which can sometimes be analysed into elements belonging both to attack and to escape behaviour, and for this reason such postures are called 'ambivalent'. The threat posture of rhesus monkeys illustrates this well. The animal sits or stands facing its opponent, the positions of its limbs revealing a compromise between advance and retreat, and alternately lunges its head forwards and pulls it back.

In some animals a detailed analysis of the different elements of a threat display has been possible. As an example we can examine the lesser black-backed gull's threat posture (Fig. 5.4a) more closely. The bird moves towards its rival with the neck stretched upward and slightly forward and the head and bill turned down. The wrist joints of the wings are lifted

(b)

(a)

Fig. 5.4 (a) The 'upright' threat display of the lesser black-backed gull. (b) The 'hunched' appeasement posture of the same species (see p. 180). It represents an almost perfect antithesis to threat posture. The head is held low on a shortened neck and the bill points upwards, whilst the wings are pressed close into the flanks so that the wrist joints—so conspicuous in threat—are completely hidden. (Photos by N. Tinbergen).

well clear of the body and the plumage is slightly raised. From a thorough knowledge of the gull's whole repertoire of behaviour it is possible to identify several elements of attack behaviour here. Gulls normally launch an attack by beating with the wings and attempting to peck down at their opponent. The raising of the head and down-pointing bill look like the first stages of actual attacks. When a gull takes off to fly, the wrists are raised only at the last moment, so the raised wrist joints in the threat posture are probably attack also. But the gull does not attack, something holds it back and elements of escape can also be seen, particularly as two rivals come very close. Now the head moves increasingly back, the bill becomes lifted and the plumage sleeker, further the bird may turn sideways on to its opponent and move parallel to, not towards it. This turning aside looks like an element of escape and gulls draw back their heads and sleek the plumage preparatory to taking off, so these may be escape elements also. In fact this 'upright' threat posture, as it is called, can take on a range of forms depending on how many attack and escape elements are shown. This range is assumed to reflect slight changes in the levels of the underlying attack and escape tendencies. Fig. 5.5 shows three gradations of the 'upright' in the herring-gull which can be called 'aggressive', 'intimidated', and 'anxiety upright' respectively, as they show increasing elements of escape.

All this analysis of the threat behaviour of gulls is derived from Tinbergen and his students, who have made gulls one of the best-known groups of birds. Tinbergen[465] provides an excellent review of this work.

Not all threat postures can be interpreted as a mixture of attack and escape elements. Some of them include elements which seem to have their origin in the physiological side effects of conflict. The secretion of adrenalin and arousal of the autonomic nervous system will, as we have seen, lead to deeper and more rapid breathing. Fish will thus have to raise their gill covers to a greater extent in a conflict, and air-breathing vertebrates fill their lungs more. Morris[355] suggests that here is the origin of threat displays which emphasize the gill covers, as in *Cichlasoma*. Other animals, including some amphibians, reptiles and birds, inflate air sacs when they display. For example, the tropical frigate-birds have air sacs on their throats. These are normally quite inconspicuous but in display they become enormous and are bright scarlet in colour (see Fig. 3.12, p. 91). In birds and mammals erection of the hair or feathers is another important result of autonomic activity in a conflict. Very many threat displays have this for a basis; birds have evolved conspicuous crests or ruffs which are raised in threat, mammals erect their hair as in the familiar threat postures of cats and dogs. Morris provides many other examples of the way in which selection has produced striking displays based upon the body's physiological response to conflict.

There is one complication to threat behaviour which is of considerable

theoretical importance. We have earlier described the different forms of a
threat posture which may correspond to changing degrees of escape
tendency (Fig. 5.5), but some animals have several distinct threat postures
which do not blend into one another. If we accept the general conclusion
that threat occurs when attack and escape balance, how can this interpretation be applied when there are several types of threat?

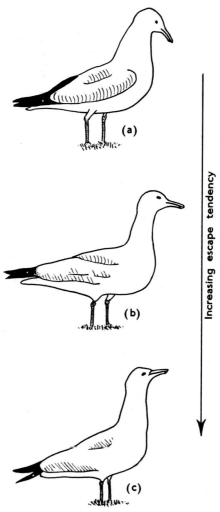

Fig. 5.5 (a) 'Agressive', (b) 'intimidated', and (c) 'anxiety' upright threat
display of the herring-gull. (From Tinbergen,[465] 1959, *Behaviour*, **15**, 1.)

One possibility is that the different postures represent different levels of the attack and escape tendencies. For threat to occur at all both must be roughly equal, but if both are low posture A appears, if both are high, posture B. On p. 169 we discussed the observation that 'pure' attack and escape were the most common sequel to threat behaviour. This fact might assist in estimating the causation of different threat postures by watching the outcome of an animal adopting postures A and B. Assuming— and it is only an assumption—a particular type of interaction between attack and escape tendencies, one might expect that if threat posture B occurs when both tendencies are high, then it would be more commonly followed by attack or escape than would posture A. Because both tendencies are low in A, other types of behaviour—feeding or preening, for example—might be able to occur because their control mechanisms would have less 'resistance' to overcome from attack and escape tendencies.

Tinbergen,[465] whose group have used this method, recognizes that it has limitations. Under field conditions it is very difficult to be sure that a bird's behaviour following threat is not affected by the responses of others. Further, as we shall discuss later, there are other types of behaviour (displacement activities) which result from a conflict, and these might appear following threat and upset the kind of analysis outlined above. Moynihan[358] has used this method with the black-headed gull which has four distinct threat postures called, 'upright' (see Fig. 5.4) 'oblique', 'forward'

(a)

(b)

Fig. 5.6 Variants of (a) the 'head-up' and (b) the 'horizontal' displays of the great tit, *Parus major*. (From Blurton-Jones,[60] 1968. *Anim. Behav. Monogr.,* **1**, 75.)

and 'choking'. He interprets their causation in the terms outlined above, with 'choking' representing the highest levels of conflicting attack and escape. Stout[160, 446-7] and his collaborators have worked with the glaucous-winged gull which has a very similar repertoire of threat postures. Like Moynihan they also rank the different postures by observing the proportion which lead to attack. They also looked at the number of attacks made by a territory owner when an intruder adopted each type of threat posture. They suggest that only the tendency to attack is varying, although presumably they would accept that some conflicting tendency inhibits its expression during posturing.

Recent work on gulls and other birds indicates that tendencies other than attack and escape may be involved in the performance of postures which hitherto have been lumped together as 'threat'. 'Choking' in gulls, which has just been mentioned, is a case in point. It is performed most commonly by birds on their territories which are very reluctant to move, and its form resembles nest building in some respects. It seems reasonable to conclude that a rather non-specific tendency 'to stay put' is involved here. Blurton-Jones[60] came to a similar conclusion from his detailed analysis of displays in the great tit where although all of the threat postures depended on the presence of a stimulus which elicited attack tendencies, other tendencies determined which posture appeared. 'Head-up', for example, (Fig. 5.6a) certainly involves an escape tendency interacting with attack but 'horizontal' (Fig. 5.6b) seems to depend, like 'choking' in the gulls, upon a strong tendency to stay put.

This ethological view of threat behaviour which we have been discussing clearly proposes that there are distinct attack and escape systems which conflict when simultaneously aroused. As we mentioned in the last chapter, ethologists have usually thought in terms of specific drives. This 'conflict hypothesis' of threat behaviour, as it has been called, has naturally attracted criticism from those who reject the idea of separate control systems and separate drives.

Some critics have suggested that threat behaviour does not require the postulation of a conflict between major control systems. Andrew[12,15] emphasizes how physiological arousal in conflict situations will lead various behavioural elements to appear some of which, such as feather raising, can often be seen in threat; we have already mentioned some other examples. He has also made a particular study of primate calls and facial expressions and suggests that many of the elements used in threatening are derived from simple reflex movements which protect the eyes and expel air from the lungs.

Andrew's suggestions are persuasive in these cases and certainly we must avoid unnecessarily complex explanations for threat displays if simpler hypotheses will serve as well. Yet in part the different types of explanation reflect the different scale of the behaviour whose causation is to be explained. If we look in detail at one aspect of a display pattern, e.g. the threat face of a rhesus monkey, then each element can probably be interpreted as the result of peripheral defensive reflexes. However, as described above, the total threat display involves a characteristic compromise positioning of the limbs and a jerking back and forth of the head and sometimes the whole body. This certainly suggests that some more central control systems are involved and competing for control of the animal's behaviour. We have also described the same co-ordinated involvement of several different parts of the animal making up the threat display of the lesser black-backed gull.

One particularly good example of two independent systems operating to determine the total form of a display comes from Wiley's[500] study of the song-spread display of grackles. The carib grackle is a large passerine bird which performs its displays both to rival males and to females (see Fig. 5.7). The song-spread varies in form and involves movements of legs, tail, head and wings, but here we are most concerned with variations in two of its components—the angle of elevation of the beak and the degree of elevation of the partly-spread wings. Wiley recorded the beak angle and wing elevation of song-spreads made to males and females and found a regular difference. Beak angles were higher in displays towards males and wings were raised more in displays towards females. The grackles have a distinct aggressive posture in which the beak is pointed upwards, so it is reasonable to suggest that beak angle in the song-spread reflects the level of the male's aggressive tendency. Another tendency, perhaps sexual, relates to wing elevation and the two tendencies can influence the form of the total display quite independently of each other. Wiley's results are certainly in accordance with the conflict hypothesis.

Another criticism which has been directed at this hypothesis concerns the results of some brain stimulation experiments. In terms of the behaviour they produce it is clear that the attack and escape tendencies are mutually inhibiting. It is perfectly reasonable to suggest that the neural mechanisms behind the overt behaviour are also mutually inhibitory. As we discussed in Chapter 1, it is by inhibition that sudden but smooth transitions from one type of behaviour to another are achieved. The behavioural evidence would indicate that attack and escape tendencies can remain in a fairly stable balance and still allow the simultaneous expression of behaviour elements belonging to both. As yet we do not know how this is achieved in neurophysiological terms. In particular we have no idea how a balance between the two struck at different levels could produce different types of threat posture. Leyhausen[293] has proposed on behavioural grounds that there is a continuous range of threat postures in the cat which represent different relative strengths of attack and escape

priscilla barrett

Fig. 5.7 A male carib grackle performs a 'song-spread' display towards a female who perches on the right. (From a photo by R. H. Wiley.)

tendencies. The facial expressions associated with these postures are illustrated in Fig. 5.8. Brown and Hunsperger[74] have studied attack, escape and threat in the cat by implanting stimulating electrodes in the hypothalamus and elsewhere. They do not consider that attack and escape are distinct systems because the areas of the brain from which they can be elicited overlap with each other and with areas which produce threat. They found it impossible ever to get 'pure' attack by electrical stimulation, it was always preceded by threat, and they could elicit threat from single points of stimulation at widely different loci in the hypothalamus and amygdala.

As we discussed in the previous chapter, there are problems in interpreting brain stimulation experiments and it is not always easy to relate the behaviour evoked through electrodes to that occurring spontaneously. Brown and Hunsperger's results cannot be said to exclude the possibility that there are distinct attack and escape systems which interact, but their suggestion that threat displays may have a more unitary basis is an important one. Delius[126] also found that he could elicit threat displays in gulls from single implanted electrodes. Baerends[20] provides a thorough survey of this and all the other problems relating to the conflict hypothesis. It still remains most useful at a behavioural level even if we must await further research before we are able to interpret threat behaviour in physiological terms.

Increasing aggressiveness ⟶

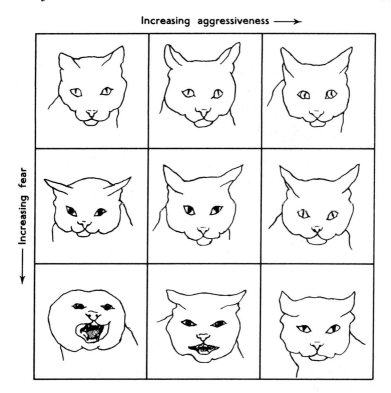

Increasing fear

Fig. 5.8 Changes in the facial expressions of the cat with their suggested motivational basis. (From Leyhausen,[293] 1956, *Z. Tierpsychol., Beiheft,* **2,** 1.)

Threat as a signal

An experienced ethologist can usually predict what an individual animal will do next from its behaviour at a given moment. The analysis of the upright threat posture of gulls is based on a full knowledge of their behaviour which enables the intention movements of attack and escape to be identified. If we can interpret intention movements, it seems highly probable that other members of the same species can do so. We have seen that many threat displays are made more conspicuous by specially evolved releaser structures which emphasize the movement or posture. Threat postures are good examples of the kinds of communication system which we discussed at the end of Chapter 3.

We learnt there that it is not always easy to measure the information conveyed by social signals and this is certainly the case for threat displays.

We can observe the general responses of rival males towards one another, but it requires rather elaborate experiments to deduce the information content of different displays. Stout's group[160, 446-7] have used models of gulls mounted in various threat postures to study communication by threat in the glaucous-winged gull. The models were placed in a pair's territory which is a rather unnatural situation in that a gull, instead of being approached by an intruder from the edge of its territory, returned from an absence to find the intruder already in occupation, as it were. They found that postures which on other grounds (see p. 174) were deemed to be most aggressive, were attacked less often and with longer latencies than less aggressive ones. The presumption was that a posture conveyed its aggressive content to the territory owners who attacked most vigorously when they had least fear of retaliation. Stout and his collaborators found that the level of the head above the ground was all-important in conveying aggressiveness—the lower the head, the more aggressive the posture. Thus dummies in the 'upright' posture were attacked with least hesitation and those in the 'choking' posture with most reluctance. Conversely, the lower the head of the model the more often territory holders made movements away from it during the course of the encounter. The angle of the bill, which varies in the natural situation (see Fig. 5.5 for example), had little effect compared with head height. Some vocalizations were important however and when a loudspeaker was used to emit calls from the dummies, the rhythmic call that accompanies the choking posture was most effective in increasing the latency of attack upon all dummies whatever their posture. No other call was so effective and clearly the choking call also functions a highly aggressive signal.

One of the most detailed analyses of the communicatory role of threat is that of Stokes[445] on the agonistic behaviour of blue tits. He observed them disputing over food in winter and recorded various elements in their threat postures, involving beak, crest, body feathers, wings, tail and orientation of the body, among others. Stokes does not choose to ascribe individual elements to attack or escape, he considers they may all reflect both tendencies. Particular combinations of elements were much commoner than others and sometimes provided a very reliable prediction of a bird's subsequent behaviour. Thus blue tits which faced their rivals with raised crests subsequently showed escape in 94% of cases and they never attacked. On the other hand, attack was much harder to predict; no combination of elements gave a probability greater than 48%.

Stokes measured the responses of one bird to a rival which adopted a particular posture, eliminating those cases when the latter actually attacked. If a bird, A, adopted an 'attacking' combination of elements this significantly increased the frequency with which its rival, B, would escape—although as we have just seen B may have escaped unnecessarily. If A adopted an 'escape' combination, B's frequency of attack did not rise but it tended to ignore A. It probably recognized that A's posture meant that the latter did not constitute a real threat to its position at the feeding dish. We badly need more quantitative information of this type to augment the wealth of descriptive data on threat behaviour which we already have for a number of species.

Appeasement displays

It is most convenient to mention appeasement displays here because they are often related to threat and are a further mechanism for reducing fighting. Functionally speaking, their role seems to be that of reducing or inhibiting attack in conditions when escape is disadvantageous. We have mentioned that an aggressively aroused male may redirect his attack upon his mate, but it may be disadvantageous for her to flee out of his territory; it is better to stay put. Young birds may also become the victims of redirected attacks and use appeasement rather than escape. Some social animals such as wolves, baboons and domestic chickens form a stable hierarchy or 'peck order' among the members of a group (see Chapter 8, p. 276). Once this is fixed, fights are rare and dominant animals have only to threaten or even move towards a subordinate and the latter defers, often with an appeasement gesture. Indeed it is often observed amongst monkeys and wolves that subordinates will approach dominant animals and go through a ritual of appeasement in the absence of any threat. Such behaviour may serve to promote the cohesion of the group, which is vital for the survival of all its individuals (see Schenkel's[418] account of submission in wolves).

At present most of our evidence on the functioning of appeasement rests on behavioural descriptions but it is none the less quite convincing. Appeasement seems to operate in two different ways. Firstly, attack can be inhibited by arousing a conflicting tendency in the attacker. A subordinate baboon of either sex turns away from the aggressor and crouches in the sexual presentation posture. Sometimes it may be mounted briefly but often it is simply allowed to move away by the dominant baboon which 'accepts' this as a gesture of submission. In Chapter 3 (p. 98) we described Chalmer's[88] observations on the effectiveness of sexual presentation in the mangabey, one of the few cases where we have some quantitative measures on appeasement (see also Fig. 8.10).

Another tendency opposing attack which appeasement may employ is that associated with parental behaviour. Among birds it is quite common to find infantile behaviour used for appeasement. Food-begging behaviour is a common preliminary to mating in gulls and some other birds and it appears to reduce a male's aggressiveness, perhaps by arousing his parental responses.

The second type of appeasement relies on being as different from threat as possible. Threat postures emphasize various releasers, but appeasement postures hide them. In the black-headed gull all of the threat postures emphasize the dark face mask. When a female alights on the territory of her mate they may threaten each other briefly, but after a second or two both lift their heads and jerk them away from each other—hiding the face mask. The kittiwake threatens with the yellow bill, often opened to display the vivid red mouth lining. The appeasement posture turns the head away from the opponent and down, hiding the bill. Fig. 5.4b illustrates the appeasement posture of the lesser black-backed gull, which is of this type and an almost exact antithesis of the threat posture shown in Fig. 5.4a. In fact, this example bridges our two types of appeasement, because this anti-threat posture is also that adopted by young birds when food begging and is used by adults in courtship.

There are examples of anti-threat appeasement postures in mammals also. Wolves threaten by snarling with bared fangs and erect ears and a defeated individual, or one who wishes to appease a more dominant wolf in the hierarchy, lays back its ears and turns away its head so as to present not its fangs but the nape of its neck; domestic dogs often do precisely the same when being punished.

Appeasement displays, just as threat, act as signals between animals. We are left with the problem of why appeasement should be effective. The appeaser is vulnerable and yet the dominant animal rarely attacks further. It is irrelevant to suggest that animals feel mercy—in fact taken out of their natural environment and crowded together they will often kill each other. We assume that selection favours the inhibition of aggression when appeasement is offered. Animals which do not are likely to be hyperaggressive and as maladaptive as the game-cock. Baboons, for example, must live in a group to survive because stragglers are usually killed by leopards or lions. Selection favours any behaviour, such as the appeasement system, which helps to keep the group together. Chance[90] has pointed out that appeasement displays may serve to reduce stress in the submissive performer because, by turning away from the dominant animal, it minimizes the 'frightening stimuli' which impinge upon it. In this way the escape tendency may be reduced sufficiently so that the submissive animal is able to stay with the group.

DISPLACEMENT ACTIVITIES

We must return to conflict situations and discuss some other types of behaviour which are observed. The name 'displacement activities' was given by Tinbergen[461] to a diverse range of behaviour patterns whose most striking common characteristic is their apparent irrelevance to the situation in which they appear.

For example, in the middle of a bout of threatening interspersed with fighting, two cockerels will turn aside briefly and peck at the ground, sometimes picking up stones or grains which they allow to fall again. A male stickleback which has been courting an unreceptive female will suddenly swim to his nest and perform the characteristic parental 'fanning' movement which ventilates the nest with fresh water even though there are no eggs present. A tern which is incubating on its nest makes a few brief preening movements just before it takes off at the approach of an intruder. A thirsty dove which is prevented from getting to its water bowl by a sheet of glass, pecks at the ground nearby.

In all these examples there is good reason to suppose that the animal is either thwarted or in a conflict between two opposing tendencies. The appearance of a brief burst of—say—feeding in the midst of bouts of threat and fighting is surprising, because it seems irrelevant to the tendencies aroused—presumably attack and escape. Normally animals are consistent and one type of behaviour continues for some time without interruption. Further the preening performed by the tern just before it leaves the nest often appears 'forced' and incomplete. So does the feeding of the fighting cockerels which, as we mentioned, may pick up grains but do not swallow them.

The obvious subjective conclusion is that the animal is under 'tension' when performing these acts. Here is one case where subjective experience, if used with caution, may help in understanding animal behaviour. We all know that when we are ill at ease we are apt to do completely irrelevant things without much conscious thought. People rarely sit peacefully in a dentist's waiting-room or outside an examination hall. They fidget, making minor adjustments to necktie or hair, light cigarettes and talk unnecessarily fast and eagerly. It would be perfectly feasible to obtain a measure of the 'stickiness' of a cocktail party from the amount of potato crisps and nuts eaten during its early stages. Nobody is hungry, often quite the reverse, but eating in some way relieves the tension of being with strangers.

Associated with these human examples there are often signs of physiological arousal with raised autonomic activity. This produces a quickened pulse, sweating and the familiar effects of psychological tension on bowels and bladder. Sometimes we can detect similar changes in animals when they perform apparently irrelevant movements.

The term 'displacement activities' refers to a particular theory of their origin which we owe to Tinbergen and Kortlandt (see Tinbergen[461]). This suggested that in the normal control of behaviour a particular quantity of 'nervous energy' is released for the performance of a behaviour pattern, A. The release of this energy through the normal A outlet can be interrupted in two ways. The correct external stimuli for A are absent or some other pattern, B, is aroused whose performance is incompatible with that of A. If this interruption occurs, the energy which pertains to A must find an alternative outlet, and it is 'displaced' through to channel C; behaviour pattern C then appears as a displacement activity. Tinbergen's theory grew out of a 'psycho-hydraulic' type of motivation model, such as we have previously discussed in Chapter 4 (Zeigler[520] provides a clear review of the whole displacement activity concept which sets this theory in perspective). It has provided a useful stimulus to research but has proved inadequate on a number of counts.

The theory set up something of a distinction between a pattern—say preening—performed normally and the same pattern performed as a displacement activity. In the former case preening would be using up its own supply of nervous energy, whereas in the latter the preening control mechanism would be receiving the displaced overflow from another pattern. This implies that preening may occur as a displacement activity without there being any previous activation of its control mechanism, either from within or without; i.e. in the absence of any tendency to preen or external stimulus to preening.

However many observations have shown that this cannot always be the case and suggest that the control of a behaviour pattern, and the influence which external stimuli have upon it, is similar no matter in what circumstances it is performed. For example Räber[387] found that turkey cocks showed bouts of displacement feeding or drinking during fights; which pattern appeared depended on whether food or water was available to the birds. Van Iersel and Bol[247] studied nesting terns which were in a conflict between the tendency to stay on their nest and the tendency to escape. They found that the amount of displacement preening increased when the terns had got wet feathers—a normal stimulus to preening. Sevenster[432] mimicked the effect of eggs developing in the nests of sticklebacks by passing through water containing dissolved carbon dioxide. This increased the amount of displacement fanning a male performed during bouts of courtship.

Sometimes the thwarting or conflict situation which leads to the appearance of displacement activities may also enhance the relevant stimuli for them. Deaux and Kakolewski[125] have shown that a brief period of stress caused by handling rats led to abrupt changes in the salt concentrations in the body fluids—one of the normal signals of water deficit—and consequently caused them to show a brief period of drinking. This might well be labelled as 'displacement drinking' by an ethological observer. The physiological arousal caused by a conflict situation will cause sweating and erection of hair or feathers, any of which is likely to produce skin irritation. It is probably not chance that preening and grooming appear so commonly as displacement activities. Andrew[12] pointed out that the feather postures of birds in a sexual conflict situation resembled those of birds which he had placed in a very hot room and which were trying to cool themselves.

Any quantitative measurements of displacement activities require strict controls. One must not label some pattern a displacement activity without good evidence. This is particularly true when something like preening is involved, because birds preen a great deal and in a variety of situations, conflict and otherwise.

Some experiments by Rowell[402] show what can be done. He worked with chaffinches in aviaries and used two methods to produce an approach/avoidance conflict. In one, a stuffed owl was placed outside the aviary which evoked the chaffinches' mobbing behaviour described in Chapter 1. The aviaries had a row of perches which allowed the birds to settle at varying distances from the owl. The second situation used hungry birds which had previously been trained to use a food dish at one side of the aviary. During these tests, when they reached the dish a bright light was flashed on inside it which provoked escape. In both cases the birds tended to make short flights between perches, resting briefly on each and gradually getting closer to the owl or food dish but subsequently retreating. After a pause on a more distant perch they gradually advanced again, and so on. With the owl we might consider the bird was in a conflict between attack and escape, whilst with the food dish, hunger and escape were conflicting.

A pause on any perch represents a brief period of stability when the bird's tendency to go forward and that to go back balance. Rowell measured the length of pauses and found that, in both conflict situations, the birds paused more often and for longer at intermediate perches, neither too near nor too far. This part of the aviary is analogous to the boundary of a territory in many ways—the region of maximum stability. It is here that we might expect displacement activities to be most frequent.

Rowell did find that chaffinches showed most preening and bill-wiping during their pauses on the intermediate perches. But one would get the same result if preening simply occurred at a fairly constant rate, wherever a bird made a pause and was doing nothing else—the longer the pause, the more preening.

In control observations of birds under no conflict Rowell found this was indeed the case, but from these controls he could calculate how much preening would be expected for pauses of a given length. The amount of preening per length of pause was significantly greater when the birds were in a conflict, which justifies labelling this extra amount 'displacement preening'. Wetting the bird's feathers tended to increase preening in both conflict and non-conflict situations, so Rowell's work agrees with that quoted earlier in showing that displacement activities are facilitated by external stimuli.

Van Iersel and Bol[247] recognize this in their 'disinhibition' hypothesis for the mechanism of displacement activities. This is represented diagrammatically in Fig. 5.9. 'Centre A' controls the performance of behaviour patterns A, 'Centre B' those of B. The two centres are mutually inhibitory, represented by the minus arrows, so that patterns A and B cannot be performed together. One or both of these centres has an inhibiting effect on a third 'Centre C'. Now suppose both A and B are aroused simultaneously, they will inhibit each other's activity and neither will be able to sustain their inhibition of Centre C which, if it is aroused, will initiate behaviour patterns C.

It is a simple matter to substitute the names of relevant tendencies from our examples for A, B and C in the diagram. We must remember that our justification for doing so is based only on behavioural observations and we cannot yet ascribe any physiological reality to the disinhibition system. In the tern example originally investigated by van Iersel and Bol, they suggest the incubation tendency is mutually inhibitory with escape. Preening, previously inhibited by incubation or escape or both, appears when they are in balance. In Rowell's experiments with chaffinches, either hunger or attack conflicts with escape and again it is preening which is disinhibited.

The disinhibition hypothesis can account only for displacement activities which occur as a result of a conflict between two incompatible tendencies. Tinbergen[461] originally suggested that the thwarting of a single tendency could also lead to displacement. More recently McFarland[313] has experimented with thwarting the drinking of thirsty doves, for example by placing a glass screen between them and water. 'Displacement pecking' occurred in this situation as well as in clear approach/avoidance conflict situations of the type used by Rowell. Both Rowell and van Iersel and Bol deny

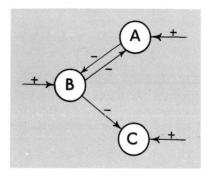

Fig. 5.9 A diagram of the 'disinhibition' hypothesis of van Iersel and Bol. See explanation in text.

that pure thwarting exists. They assume that thwarting is punishing in itself and therefore leads to avoidance, but there is no real evidence that this is always so. McFarland[314] discusses this problem more fully and suggests a rather different type of mechanism for displacement activities which attempts to draw together conflict and thwarting situations alike. In brief outline, McFarland's theory proposes that the brain is constantly comparing the results of the various behaviour commands that it produces, with an internal 'expectation' of their results. (This expectation may be built in to the nervous system during development or it may have to be learnt—it will certainly be subject to modification by experience.) We have already met the concept of a behavioural output being compared with an internal expectation when we discussed the development of bird song, p. 56. There the singing bird compared the sound it produced with an internal 'template' which effectively functions as an 'expectancy' of what the correct song should sound like. McFarland suggests, for example, that a dove approaching a situation in which it has previously fed has an analogous 'expectancy' of what the result should be. Normally the result and the expectancy match up—the bird pecks in a familiar food bowl and its crop fills with grain etc. There are two different ways in which this situation can go wrong. Firstly the food bowl may be covered by a glass plate—a thwarting situation. Secondly the dove may have another simultaneous behaviour command to avoid the area near the food bowl because of previous unpleasant experience there. This second command will also produce a second expectancy and a conflict situation develops. In either case there will be a mismatch between the results of the animal's behaviour and the expectancy or expectancies. Then, according to the theory, a higher brain centre which determines the behaviour commands that are issued comes into play. So long as expectancy and results match, the behaviour command is stable and changes only when external or internal factors change; when expectancy ceases to be matched there is an increasing tendency for the higher centre to switch attention to another command and thus another type of behaviour—the displacement activity—appears.

McFarland's theory involves a type of disinhibition, because that is effectively what the switch of attention does, but it allows for disinhibition in thwarting situations as well as in conflict. It also allows for the accessibility of displacement activities to the normal external and internal factors that affect their performance in other situations. McFarland claims to be able to detect behaviourally the switch of attention that accompanies the performance of displacement activities. In a conflict or thwarting situation his doves either stand still, fixating their objective, e.g. the food bowl, or stand in a relaxed fashion, looking around. It is during the latter 'attentive' posture that they turn to preening or pecking at the ground.

Several questions remain concerning the outcome of conflict situations.

For example, when there is mutual inhibition between attack and escape, what determines whether threat or, say, displacement preening occurs? Does it depend upon the absolute levels of attack and escape, or upon how closely they are balanced, or on some other factor? Some clues are suggested by Kruijt's[278] observations in the jungle fowl where several different behaviour patterns appear during fights between males. Kruijt found that the frequency with which the patterns were performed varied greatly depending on whether a bird was winning or losing. Thus to take the two extremes, pecking at the ground was more than four times as common in winners than losers, whilst the converse ratio was found for preening. It is reasonable to suggest that ground pecking results from a higher level of aggressiveness than preening and that birds which preen are more motivated by escape. Should we regard these behaviour patterns as displacement activities resulting from different levels of conflict between attack and escape, or are they more direct reflections of single motivational states? Ground pecking might arise from redirected aggression in victorious birds and preening from the more stressed condition of a loser. The behavioural evidence is so far inadequate to provide an answer, nor can physiology help yet.

If a displacement activity is to appear it seems probable that the thwarting or the equilibrium between two conflicting tendencies must persist for long enough to allow its performance to 'break through'. For example, if one tendency is rising very rapidly with respect to another with which it conflicts, then there will be no time for a displacement activity to appear during the brief period in which equilibrium exists.

In this connection Rowell makes an important point about the role which disinhibition may play in the change over from one type of behaviour to another. He describes how chaffinches often show a cycle of behaviour, lasting about 20 minutes, in which periods of sleep or rest on a perch alternate with periods of feeding on the ground. Quite long bouts of preening occur, especially during 'waking up' when, presumably, the feeding tendency is slowly rising with respect to sleeping. Preening is probably the result of disinhibition here just as in the conflict situations we have been considering.

In other words, preening is a rather 'low-priority' activity. Plumage can be kept in order even if preening occurs only when a bird has no other strong tendencies. Consequently it will appear between periods when one or other of the stronger tendencies has control of behaviour. If a period is short the preening may have the characteristics of a displacement activity but if it is long, as in the example just quoted, we would usually call it 'normal preening'.

Displacement activities are probably just one example of a general feature of behavioural organization. There will be 'competition' for control of an animal's behaviour and various factors, such as mutual inhibition between two tendencies, the external stimulus situation and perhaps even the posture or movement which the animal is performing,[296] may facilitate another tendency taking over. If a tendency is increasing, it is more likely to succeed in taking over, whether there is conflict or not. McFarland[313] has shown that when doves are hungry they show more 'displacement feeding' behaviour in a thwarting situation and their behaviour will, of course, become increasingly 'food orientated' in all respects.

In conclusion it may be noted that, just as with ambivalent postures, natural selection has modified some displacement activities to serve as social signals. These are dealt with in the next section because all the best examples come from courtship behaviour.

COURTSHIP AS CONFLICT BEHAVIOUR

One of the most important contributions which ethology has made to the study of behaviour has been in the field of animal courtship. Courtship may be defined as specialized behaviour patterns which form the normal preliminaries to mating. It occurs rather sporadically through the animal kingdom and is commonest in the arthropods and the vertebrates.

One function of courtship is to synchronize the reactions of male and female so that copulation can take place. Overall synchrony of reproductive behaviour is achieved by day length or temperature via the endocrine system, but a much more precise synchrony is necessary. This is because the close presence of a potential mate arouses not only sexual tendencies, but also others which may be incompatible with sex. In some arthropods, such as spiders, this is seen at its most dramatic because a male is apt to be treated as prey by the larger female! His courtship is in fact a signalling system designed to establish his identity and, finally, to immobilize the female so that he can mate with her. The courtship of arthropods is, so far as we know, very different in origin and detailed function from that of vertebrates with which we are most concerned here. More detailed discussions of arthropod sexual behaviour can be found in Evans[148] and Manning.[326]

In territorial vertebrates a male's first response to a female may show elements of attack and escape in addition to those of sex. This holds true even when there is strong sexual dimorphism and it would seem easy to distinguish a female from a trespassing male (see p. 281). Sometimes the female is attacked, although often she does not flee but remains still and may take up an appeasement posture. We have already discussed an example of this type in the pairing behaviour of the highly aggressive cichlid fish *Etroplus* (see p. 139). In some birds such as chaffinches the female becomes dominant after pair formation and the male's subsequent behaviour shows more overt signs of fear.

Attack and escape behaviour can also be seen at the time of courtship. The zig-zag courtship dance of the male stickleback is commonly interspersed with actual attacks on the female. Similarly the waltzing display performed by a cockerel to a female chicken is the same as that which he performs towards males in the preliminaries to fighting. Again as we described on p. 175, the song-spread display of male grackles does indicate a certain level of aggressiveness even when displayed towards females in a courtship situation. Conversely, copulation in chaffinches is sometimes broken off when the female turns and attacks the male who flees. In animals which pair for any length of time, aggression and fear tend to die down as they get to know each other but it rarely disappears completely.

Lorenz[309] has stressed the importance of such aggression in the formation of the 'pair bond' in many vertebrates. This aggressiveness, although it may be aroused between the male and female, is largely redirected outwards towards neighbouring animals. Often male attacks male and female attacks female; their co-operation in the defence of the territory appears to strengthen the bond between them. Some birds have evolved elaborate mutual courtship displays to fulfil the same function. The elegant displays between male and female great-crested grebes, first described and analysed by Julian Huxley[245] more than fifty years ago, are of this type (see Fig. 5.10); their displaying reaches a peak in the early part of the breeding cycle when the pair are settling down together, although it persists to some extent all through the summer. Feeding of the female by the male is also a common feature of bird courtship, and obviously derives from the appeasement gestures resembling the food-begging of young birds which the female makes when she enters the male's territory (see p. 179).

The reduction of aggressiveness is certainly an important aspect of courtship in vertebrates. Coulson[104] has shown that in the kittiwake—a cliff nesting gull—the nesting success of birds paired with their old mate is greater than that of birds of the same age which have changed mates. Significantly the former lay their eggs sooner and this may be because they settle down together more rapidly. Kittiwakes are highly aggressive and there is a high level of aggression within a pair at the beginning of the season (see p. 198).

We mentioned that courtship feeding obviously has its origins in the male's aggressiveness towards the trespassing female and her appeasement gesture. It is not surprising that we can identify other elements of attack and escape which have become incorporated into courtship behaviour. The courtship of male chaffinches and snow-buntings resembles escape behaviour because they do not face the female but run away from her with spread wings. The zig-zag dance of the stickleback may have evolved

Fig. 5.10 Male and female great crested grebes, *Podiceps cristatus*, performing one of their elaborate mutual displays. This one involves presentation of nest material. (After Huxley,[245] 1914. *Proc. zool. Soc. Lond.*, **1914**, (2), 491.)

from an alternation between attacks on the female and leading to the nest; sometimes, as mentioned above, the 'zig' part of the movement still finishes as an actual attack whilst the 'zag' may finish as leading.

From the descriptions given earlier there are reasonable grounds for suggesting that the courtship situation may involve a conflict between sex and attack or escape and courtship is often interspersed with short bouts of preening or other patterns which resemble displacement activities. In some animals natural selection has apparently so modified movements which originally occurred as displacement activities that they now form a normal part of courtship. Morris[357] describes movements in the courtship of grass-finches which probably represent different stages in

Mallard   Garganey

Shelduck   Mandarin

Fig. 5.11 Courtship preening in four species of duck ; in every case the move-ment serves to emphasize bright markings on the wings. (From Lorenz,[304] 1941, *J. Orn. Lpz.*, **89**, 194.)

the modification of displacement bill-wiping (see further discussion in Chapter 6, p. 219 and Fig. 6.9). Some of the most beautiful examples are in the ducks where Lorenz[304] and his pupils have been able to trace the evolution of displacement preening and drinking movements into displays. Some species have evolved further than others in this respect, but in all cases releasers have developed which emphasize the effect of the move-ment. In the courtship preening movement, a drake's bill is drawn along brightly coloured feathers on the wing which is usually elevated (Fig. 5.11).

We do not fully understand how this or any other type of courtship display functions. In the context we have just been discussing, presumably selection has modified those patterns which most stimulate the female sexually or at least inhibit her from moving away. However there are at least two other functions of courtship which may impose other types of selection pressure. There is the requirement for sexual isolation (discussed more fully on p. 224 which will favour courtship which is specifically very distinct. Secondly, a male's courtship displays may have been influenced by sexual selection (see p. 198) which often leads to the evolution of flamboyant releasers and displays in order to attract the maximum number of females.

Probably all courtship has evolved by some combination of these and other factors and each species will have its own particular specializations. Courtship behaviour is a fascinating study and it is important for illustrat-ing a number of behavioural principles. The reader is recommended to more complete accounts in books by Tinbergen[463] and Bastock.[32]

PROLONGED CONFLICTS AND 'EXPERIMENTAL NEUROSIS'

We mentioned at the start of this chapter that experimental psychologists have studied conflict in animals because of its possible relationship to human neurosis. The 'natural' conflicts we have been discussing rarely persist for long. When two tendencies conflict one usually gains fairly rapid precedence over the other and territorial or courtship conflicts are usually resolved one way or the other within a few minutes. By contrast some laboratory-staged conflicts are prolonged indefinitely and generally allow the animal no chance to escape from the situation.

A common technique has been to set up a stable approach/avoidance conflict of the type Rowell[402] used with chaffinches. Animals are given an electric shock when they approach food, for example. Masserman[337] trained cats to open a box for a food reward when a signal light flashed. Later when the food box was opened the cats sometimes received a strong blast of air. Under these conditions the animals' behaviour often became severely disturbed. Some of the cats became hyperexcitable, others moped in corners for days on end. They nearly all showed signs of acute stress, with raised blood pressure, hair erection and gastric disorders.

A similar kind of disturbance can be produced by presenting animals with an unsolvable problem. Pavlov[378] described various 'experimental neuroses' in dogs he was using for his conditioned reflex experiments (see Chapter 7). These developed when a dog was rewarded for responding with—say—a leg movement to a circular patch of light but punished it if responded to an elliptical one. Once this discrimination had been learnt, the elliptical patch was gradually made more circular over successive trials. There came a point at which the dog could no longer distinguish between the two. Faced with this situation some dogs became violently agitated, others responded to every stimulus regardless of shape, others ceased to respond at all and fell asleep.

None of the behavioural disturbances just described has been analysed in ethological terms. Undoubtedly some of the behaviour patterns could be considered as displacement activities and are similarly produced, but the disturbance goes far beyond this. Masserman's cats if forced to drink milk containing alcohol during the stressful period would come to prefer it to unadulterated milk and obviously sought the temporary oblivion it produced. It was often months before they recovered normal behaviour (and normal aversion to alcohol) and then only if they were kept away from the laboratory.

We know that the stress which results from a prolonged conflict is often severe enough to produce physical damage. Animals develop gastric ulcers ır tumours in the pituitary gland. In such cases it is impossible to regard he response as being in any way adaptive. The unfortunate animal finds itself in a situation to which there is no solution. It is not surprising that no mechanism exists to deal with chronic conflicts. It could provide little selective advantage in the natural world where escape from the conflict situation is almost always possible as a last resort.

6

Evolution

All through this book we have been using a biological approach which emphasizes the adaptive role of behaviour in an animal's life. The concepts of evolution have been implicit and often explicit, and consequently there is bound to be some overlap between this chapter and the others.

Here we shall concentrate on some selected aspects of behavioural evolution—in particular its relationship with genetics—and also on some of the ways in which behaviour can influence the course of evolution. By their behaviour even the simplest animals can change their environment to suit themselves and thus modify the selective forces which affect them. Wood-lice move out of dry places into moist ones where they survive better; blue tits select an area of deciduous woodland for breeding because it provides the most food, and so on.

Of course animals do not simply rely on choosing from a range of environments that are already available; many can modify external conditions to match their requirements. Beavers dam up streams to provide deeper water, termites avoid desiccation whilst foraging by constructing covered paths over dry terrain; both termites and honey-bees use a complex range of stereotyped behaviour patterns to construct around themselves a nest or hive whose physical characteristics—temperature, light and humidity for example—are kept rigidly controlled.

THE ADAPTIVENESS OF BEHAVIOUR

We are familiar with the overall adaptiveness of animal behaviour and the way in which it is correlated with morphology but it often requires

* A fuller treatment of many topics covered in this chapter can be found in Manning[327] and in Brown.[73]

detailed study to reveal just how perfect such adaptation is. For example, there are two or more species of mole cricket (*Gryllotalpa*) living in Europe. They are large burrowing insects and the males sing from their burrows below ground. As with many crickets and grasshoppers the song of each species is highly distinctive; (they certainly play a role in the sexual isolation between species, see p. 224). The character of the song is determined in part by the structure and size of the forewings which are rubbed rapidly together to produce sound. A scraper or plectrum on the hind margin of the left wing rubs along the toothed under surface of a vein on the right wing (the 'file'). In two species studied by Bennet-Clark[47-8] the fundamental sound frequency within the pulses of the song (see Fig. 6.1) was about 1,600 Hz in *G. gryllotalpa* a species with small wings,

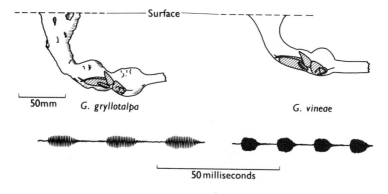

Fig. 6.1 Side views of males of two species of the mole cricket (*Gryllotalpa*) sitting in their burrows, head downwards in the singing position. The shape of their burrows approximates to an exponential horn. *G. vineae*, on the right, has a smoother tunnel and a much louder song, but in both cases the shape of the burrow is adapted to their song frequency (oscillograms of which are illustrated below) so as to emit with maximum efficiency. (Adapted from Bennet-Clark,[48] 1970. *Proc. R. ent. Soc. Lond.* (B), **39**, 125–32.)

shallow teeth on the file and a quiet song, but 3,500 Hz in *G. vineae* which has large wings, deep teeth on its file and a much louder song.

Here then we can see the direct correlation between behaviour and morphology, but the mole cricket's adaptiveness goes much further than this. The male of each species excavates its burrow in a different way, illustrated in Fig. 6.1. Bennet-Clark has shown that the burrow forms an exponential horn with a bulb at its base, whose physical properties are matched to the frequency of the song, emitting it with the maximum efficiency. When singing the male positions himself head downwards so that sound is produced just at the origin of the horn, as shown in Fig. 6.1. Thus the wing structure, singing and digging behaviour of the mole crickets are all precisely co-adapted each with the others so as to provide the most efficient sound production, 'aimed' at the females who are attracted to the male burrows as they fly above.

As a further example of the precision with which behaviour has evolved to meet the demands of the animal's environment, we may refer again to the habit of black-headed gulls removing the empty eggshells from their nest. (The stimuli to which they respond in this situation were described on p. 73.) Gulls breeding for the first time carry away eggshells promptly although they never pick up eggshells in any other situation. It is an activity which occupies them for about 5 minutes each year and might, for this reason, seem of trivial importance. Nevertheless Tinbergen and his collaborators[400] have clearly shown that eggshell removal is important for camouflage of the nest. Predation of chicks by crows, stoats and foxes is rife in gull colonies and such predators discover nests and chicks more easily if the shells with their smell and white shell membranes (see Fig. 3.6) are nearby. Natural selection will have operated to ensure that parent gulls never fail to perform this brief but vitally important behaviour. Clearly we must never write off as functionless—however trivial it may seem—any piece of behaviour which we observe regularly in the natural situation.

The functioning of many behaviour patterns requires that they match very precisely with those of other individuals. Most of the examples of communication which we discussed in Chapter 3 involve such mutual adaptation. In a courtship display, for example, the responses of one partner will act as one of the selective forces which shapes the behaviour of the other and *vice versa*.

Tschanz and Hirsbrunner-Scharf[472] have been able to demonstrate most beautifully how the responses of parent and young auks are adapted to each other. They studied two species, the guillemot and the razorbill, which nest on the same sea cliffs but use very different sites. Guillemots nest packed closely together on flat ledges whilst razorbills nest singly in crevices or small caves. Young guillemots tend to rest in a rather upright posture, keeping their bodies clear of the ledge which is usually wet and covered in guano. Razorbill chicks rest more horizontally; their nest crevices are drier and more sheltered. This behaviour is independent of experience and does not alter when chicks of both species are given the same surface on which to rest, or when they hatch into a foreign environment after eggs are exchanged between guillemots and razorbills.

The parents of both species bring back fish for their young but their manner of presenting them is strikingly different, see Figs. 6.2 and 6.3. The razorbill brings several fish and stands rather passively as its young

Fig. 6.2 A razorbill brings food to its chick, who pecks at the fish presented openly in the parent's bill. (From Tschanz and Hirsbrunner-Scharf[472] in *Function and Evolution in Behaviour.* Oxford University Press, London.)

peck at them. Parent and chick have plenty of space and are not disturbed during feeding. Guillemot parents present a single fish tail first, sheltering the feeding chick with wings and feet. The chick lets the fish slide through its bill, and then swallows it head first. This feeding ceremony gives some protection and prevents poaching on the densely packed ledges.

When Tschanz and Hirsbrunner-Scharf transferred eggs between guillemots and razorbills (a process which could be accomplished only by the gradual obliteration of their markings over a period of days before the transfer), they could observe the responses of chicks and foster parents. In every case each first behaved in the species-typical manner and this caused great problems during feeding, particularly for young razorbills whose responses to the guillemot's feeding ceremony were quite inappropriate. It took many days before foster-parents and chicks learnt partially to adjust to one another and many chicks died. These results indicate the strong survival value of the mutually evolved responses of parent and young. We may not always be conscious of such evolved co-adaptation when we observe the smooth flow of interactions between parents and

Fig. 6.3 A guillemot presents food to its chick : a single fish is presented tail first. The chick seizes it, runs it through its beak, turns the fish round and swallows it head first. (From Tschanz & Hirsbrunner-Scharf[472] in *Function and Evolution in Behaviour*. Oxford University Press, London.)

offspring, yet it is obvious that for many animals learning and practice alone will be inadequate to ensure survival.

The precision with which natural selection can equip animals with a beautifully adaptive behavioural repertoire is most impressive. Nevertheless we must remember that evolution can rarely affect any character and especially behaviour in isolation. In the end it is whole organisms that count and selection can only make use of the inherited variations that are available to produce not the maximum possible response from the character, but the best compromise with all the others.

We noted in Chapter 3, p. 73, that it was often possible to make an artificial releaser which was more effective than the natural one. The latter may be part of a butterfly's wing, for example, which also has to serve for flight and selection has to compromise at a level lower than maximum. We can see a similar situation with the nuptial colours and songs of male birds. The bright plumage and loud song which attract a mate are also likely to attract a predator. Most males stop singing as soon as their mate begins to incubate and some lose their breeding plumage at the autumn moult. There must obviously be a compromise between reproductive success and danger from predation. This will be struck at different levels depending on other circumstances. Polygamous species like the pheasants where not all males are successful and one may mate with several females, have developed very extravagant reproductive plumage and keep it all the year round. However, male pheasants have nothing to do with incubation or rearing young which is the task of the cryptically coloured females. A brilliantly coloured male who attracts and mates with six females in one season before falling prey to a fox, will leave more descendants than a more sober male who lives for 10 years but never attracts a mate!

The work of Tinbergen and his group on gulls has revealed several beautiful examples of the way in which their behaviour has been forced to change to fit in with other environmental circumstances. For example, Cullen's[112] study of the kittiwake (*Rissa tridactyla*) has revealed numerous differences from the other gulls. The kittiwake has taken to nesting on narrow cliff ledges and by this means protects its eggs and young from predators which take a considerable toll from the ground-nesting gulls. However, suitable ledges are in short supply and there is intense competition for nest sites. Prolonged fights involving both males and females take place on the ledges and, beaks locked together, kittiwakes often fall into the water where they continue to beat one another with their wings! Fights of this severity and persistence are virtually unknown in the other gulls. The high level of aggressiveness in kittiwakes affects not only their fighting but also their courtship displays. These show a high proportion of aggressive elements and overt aggression between male and female persists late into the breeding season. To counteract this aggressiveness, appeasement displays are also well developed (see p. 179). Here they consist of turning the head away from the opponent and depressing the beak—an example of the 'anti-threat' type of appeasement we discussed in Chapter 5. The commonest form of threat is 'choking', also described in Chapter 5, where we considered evidence that when this posture is adopted by other gulls—in which it is far less common—it indicates that the bird is strongly motivated to stay put. The kittiwake's attachment to and defence of its nesting ledge has to be most intense. Accordingly selection has favoured a compromise between attack and escape tendencies which is more biased towards attack than in the other gulls.

Cullen's study revealed a number of other behavioural modifications both in adult and young which could all be traced back to the original shift to cliff nesting. More recently Cullen and Ashmole,[115] and Hailman[191] have shown a similar series of adaptions in a cliff-nesting tern, the black noddy, and a second cliff-nesting gull, the Galapagos swallow-tailed gull, respectively.

CULTURAL TRANSMISSION OF BEHAVIOUR

In Chapter 2 we discussed development and the manner in which both genetic and environmental factors contribute to the final form which behaviour takes in the adult animal. Behaviour can evolve only if it varies and if such variations are transmitted from one generation to the next. Then natural selection can come into operation, favouring some variants—which thus spread—and eliminating others. With most morphological and physiological characters there is only one way in which such inherited variations can be produced—by genetic changes which will involve new mutations or recombinations of genes already present.

Behavioural evolution has another dimension because an animal may learn certain patterns of behaviour from its parents or others in its group. Later it may in turn act as a model from which its own offspring will modify their behaviour. Suppose now one animal learns a completely new behaviour pattern which proves more successful for some purpose than the hitherto typical behaviour. This pattern may be passed on to succeeding generations, and gradually replace the old pattern, without any genetic changes being involved.

We can begin our discussion of evolution by considering some examples of this relatively rare, but potentially very important, manner of behavioural transmission in animals. We are familiar with such 'cultural' evolution in human behaviour, and many would argue that almost everything of importance in human behaviour is transmitted in this way from one generation of a society to the next. The various human languages are an obvious example of a continuous cultural tradition which maintains different types of behaviour in different populations.

Clearly cultural evolution is possible only among animals with a considerable ability to modify their behaviour by copying and practice. We might not be surprised to find examples from the other primates, our closest relatives. Intensive, long-term observations on the Japanese monkey, *Macaca fuscata*, has shown that some differences between the behaviour of monkey troops are indeed of cultural origin. The monkeys are attracted to feed on corn, sweet potatoes and other foods artificially put down for them. The potatoes are often caked with earth which monkeys usually rub off with their hands. One day observers saw a young female dip her potato into a stream and wash it. She persisted in this habit, which was copied, first by one of her infants, and subsequently by nearly all the younger members of her troop, which is now clearly to be distinguished as a 'washing subculture'. A wide range of feeding differences between troops have become established in a similar manner (Frisch[155] reviews some of the Japanese work). There seems little doubt that close observation of other primate groups will reveal a similar story, not just for feeding habits but for the use of simple tools which is certainly learnt afresh when each new generation of young chimpanzees copies from its parents. There may also be cultural transmission of some details of social behaviour patterns (see Guilmet[186] and McGrew and Tutin[318]).

We can readily accept cultural transmission in primates but there is growing evidence that it occurs, though perhaps sporadically, in other vertebrates, and Galef[159] provides a useful review. It is certainly cultural transmission that maintains the different song dialects in white-crowned sparrows as described on p. 55. The young male models the finer details of his song on that of his father and neighbouring males singing close by. We know that he would pick up a different dialect if he grew up in another area.

Recently Norton-Griffiths[370-1] has discovered a remarkable case of cultural transmission in the feeding habits of the oyster-catcher, *Haematopus ostralegus*. These shore birds feed extensively on mussels and they have two distinct methods of opening the shells. One is by hammering on the ventral surface of mussels which are carried from the rocks to an area of firm sand suitable for the purpose. The second method involves stabbing the bill into the gape of slightly open mussels which are still attached to rocks and covered by sea water. Norton-Griffiths found that individual oyster-catchers use only one method to open shells. Young birds follow their parents down to the mussel beds and gradually acquire the same technique which they use exclusively thereafter. Different mussel beds vary in the degree to which they are suitable for hammering or stabbing and oyster-catchers select areas which suit their own technique. There is certainly the basis here for real 'cultural isolation' between hammering and stabbing birds but, as Norton-Griffiths points out, oyster-catchers have many other foods besides mussels. Cultural transmission of novel feeding habits in birds has also been described for the familiar plundering of milk bottles by titmice,[220] and for the way greenfinches (*Chloris chloris*) acquired the habit of extracting seeds from the unripe fruits of *Daphne mezereum*, a plant whose ripe fruits have inaccessible seeds.[381]

The cultural transmission of behaviour can certainly affect its evolution because it may change the selective forces acting on the animals. Hardy[200] has illustrated this with a vivid—if hypothetical—example. Suppose a group of primates lives on land which adjoins the sea and begins to forage for food on the sea shore. One primate begins to explore beyond the water line and then to swim whilst gathering its food. The habit spreads, as has food washing (and one troop of Japanese macaques *does* swim in the sea), and so through cultural transmission a group of primates has begun to gather sea food. This will immediately change the course of selection—loss of hair, development of sub-cutaneous fat, webbed feet, physiological adaptations to diving—any of these and other aquatic adaptations which hitherto have been totally irrelevant and probably highly disadvantageous may suddenly become of high selective value simply because of changed behaviour.

We do not know for certain of any examples with effects as large as this but such effects must be there at least among the primates although they may go undetected. However, despite its occasional importance cultural transmission remains limited in animals and as a rule the evolution of behaviour proceeds by selection operating on inherited variations. Consequently before we can profitably discuss what types of change have occurred during evolution, it is necessary to consider what is known about the genetics of behaviour.

GENES AND BEHAVIOUR

Behaviour genetics is not a straightforward field of study. Genes contain the information which determines the proteins a cell makes. If one is interested in biochemical genetics then the causal links between the genes and the biochemical variations under study are close and often quite clear. It is more difficult to work out the links between genes and morphological variations but with the genetics of behaviour the 'distance' between the genes and the end product under study is at its greatest. So many different things can affect behaviour, the general metabolic state of the animal, rates of hormone secretion, the physiology and morphology of the muscles and nervous system—the list can be made both long and complex. Through all this maze one has to try to relate genetic differences to behavioural differences in a way that makes sense. The diversity of behavioural phenomena adds a further complication for we know of genetic effects on motivation, learning, general reactivity, sensitivity to particular stimuli, performance of fixed action patterns and others. We cannot expect to find any common genetic pathway in the development of such a range of different measures. Such difficulties have not prevented a great deal of interesting progress in behaviour genetics (see Fuller & Thompson[157] and Ehrman & Parsons[142] for good surveys) but it cannot be claimed that we have a coherent body of knowledge.

Behaviour involves the co-ordinated control of a number of different body systems. The fruit fly, *Drosophila*, which is better known genetically than any other animal, provides good examples of the various ways in which genes can affect behaviour. Many hundreds of mutant genes are known in *Drosophila* and their position on the chromosomes has been mapped. For example, the gene called *Bar* reduces the number of facets in the compound eyes, *white* reduces pigmentation in the eyes, *forked* and *hairy* affect the number and structure of the bristles, *vestigial* and *dumpy* alter the form of the wings, *yellow* and *black* affect the general pigmentation of the body, and so on.

All these genes have been given names which roughly describe the most conspicuous effect they produce. They certainly have other effects too, which are sometimes detected in tests of metabolism, fertility, longevity and the like. Most of the genes mentioned above have also been shown to alter at least one aspect of behaviour—males which carry them are less successful than normal males in stimulating females to mate with them. However, although this superficial description lumps all of them into the same behavioural category, it is obvious that they produce their common end result in quite different ways. *Bar* and *white* affect vision: mutant flies cannot see as well as normals and will have trouble both in locating females and in receiving visual stimulation from them during courtship. Bristles are tactile sense organs, and males carrying *forked* or *hairy* genes will be defective in this respect. An important part of the normal *Drosophila* male's courtship is wing vibration in which he brings out one wing laterally and vibrates it in the horizontal plane. This stimulates sense organs at the base of the female's antennae. *Vestigial* and *dumpy* males have grossly malformed wings which cannot be vibrated properly and it is not surprising that they have poor mating success. It is less obvious why *yellow* and *black* males take longer to mate than normal males. It is not their abnormal colour that puts the females off, because they are equally at a disadvantage when competing for mates in total darkness. There are no obvious deficiencies in the sense organs or wings, though the elaborate tests needed to prove this have not yet been made. The genes could be operating on the nervous system, the muscles or the general metabolism of the insects.

Until recently it has not usually been possible to identify the pathway by which a gene affects behaviour unless the effect is large. The genes for microcephaly and phenylketonuria in man produce gross defects in the structure and biochemistry of the brain with resulting severe mental defect. In mice there are a number of genes known which upset their posture and balance. The 'waltzing mice' spin round in circles when disturbed, others show continuous head shaking or circling. These behavioural results can be correlated with structural defects in the middle ear and brain centres responsible for balance. Such changes are both large and deleterious, and on both counts they have little relation to the kinds of genetic change—presumably small and favourable—which have been important in the evolution of behaviour.

A less drastic change and a more precise analysis has been provided by Ikeda and Kaplan.[248-9] They have studied a number of *Drosophila* mutants called collectively 'hyperkinetics', because the affected flies show a particularly exaggerated type of leg twitching when under anaesthesia and tend to make sudden jerks in their walking. Using refined neurophysiological techniques they have located groups of motor neurons in the lateral part of the thoracic neural complex which have abnormally high rates of spontaneous discharge. Sudden bursts of firing from these neurons can be completely independent of any external inputs: it really looks as if Ikeda and Kaplan have located the neural sites of gene action: presumably the membrane physiology of the affected cells is changed and results in the abnormal rates of discharge. In this case we really can explain the behavioural changes—albeit still fairly gross ones—in terms of the neurophysiology of gene action.

A considerable number of genes have now been identified and mapped purely on behavioural characteristics. A batch of *Drosophila* are exposed to a powerful mutagenic chemical and their progeny screened for some aspect of behaviour. Mutations are rare, so large numbers have to be measured and thus the screen has to be a relatively simple one. For example, normal flies run towards light when disturbed, the occasional mutant fly which does not can be easily picked out. Normal male flies succeed in mating with females within a few minutes, those few who do not can be isolated for further checking.

Benzer[50] and his co-workers have used ingenious screening techniques to isolate genes which affect phototaxis, chemotaxis, circadian rhythms, courtship, learning ability, etc. Such mutants are potentially most valuable material for studying the way genes affect behaviour and it is also possible to discover the site of gene action in some cases.

This can be achieved by making use of some of the remarkable genetic manipulations which are possible in *Drosophila* to produce 'mosaic' individuals some of whose cells are affected by the mutant to be studied whilst the rest are normal (see Hotta and Benzer[233-4]). By studying a number of mosaics with different regions of the body affected by the mutant one can discover which parts of a fly must carry a gene in order for its effects to be shown. The genes affecting circadian rhythms, for example, can express their effect only when carried by the insect's brain. (The same method can be used to study which parts of the nervous system must be male or female in order to produce sex specific behaviour patterns.[234])

The study of single gene effects is one important approach to behaviour genetics, but we also need to study the inheritance of behaviour patterns themselves, especially if we want to understand the mechanisms of behavioural evolution. As far as possible the methods to use are those of classical genetics. Crosses are set up between animals whose behaviour differs and the effects are observed in the F_1 hybrids, in the F_2, in backcrosses and so on.

One immediate difficulty is the choice of suitable 'units of behaviour' for genetic analysis. This is one example of a problem which has always faced geneticists. Mendel's stroke of genius was to choose characters in his pea plants which were clear-cut and could be counted. In this case, each 'unit' —tallness, seed colour, texture of seed coat, etc.—corresponded to a single gene, but there is little probability of getting such a simple correlation with behaviour.

It is impossible to lay down rules for the selection of behaviour units; the genetical tests themselves will usually suggest the best ones to use. Fixed action patterns immediately suggest themselves and certainly they are clear cut and reasonably easy to count. Further we know their development is well-buffered against many types of environmental disturbance. They are phylogenetic units, but this does not mean that they will necessarily have a simple genetic basis. Indeed there is every reason for believing that they will depend on the action of many different genes for their development, but simpler forms of inheritance may control the frequency with which such patterns are performed or the ease with which they are elicited and show us a unit structure.

Rothenbuhler's[400-01] work with honey-bees is one of the few examples. Certain strains of bee are called 'hygienic' because, if a larva dies inside a cell, the workers uncap the cell and remove the corpse; 'unhygienic' strains do not respond in this way and the workers leave the cells alone. The hybrids between these strains are all unhygienic which thus appears as a dominant character. Rothenbuhler back-crossed the hybrids to the recessive, hygienic strain with the following remarkable result. From 29 back-cross colonies he found:

1. Nine of the colonies uncapped cells containing dead larvae, but left the corpses untouched.
2. Six did not uncap cells, but would remove dead larvae from cells which were uncapped by the experimenter.
3. Eight neither uncapped cells nor removed larvae, i.e. were unhygienic.
4. Six uncapped cells and removed larvae, i.e. were hygienic.

The proportions of the four classes do not differ significantly from equality and it is possible to explain this result in terms of two pairs of alleles, one of which controls the expression of the uncapping pattern, the other that of removing a dead larva. The 'unhygienic' alleles are dominant, we may call them 'U' for uncapping and 'R' for removing; then their 'hygienic' alleles are 'u' and 'r'. Worker bees of the hygienic strain must have the genetic constitution uurr, unhygienic workers have UURR. All the worker and queen F_1 hybrids will have UuRr and be unhygienic. Since the male bees or drones are haploid carrying only one set of chromosomes (see p. 211), Rothenbuhler had to mate 29 different F_1 drones to double recessive, hygienic strain queens in order to produce the backcross colonies. These drones must have included all the four possible types, UR, ur, Ur, and uR, and the outcome was as follows:

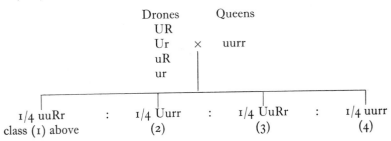

	Drones	Queens
	UR	
	Ur	× uurr
	uR	
	ur	

1/4 uuRr : 1/4 Uurr : 1/4 UuRr : 1/4 uurr
class (1) above   (2)   (3)   (4)

The simplicity of this result does not imply that a single locus controls the development of whatever complex neuronal network organizes the fixed action pattern of uncapping or that of removing corpses. Rothenbuhler has evidence that unhygienic workers do perform these activities but at a very low frequency and they require a more powerful stimulus. The alleles U and u act as 'switches' determining the threshold of the uncapping pattern in a virtually all-or-nothing fashion.

We have other examples of fixed action patterns which inherit completely or not at all. The courtship display of ducks consists of a series of patterns most of which can be observed, with slight modifications, throughout the family. One such pattern Lorenz[304] calls the 'down-up' in which a drake dips his bill into the water and then suddenly lifts his head, raising a plume of water. This pattern is not seen in the yellow-billed teal (*Nettion flavirostre*) or the pintail (*Dafila acuta*), although found in a number of their relatives. However, the 'down-up' reappears in its typical form in the F_1 hybrids between these two species.[308,481] The most likely explanation for this is that the block of genes necessary for the organization of the 'down-up' are still present in both parent species but natural selection has eliminated its performance by raising its threshold. This threshold change has been accomplished by a different set of genes in each species. When they combine in the F_1 hybrids, their effect is reduced so that the threshold is lowered and the down-up is once more performed in the normal situation. We should need to study F_2 hybrids and back-crosses to confirm this hypothesis, but unfortunately, as is usually the case, the hybrids are very infertile.

We have described two examples where genes affecting thresholds may be operating rather like a 'switch' determining whether or not a pattern appears. 'Switch genes' are known in other circumstances[434] controlling the operation of a group of genes which inherit as a block. Some butterflies have evolved elaborate wing patterns which mimic those of distasteful species and thereby gain some protection from predators. The development of the mimic patterns is known to be under the control of many genes but it is a single switch gene which determines whether they operate or not. Similarly for behaviour it is reasonable to expect that natural selection will favour linking together genes which all contribute to a common feature which must appear fully formed if it is to be of any use. The closer such genes are linked the less the danger that individuals will lack some of them and develop imperfectly.

We have a possible example of this from the work of Ewing[150] who found that the specific form of the courtship 'song' (produced by the male's wing vibration display) in *Drosophila pseudoobscura* and *D. persimilis* is determined by the X chromosome. F_1 hybrid males produce 'songs' which resemble that of their mother's species since they receive their X chromosome from her. A number of genes must be involved in determining the song characters and the fact that they are all linked together on the X chromosome can hardly be fortuitous. It is tempting to speculate that such gene blocks may be a feature of the inheritance of fixed action patterns. We need many more such studies but it is not easy to find suitable material for genetic work of this type.

It is most interesting that when a number of fixed action patterns normally occur together in a particular sequence, often it is the sequence which becomes disrupted in hybrids, although the patterns themselves remain intact. The extraordinary nest building behaviour of hybrid love-birds described on p. 53, is a case in point. Lorenz[307] and Ramsay[389] have described striking examples from the courtship behaviour of ducks. The drake mallard (*Anas platyrhynchos*), in common with other surface-feeding ducks, has a repertoire of some ten highly stereotyped courtship patterns some of which are illustrated in Fig. 6.4. The sequence shown—'bill-shake, grunt-whistle, tail-shake'—is called by Lorenz an 'obligatory sequence' although Ramsay mentions that occasionally he observed 'tail-shake, bill-shake, grunt-whistle' both in the mallard and in the closely related black duck (*Anas rubipres*). In hybrid drakes between these two species, he saw 'grunt-whistle, tail-shake, bill-shake', and other combinations which were never observed in either parent species. Clearly the genetic control of sequences is independent of that for the fixed action patterns themselves, but we know too little of the behavioural mechanisms involved to be able to comment further. Franck[153] discusses a number of other examples of abnormal combinations of fixed action patterns from his work on hybrid swordtail fish (*Xiphophorus*) and also provides a useful review of observations on hybrids generally.

Whilst the 'switch genes' discussed above produce dramatic changes to the thresholds of behaviour patterns, it is far more common to find genes producing relatively small quantitative changes. We have mentioned the reduced mating success of male *Drosophila melanogaster* which carry the

Fig. 6.4 One common sequence of stereotyped fixed action patterns in the courtship of mallard drakes. Hybrid drakes show these typical patterns but their sequence is abnormal (see text). (From Lorenz, K.,[307] 'The Evolution of Behavior'. *Scientific American.* © 1958 by Scientific American. All rights reserved.)

gene *yellow*. Bastock[31] has shown that their courtship is less stimulating to females because it contains a smaller proportion of wing vibration. Yellow males perform this fixed action pattern in the same way as normal males, but do so less often. A number of other examples of this type are known from *Drosophila* where it is easy to produce stocks which, as near as possible, differ by only a single gene.[325]

In other animals most studies have used either inbred lines, which will differ by many genes, or lines which have been deliberately selected for differences in behaviour. The same kinds of quantitative differences are recorded with the same fixed action patterns being performed at different frequencies. Thus different inbred strains of mice[316] and guinea pigs[175] show variations of this type in their sexual behaviour. Scott and Fuller[428] give numerous examples from domestic dogs, where the different breeds are the result of a combination of selection and inbreeding. It is a remarkable testimony to the stability of fixed action patterns that some 8,000 years of domestication with at times intense selection, has produced so little modification of the ancestral patterns. All the behaviour repertoire of the domestic dog breeds is represented in the wolf but frequencies vary. For example, the terrier breeds bark frequently, spaniels do so less often, and wolves and basenjis scarcely ever bark. Similar frequency differences are found for all the other patterns.

Deliberate selection for behavioural characters has nearly always been successful in changing them in a quantitative fashion. We have already described how selection changed levels of aggressive behaviour in mice (p. 136). In a similar way and with remarkable speed—only three generations of selection—Wood-Gush[511] was able to alter substantially the frequency with which chickens performed sexual behaviour patterns. Manning[324] selectively bred *Drosophila* for fast and slow speed of mating over a number of generations; see Fig. 6.5. 'Mating speed' is a complex character which involves the interaction of male and female, but analysis showed that the behaviour of both sexes had been altered in a quantitative way. The males from fast-mating lines perform high intensity courtship movements more frequently than those from slow lines. Conversely, females from fast lines are more easily stimulated to accept males of their own or from other lines than are the slow mating females.

As well as changing the frequency of fixed action patterns, selection has been equally successful for altering more 'general' aspects of behaviour, such as locomotor activity and 'emotionality'. For example, Rundquist[408] selectively bred two lines of rats which showed respectively high and low activity in a running wheel attached to their home cages, which they could enter at will. He could produce little change in the direction of increased activity, but after 12 generations rats of his low activity line were averaging 6,000 turns of the wheel in a 15-day period as compared with over 100,000

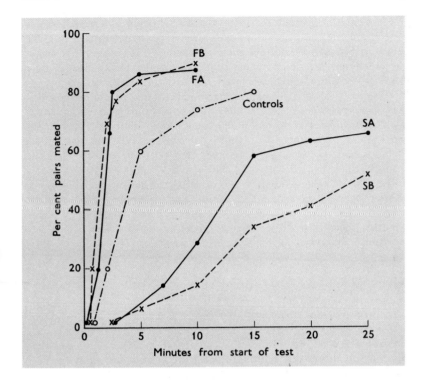

Fig. 6.5 The mating speed of groups of 50 pairs of *Drosophila melanogaster* from two lines selected for fast mating, FA and FB, and two selected for slow mating, SA and SB, compared with unselected controls. These samples are from the 18th selected generation. Some 80% of the fast lines have mated before the first of the slow lines begin. (From Manning,[324] 1961, *Anim. Behav.*, **9**, 82.)

for normal rats. Though we have no details, a change of this type will probably have effects on many different aspects of the rats' behaviour. So do changes to 'emotionality' which have been extensively investigated by Broadhurst.[68-9] He has selectively bred rats which show 'emotional' and 'non-emotional' (or as he terms it, 'reactive' and 'non-reactive') responses to a mildly frightening experience—being placed in an arena under bright light. We have already mentioned this type of 'open field' test when discussing the effects of early experience on emotionality (see p. 29). 'Reactive' rats are defined as those which show a lot of defaecation and urination and stay still, often crouching in one place. These animals are likely to be those which have a high degree of physiological arousal (see p. 162) as compared with the 'non-reactive' rats which do not defaecate

and move around more in the arena. Broadhurst obtained a large response
to selection and the reactive and non-reactive lines have proved to differ
in several behavioural and physiological measures. For example, reactive
rats are slower to learn a simple response in order to avoid an electric
shock[70]—their intense arousal 'interferes' with their learning (see p. 245).

The success of selection experiments indicates that natural populations
carry numerous genes which affect behaviour quantitatively. Natural
selection will maintain the optimum level of response but there is plenty of
variability for selection to act upon. Such variation is the raw material for
evolution and, as we shall discuss later, it is relatively easy to relate the
changes which have occurred in the so-called 'micro-evolution' of behaviour (those that take place at the earliest stages of the divergence
between species, for example) to the effects of mutation and artificial
selection which we have just been describing. Before considering micro-
evolution there is another aspect of genetics and evolution which has
profound implications for behaviour.

KIN SELECTION AND INCLUSIVE FITNESS

In *Erewhon* (1872) Samuel Butler wrote 'A chicken is an egg's way of
producing another egg', a statement which may seem merely eccentric but
in fact contains an important biological idea. It is possible to argue that
it is only our genes that reproduce and leave descendants and our adult
bodies are simply the elaborate packaging that protects them and maxi-
mizes their chance of success. This viewpoint is argued with vigour and
wit by Dawkins in his book *The Selfish Gene*.[122] We do not have to accept
the argument *in toto* to recognize its usefulness. If we take the genes as
units of selection, rather than the individual organism, then we can regard
reproduction and parental behaviour as a means of perpetuating genes.
Each offspring will share half its genes with its parents and leaving
two offspring will—on average—ensure that any gene's frequency stays
the same from generation to generation. Up to this point the argument
may seem rather academic since we could equally well think in the
familiar terms of individuals reproducing themselves. But the gene
selection idea suggests other possibilities for it is not only one's immediate
offspring who share one's genes, so do other relatives. Hence J. B. S.
Haldane's after-dinner remark following some brief calculations on the
back of the menu card, 'I am prepared to lay down my life on behalf of four
grandchildren or eight first cousins!' In each case, if one carries genes for
self-sacrificing behaviour then such groups of relatives will, on average,
include the equivalent of one's own gene complement.

The important concept to emerge from this approach is that selection may favour 'altruism'—sacrificing one's own immediate interests, even one's survival—if thereby sufficient animals related to oneself have their chances of survival increased. We are familiar with this concept when it involves parents sacrificing their own interests to preserve their offspring but we can now extend it to include relatives. The fitness of an individual can be roughly equated with the number of descendants it leaves. The fitness of a gene (or a genotype) can be extended to include all survivors which incorporate copies of itself, and this we call its 'inclusive fitness'. Smith,[439] whose book gives an excellent modern account of evolutionary theory, suggests the term 'kin selection' to describe selection which takes account of other relatives as well as immediate descendants.

Kin selection will often operate on genes which affect behaviour, especially towards other animals and there has recently been a great upsurge of interest in its role in the evolution of all types of social behaviour. Much of this interest can be traced back to the publication of two important papers by W. D. Hamilton.[199] Hamilton addressed himself to a problem which had long puzzled zoologists. True social life among the insects, i.e. societies with overlapping generations, sterile workers and communal feeding of the young (see p. 269) has evolved in only two orders, the Isoptera (termites) and the Hymenoptera (ants, bees and wasps). The remarkable thing is that in the Hymenoptera all the evidence suggests that it has evolved independently at least *eleven times* and perhaps more. Clearly it is not easy to evolve social life or more different types of insect would be expected to have done so, yet somehow this one order of Hymenoptera seems predisposed to it.

Hamilton draws attention to the significance of the Hymenoptera's unique form of sex determination. They exhibit 'haplo-diploidy'; males are haploid and develop from unfertilized eggs, females are diploid and develop in the normal fashion from fertilized eggs. A diploid organism gets half its chromosomes from its mother, half from its father. When it forms gametes the cells which develop into sperm or eggs undergo a 'reduction division' (meiosis) whereby the chromosome number is halved. The homologous maternal and paternal chromosomes pair up and exchange some genetic material. Then the members of each pair move at random to either daughter cell. This means that although each gamete has a complete haploid set, the proportion of maternal and paternal chromosomes varies, although on average it will have 50% of each. When a haploid male Hymenopteran forms gametes there is no reduction division to halve the number of chromosomes and consequently all his gametes are identical, containing simply copies of his complete set of chromosomes.

One of the most striking features of a Hymenopteran colony is that it usually consists of a single family—a queen with her large progeny, most of them females. Hamilton examined the effects of haplo-diploidy on genetic relationships within such a family. Since normal offspring get a haploid set of chromosomes from each parent we can say that its genetic affinity to either mother or father is $\frac{1}{2}$. As we described above, the gametes each offspring receives vary in their genetic makeup but on average they contain 50% of the same genes. Different offspring of the same parents receive, on average, half their genes in common from their father and similarly half in common from their mother. Thus the average genetic affinity of one sibling to another is $\frac{1}{2} \times \frac{1}{2}$ (from father) $+ \frac{1}{2} \times \frac{1}{2}$ (from mother) $= \frac{1}{2}$. So, as far as genes go, one is as closely related to a sibling as one is to a parent.

A honey-bee daughter develops when her mother—the queen bee—fertilizes an egg and her relationship to her mother is $\frac{1}{2}$ as normal, but the same is not true between daughters. Since all the gametes from the father—the haploid drone—are identical, all daughter bees share 50% of their genes in common derived from him. They also share 50% of the genes they inherit from their mother in common, so their average affinity to each other is $\frac{1}{2}$ (from father) $+ \frac{1}{2} \times \frac{1}{2}$ (from mother) $= \frac{3}{4}$. Hymenopteran females share more genes with their sisters than they do with their mothers.

In the family of the Hymenopteran society only very few of the daughters actually reproduce, most are sterile workers who help their mother to rear these few privileged young queens. But as Hamilton points out, in terms of replicating one's genes this is to their advantage. If a young female has a 'choice' in evolutionary terms between leaving the colony and rearing daughters of her own or staying put and helping her mother to rear more sisters for her, she should choose the latter.

Whilst there are certainly other pre-requisites to the evolution of social life in the insects (see Kennedy[267]), there can be little doubt that Hamilton was correct in suggesting that haplo-diploidy must have played an important part in its remarkable development in the Hymenoptera. Kin selection can operate with particular force because of the sex determination mechanism. (There is a good discussion of Hamilton's hypothesis in Wilson.[506])

With kin selection in mind it is possible to look at the evolution of social behaviour, and particularly acts of apparent 'altruism' or self-sacrifice in a different light. Selection cannot favour sacrificing one's own fitness for unrelated animals so if kin selection is to work there must be some way of identifying relatives. This is no problem for social insects nor in long-lived animals which live in small groups, e.g. primates (see Chapter 8). It is much more difficult for animals which disperse after breeding and lose touch with one another.

Birds offer some interesting comparisons here. Many tropical and sub-tropical species live a relatively sedentary life and offspring do not disperse much. This is in strong contrast to temperate zone species which com-

priscilla barrett.

Fig. 6.6 Three adult Florida scrub jays at the nest. All are colour ringed and can be identified as the two parents and a yearling bird from a previous brood. This latter is now helping to rear its young siblings in the nest. For further explanation see text. (From Wilson,[507] 1975, *Sociobiology*. Harvard, The Belknap Press, Cambridge Mass.)

monly disperse and often migrate to warmer areas in winter. In view of these differences it is particularly interesting to find that the phenomenon illustrated in Fig. 6.6 is common amongst tropical birds. There are *three* adult jays at the nest helping to raise the young and not all of them can be the true parents. The extra 'helpers' sometimes prove to be young birds raised in a previous brood but sometimes they are simply neighbours whose own nest has been destroyed (nest mortality is very high in many tropical species). By contrast we may note that helpers are scarcely ever observed in temperate birds.

It is quite easy to explain the behaviour of helpers in terms of kin selection. The young birds are helping their parent to rear more siblings to which they are just as closely related as they would be to their own offspring. Very often young birds are relatively unsuccessful at breeding and it might make good sense even in terms of individual advantage to stay around the parental nest and gain experience in food seeking etc. whilst helping to feed the next brood. Since we know that dispersal is slight in such species, even neighbouring birds will be likely to share some genes in common. If your own nest is lost it may be too late to start again that season, better to cut your losses and help your partly related neighbours raise their young. Neither argument applies strongly to temperate species who disperse more widely and often have time to rear only one brood. Such explanations are quite simple to construct but they have yet to be proven, nor are they uncontested. Zahavi[518] has argued against kin selection and claims that in the Arabian babblers he has studied, the helpers at the nest of their parents do more harm than good, creating disturbance and attracting predators. Certainly it is important for the kin selection hypothesis to be able to demonstrate that nests with helpers are more successful than those without and at present we have very little good data bearing on this question. (There is a very full discussion with more examples in Brown.[73])

It could be claimed that 'helping' is not an especially strong example of the action of kin selection for rather little self-sacrifice need be involved. There are several examples where sacrifice is more immediately obvious. One of the most striking concerns the sexual behaviour of wild turkeys which was studied by Watts and Stokes.[489] The males of each brood stay together in groups of two or three and display synchronously to females on the communal mating ground. Only the most dominant (see p. 276) male of the group of brothers mates with the females which are attracted by their displays. The others defer to him and are thereby effectively sterile. However, as with many communally displaying birds (see the description of blackgrouse on p. 279), very few turkey cocks actually succeed in attracting females to mate. Kin selection will favour less dominant males supporting their brother and helping to increase his mating success by their displays. In this way they will greatly increase their inclusive fitness.

Kin selection is a useful concept which has extended our thinking on the evolution of behaviour. Perhaps its chief drawback is the relative ease with which one can construct all kinds of untestable hypotheses to explain little understood types of social interaction. Further, as we have seen when considering 'helpers', it is not always possible to separate out advantages which accrue to self from those to kin. However if used with caution kin selection can generate hypotheses for testing and, as we shall discuss in Chapter 8, it is an important component of recent developments in social behaviour.

THE MICRO-EVOLUTION OF BEHAVIOUR

The term 'micro-evolution' is used to describe the earliest stages of divergence between populations leading up to the origin of species themselves. It is relatively easy to relate the behavioural changes involved in the micro-evolution of a group to the genetic changes we have just been considering.

If the behaviour of a group of related species is compared, it is usually possible to identify the same fixed action patterns throughout. These patterns are truly homologous, just as the skeletal elements of the different vertebrate groups are homologous—they have descended to each species from a common ancestor. However, the patterns themselves have been modified by selection in a number of ways. The commonest types of change affect (2) the frequency of performance of patterns and (b) their form or 'emphasis'. The former category needs little further comment, it is exactly comparable to the types of inter-strain differences we have just been describing. To give just three examples from many, the ducks have a common repertoire of courtship patterns which appear at different frequencies in each species. Some, as we have noted, do not perform certain patterns though carrying the genetic potentiality to do so. In *Drosophila*, *D. melanogaster* and *D. simulans* show similar differences; in the former the commonest courtship pattern is vibration, in the latter 'scissoring'—a pattern involving synchronized movements of both wings.[323] Clark *et al.*[95] have described frequency differences between homologous courtship patterns in two tropical freshwater fish, the swordtail and the platyfish.

Micro-evolutionary changes to the form of patterns are equally universal; they produce different kinds of 'emphasis' to common display patterns. Two examples can illustrate this, more are given by Manning.[325] Figure 6.7 illustrates a typical case; the 'long call' of gulls is equivalent to the territorial singing of passerine birds. The two species shown go through the same sequence of head movements when calling but 'emphasize' different parts of the sequence and this gives a characteristic 'look' to each display. Crane[106] has made a thorough comparative study of the courtship displays of the fiddler-crabs (*Uca*). The males dig burrows on the beach and stand at the entrance; one of their claws is much larger than the other and is brightly coloured. When a female comes near they rhythmically wave this claw and, if the female is receptive, she eventually approaches. The exact form of the male's wave is distinct for each species but the genus *Uca* can be divided into two main sub-genera on both behavioural and morphological criteria. The one, which Crane calls the 'vertical wavers', extend their claw only slightly and merely lift it up and forward. The 'lateral wavers' extend the claw out sideways and lower it to complete a more or less circular path.

For convenience we have isolated two types of change—frequency of performance and form of pattern—but clearly they normally occur together and combined with other types of change. For example, a great variety of changes are involved in the radiation of different types of cricket song from a common ancestral type—Alexander[3] discusses this with

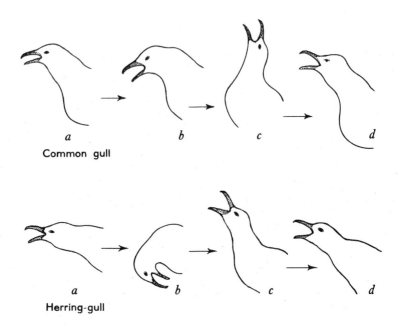

Fig. 6.7 Variations of emphasis in the 'oblique long call' sequence of two gull species. The sequence reads from left to right. At the start the head is in the oblique posture (a), it is jerked down as the bird begins calling (b), and is then thrown back (c) as the calls continue. The head is gradually lowered (d) as the calls die away. The common gull shows little emphasis of (b) and much of (c); the converse is true of the herring-gull. (From Tinbergen,[465] 1959, *Behaviour*, **15**, 1.)

numerous examples. Cricket songs vary in the amplitude, length and frequency of pulses and thereby produce a wide range of trills and chirps, each of which is quite distinctive. The same kind of factors must have been involved in the evolution of bird song. The song thrush (*Turdus ericetorum*) characteristically repeats each phrase of its song two or three times before passing on to the next.* Its close relative, the blackbird (*T. merula*), does not do so.

* 'That's the wise thrush; he sings each song twice over,
    Lest you should think he never could recapture
    The first fine careless rapture!'
                                    Robert Browning.

THE NATURE OF GENETIC AND MICRO-EVOLUTIONARY CHANGES

It is worth while trying to discover if there is a common thread running through the diversity of genetic and micro-evolutionary changes just described. We have emphasized the quantitative nature of the changes and many of them can, in behavioural terms, be directly related to changes in thresholds. If, in the same situation, two species or strains differ in the relative frequency with which they perform patterns A and B from a common repertoire, it is reasonable to interpret this as due to threshold differences in the mechanisms controlling A and B in each case. There are numerous sites at which the control mechanisms could be affected. There might be changes of threshold in the sensory system, in systems affecting motivation, in the motor system, or in any combination of these. Threshold changes could be produced by genes acting on the nervous system itself as in the hyperkinetic mutants of *Drosophila* described on p. 202—or more indirectly, by genes affecting metabolic rate or hormone secretion which will in turn affect the nervous system. In no case do we have proof, but sometimes the circumstantial evidence for direct action on the nervous system is strong. Thus the differences in the performance frequency of various sexual patterns between inbred lines of guinea pigs remain even after injections of sex hormones[185] or thyroid hormones, which increase the basic rate of metabolism.[395] It seems most probable that the lines differ in the threshold properties of the sexual mechanisms in the brain.

It is not difficult to explain changes in the qualitative form or speed of patterns in terms of threshold changes also. Each fixed action pattern must have a co-ordinating ' centre ' of some kind whose structure and properties are inherited and which, when activated, produces a stereotyped pattern of output to lower centres controlling muscles and groups of muscles. The centre calls into play each muscle group in the correct order, at the correct time and for the correct duration. This result must depend on a subtle series of threshold relationships both within the co-ordinating centre itself and between the various motor centres which control the muscles. The output of the co-ordinating centre may be completely ' pre-set ', or it may be modified by feed-back from the muscles as the fixed action pattern is actually being performed. In insects we know centres, probably small groups of neurons, which control singing in crickets[238-9] and flight in locusts.[504] These produce quite normal output when isolated, although they can be modified by feed-back. Hoyle[235] discusses the different types of control in more detail in relation to his own work on insects, which have proved excellent material for such studies (see also Chapter 2, p. 43).

Genetic changes affecting thresholds either within the co-ordinating centre or one of the subordinate muscle groups will change the form of a fixed action pattern. For instance, muscles might be brought into action earlier or later in a sequence; they might be held on for a longer or a shorter time, or they might change the intensity and speed of their response as more or less of their constituent fibres are activated. Figure 6.8 shows a beautiful

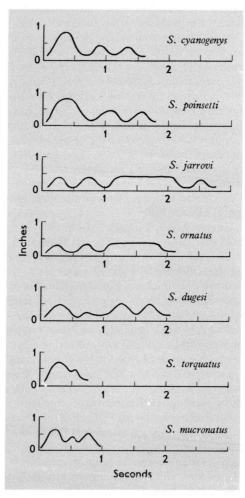

Fig. 6.8 The specific head-bobbing movements of some *Sceloporus* lizards. The movements of the head are represented as a line with height on the vertical axis and time on the horizontal axis. Changes to amplitude, speed and length of movements are all clearly shown. (From Hunsaker,[240] 1962, *Evolution*, **16**, 62.)

example of the behavioural results of such changes from Hunsaker's[240] work on lizards. Males of the genus *Sceloporus* show rhythmic head bobbing movements during courtship and also whenever they meet other lizards as a kind of species-identity signal. Each species has a characteristic pattern of bobbing which results from a series of contractions in the muscles which extend the front legs and thereby raise and lower the head and shoulders. Micro-evolution has produced several distinct versions by a series of changes of the type outlined above. This example is particularly clear because only two main groups of muscles are involved, but it is not difficult to envisage the differences in emphasis illustrated in Fig. 6.7, evolving in just the same way.

We may conclude that genes affecting thresholds within the nervous system have been most important in the evolution of behaviour. Most of the amazing diversity of fixed action patterns we observe will have evolved by the accumulation of small quantitative changes, in just the same way as the body form of animals has evolved.

RITUALIZATION

Many of the fixed action patterns whose evolution we have been considering function as social signals and form part of threat, courtship or appeasement displays. We have already discussed the origins of such patterns in Chapter 5; they do not usually evolve *de novo*, but develop from 'intention movements'—especially of attack and escape—or displacement activities, of which breathing movements, preening, drinking or eating are amongst the most common. Such displays are called 'derived activities' by Tinbergen[461] to signify their origin from other types of behaviour, and the evolutionary process whereby they become modified to form social signals is called 'ritualization'. This term we owe to Julian Huxley[245] who first introduced it to describe the evolution of the elaborate mutual displays of male and female great-crested grebes, which we have already discussed (see p. 188 and Fig. 5.10). The display illustrated there has obviously been derived from elements of the grebes' nest-building behaviour, but some courtship displays have been so modified during the course of evolution that without comparative evidence from less modified species we would be hard put to suggest their origin. This is certainly the case with some of the movements used in the courtship of the tropical grass finches (Erythrurae) illustrated in Fig. 6.9. Male zebra finches (a) often perform beak-wiping on the perch during courtship—probably this is an un-ritualized displacement activity. In the related striated finch (b) and spice finch (c) the male performs a bow and holds himself with head lowered close to the

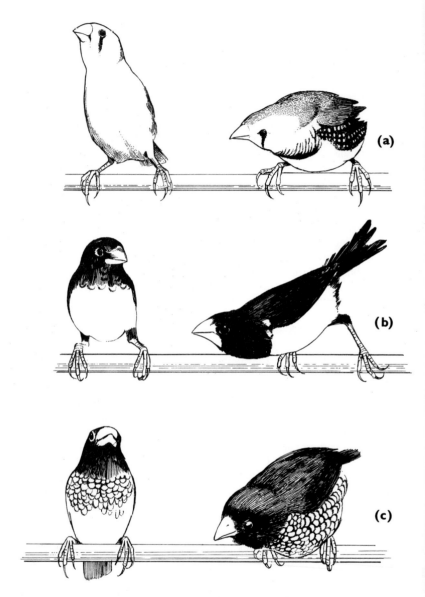

Fig. 6.9 Beak-wiping in the courtship of three grass finches: (a) the zebra finch, in which the movement is unritualized and we see here the male just about to wipe his bill across the perch ; (b) the striated finch and (c) the spice finch, in both of which what is very probably the same movement has become ritualized. In these two species the male remains stationary in this position for some seconds and ritualized beak-wiping now looks rather like a bow. In all three cases the bird on the left is a female (Redrawn from photographs in Morris,[357] 1958, *Proc. zool. Soc. Lond*, **131**, 389).

perch beside the female. The similarity of this posture to a phase of beak-wiping is quite striking and Morris[357] suggests that the bow is a highly-ritualized pattern derived from the former.

Ritualization involves all the threshold and frequency changes we have just been discussing and, in particular, changes in form and emphasis. Social signals must be conspicuous and most are extremely exaggerated in form and are accompanied by specially evolved releasers which enhance the effect of the movement (see the examples in Figs. 5.3 and 5.11). Good reviews of the ritualization of intention movements are provided by Daanje[116] and Andrew[13]; Tinbergen[461] and Morris[355] include other types of derived activities.

There are only two points connected with ritualization that require further mention here. Firstly, the concept of *typical intensity*, which term we owe to Morris.[356] If they are to be maximally effective, social signals must be clear-cut and unambiguous. For this reason many of them do not vary the form of the pattern to varying strengths of stimulus. A male cutthroat finch has a characteristic 'fluffed' courtship display posture. With weak stimuli he holds this posture only briefly but with strong stimuli he persists and begins his courtship dance. In either case, as shown in Fig. 6.10, his display posture is almost identical, i.e. it shows a typical intensity. This situation may be compared with the threat posture of the cat (Fig. 5.8, p. 177) where the intensity of the response matches that of the stimulus. Postures or movements which have a typical intensity are more easily recognized but correspondingly convey less information about the sig-naller's motivational state. Both types of organization must have advant-ages and which of the two evolves presumably depends on details of the context in which the display has to function.

The second concept we must mention is the *emancipation* of ritualized patterns. The courtship of ducks, for example, includes patterns which are clearly derived from displacement preening (Fig. 5.11) and displacement drinking. Following the arguments developed in Chapter 5 (p. 190) it is possible that these patterns originally occurred by disinhibition as a direct result of a conflict between sexual and attack or escape tendencies when males courted females. One supposes that in ancestral males the appearance of preening or drinking depended on the conflict being suffici-ently balanced. Those males which performed these displacement activities most regularly were more successful in mating. In some way the appearance of these patterns served to arrest the female's attention and perhaps stimulate her sexually. However, in their descendants conflict may no longer be a prerequisite for the performance of preening and drinking patterns in the sexual situation. They are now almost invariable parts of courtship and highly modified in form. Tinbergen[461] suggests that they have become 'emancipated' from their original controlling mechanisms and are now controlled by the sexual mechanisms alone.

(A)

(B)

Fig. 6.10 The ruffled courtship posture of the male cutthroat finch. These two photographs illustrate the very small difference between the low intensity form (A), and high intensity form (B). The posture has developed a 'typical intensity' and, if shown at all, is shown nearly at its maximum, The small extent of the intensity differences can best be judged by comparing the two displaying males with the posture of the non-courting male on the lower left. The right-hand birds in both photographs are females (see p. 221). From Morris, [356] 1957, *Behaviour*. **11**, 1.)

The two chief criteria for judging that a pattern is emancipated are (a) that it occurs out of the context which is appropriate for the presumed ancestral pattern—it is no longer evoked by the previously effective stimuli, and (b) that fluctuations in the tendency to perform the presumed ancestral pattern have no effect on the tendency to perform the emancipated pattern. The first criterion is the same as that applied to justify any behaviour pattern being called a displacement activity, whilst the second predicts that, for example, the state of thirst should not influence the amount of display drinking a drake performs in courtship. We have little firm evidence that these criteria are met but even so the reality of emancipation is in little doubt. Comparative evidence shows clearly the derivation of display drinking but the pattern is greatly modified—in some species it is little more than a bob of the head—and it obviously has nothing to do functionally with water intake. At present we have little idea of what the emancipation process involves (see Blurton-Jones[60] for a discussion of emancipation in threat displays).

On p. 186 in our discussion of displacement activities, we mentioned the need for a relatively stable balance between conflicting tendencies if a displacement activity was to have time to 'break through'. Such a balance can never stay perfect for long and presumably there is a range of conflict states in which equality of the two tendencies is sufficiently close to allow another behavioural system to manifest itself briefly—Rowell[402] refers to this range as 'effective equality'. One initial step in ritualization might be to widen the range of effective equality between two conflicting tendencies and thereby increase the frequency of a displacement activity. Blest[57] discusses in more detail some problems of emancipation and the ritualization concept generally.

So far in this discussion of ritualization we have applied the term solely to the evolution of courtship and threat displays, for which it was used originally. It is worth noting however, that the opportunistic nature of selection whereby an element of behaviour occurring in one situation serves as raw material for the ritualization of another pattern is also seen in other contexts.

Many tropical fresh-water fish, such as gouramis and fighting-fish, build nests of air bubbles. A male takes a mouthful of air at the surface and blows bubbles coated in saliva, which he builds up into a cluster clinging to the surface film amongst water weed. He deposits the fertilized eggs into the bubble nest after courtship and mating. Braddock and Braddock[64] suggest that the construction of bubble nests has followed on the evolution of air breathing. This habit evolved in quite another connection, as a response to the low oxygen content of water in tropical rivers and lakes, but it meant that the fish was constantly taking mouthfuls of air at the surface. Usually when it does so a few bubbles escape from its mouth and remain intact at the surface film for a time. From this point onwards it is not difficult to imagine the stages whereby bubble nests became well-oxygenated shelters for the developing eggs.

Another remarkable example of evolutionary opportunism is the dance language of the honey-bee which we discussed in some detail in Chapter 3, p. 99. The basis of this communication system is the honey-bee's capacity to convert its orientation path from an angle with respect of the sun, to one on the vertical comb at the same angle with respect to gravity. This capacity seems at first sight quite extraordinary and requiring some unique evolutionary event. However we know that light and gravity are the dominant features in the orientation of most insects and there is nothing exceptional about transposition between them. For example ants which have been running at a particular angle with relation to a light on a horizontal board, change course abruptly to take up a related angle with respect to gravity, when suddenly the light is switched off and the board tilted into the vertical plane (see Vowles[478]). Certain beetles behave in an analogous way. It is probable that honey-bees have modified an orientation system which is common to all insects. So far as we know no other insects use this ability for communication, but the potential was there. Given the demands for maximum efficiency in foraging by honey-bee colonies, natural selection has been able to elaborate this remarkable means of communication from relatively mundane and inconspicuous origins.

SEXUAL ISOLATION

Related species do not normally hybridize under natural conditions because it is disadvantageous for them to do so. Hybrids, although they may show some signs of 'hybrid vigour', are usually sterile or sub-fertile. Even if they are fertile they are rarely as successful as either parental type. The latter each have a set of genes which has been selected over many generations as the best for their own environment. The hybrids inherit a compromise set of genes which does not equip them so well in either parental environment. We know several cases in plants where hybrids between two species have invaded new environments and been very successful but this is very rare, if it occurs at all, among animals.

Consequently there is a strong selective advantage to choosing a member of one's own species for a mate. This is particularly true for females; males can usually mate several times in their lifetime but some female insects, for example, mate only once. If they mate with the wrong type of male they are effectively sterilized. Accordingly it is common to find that discrimination is stronger by females than by males. Females cannot afford to take risks but males can, and indeed too much discrimination may be actually disadvantageous for them. It will usually be better for a male to risk the occasional infertile mating than ever to miss the chance of a fertile one. In *Drosophila*, for example, females are more selective than males and it is often their active rejection of foreign suitors which prevents interspecific hybridization.

Sexual isolation may be defined as 'behavioural barriers to hybridization between species or populations'. It forms one aspect of the more general phenomenon of reproductive isolation, which is fully discussed by Mayr,[339] and is one of the most important ways in which behaviour affects the evolution of animal populations.

The degree to which two related species will meet under natural conditions varies greatly. Even if they co-exist in the same general area, two species will rarely live in exactly the same part of the habitat; competition is bound to drive them into specialization. In Britain, chiff-chaffs (*Phylloscopus collybita*) and willow warblers (*P. trochilus*) inhabit the same woods and both nest on the ground. However, for feeding the chiff-chaff moves amongst the high trees and the willow-warbler in the lower branches of trees and bushes. Clearly they are specializing, and in the Canary Islands where the willow-warbler is absent, the chiff-chaff occupies the willow-warbler's niche as well.

Obviously the selection of a habitat characteristic for the species will form part of normal reproductive isolation. *Drosophila pseudoobscura* and *D. persimilis* are two closely related species whose ranges overlap extensively in the south western part of the United States. They can be hybridized in the laboratory, particularly at low temperatures, but although many tens of thousands of flies have been caught in the wild and examined, no natural hybrid has ever been found. The two species seek out different micro-habitats within an area, *pseudoobscura* drier and lighter places, *persimilis* cooler and moister, and their preferences keep them effectively isolated. Even in the laboratory, *pseudoobscura* females will not readily accept *persimilis* males or *vice versa* because their courtship songs are so different, (see p. 206). In most animals isolation by habitat is supported by the evolution of highly specific signals which enable animals to detect their own species, sometimes from a considerable distance.

In the crickets, grasshoppers and cicadas among insects, and the frogs and toads among vertebrates, the males have 'assembly calls' which attract females in the mating season (see Alexander[2] for insects and Blair[53] for Anura). A number of different species may live in the same area and with frogs males of several species may be calling from the same small pond or ditch. It is remarkable how distinctive is each species' call and females respond only to the call of their own males. There can be little doubt that the diversification of calls is a direct result of selection in many cases. For example, Blair[52] describes how two species of frog, *Microhyla olivacea* and *M. carolinensis*, have a wide range in the southern United States, the former more to the west and the latter more to the east. At the extremes of their ranges where only one species is present their assembly calls are quite similar. More centrally where both species overlap and use the same ponds for breeding the calls have diverged and are quite distinctive. Hybrids are occasionally found in nature and Blair has shown that, apart from calls, there is little other barrier to hybridization. The males, like those of all Anura, will clasp and attempt to mate with almost any object of the right size and the females, if placed with males of the wrong species, will accept them readily.

Perdeck[380] describes a similar example in the two closely related grasshoppers, *Chorthippus brunneus* and *C. biguttulus*. These are very similar morphologically and occur in the same areas but their songs are very distinctive and, once again, females are attracted only by the song of their own males. If they are lured to foreign males by tape recordings of their own species, courtship and copulation proceed normally and hybrids are produced.

In cases such as these just quoted it seems likely that two populations with similar assembly calls came to overlap after many generations of geographical isolation, during which time they had diverged in a number of other ways. Selection then favoured the divergence of their calls because hybrids were at a disadvantage. Alexander and Bigelow,[5] describing the calling songs of the large variety of crickets and grasshoppers which live together in the eastern United States, find that—with one exception—they all have different calling songs. (Sometimes species are so similar morphologically they have been first distinguished by song.) The exception is the pair of cricket species *Acheta pennsylvanicus* and *A. veletis* which live in exactly the same area but have identical calling songs. However this is the exception that proves the rule, because *veletis* and *pennsylvanicus* mature in spring and autumn respectively, and therefore males are calling and attracting females to mate at completely different times of the year. These two species have diverged from a common ancestor by specializing in the early or late timing of their breeding cycle. The fact that during this evolution their songs have remained identical indicates that unless there is active selection for divergence calls will remain very constant, presumably because variants are likely to be 'misunderstood'. The most compelling evidence for this conclusion is provided by three crickets of the genus *Gryllus*, *G. campestris* from Europe, *G. bermudiensis* from Bermuda and *G. firmus* from the eastern United States. They must have been isolated from each other for many thousands of generations and they have diverged considerably in structure, but their calling songs remain almost identical.[4]

There is mutual adaptation between a male's call and a female's response. If selection favours the divergence of calls then changes to the behaviour of one sex will impose selection on the other to respond accordingly. Sudden, large changes will tend to be disadvantageous because they will be too 'difficult' for the opposite sex to follow. Assembly calls are only one type of sexual isolation mechanism. Other groups of animals discriminate their own kind by scent, colour pattern or some combination of features. Colour pattern is obviously important in some birds; two or three species of ducks may gather and court on the same pond but the males each have highly distinctive patterns. Scent differences are of great importance for sexual isolation in insects. Some virgin female moths 'assemble' males to their highly specific scents.[423] In *Drosophila* stimuli received when a male approaches and taps the female with his fore-tarsi provide chemical evidence of identity for both sexes.[325]

Courtship behaviour itself, if we exclude song patterns, is probably used less often for identification. The differences between close relatives are, as we have seen, mostly quantitative. It seems that selection for divergence can operate most efficiently to change colour or scent in a clear-cut distinctive way. Courtship patterns may help in sexual isolation in that the movements and postures used may serve to display specific colours or scents in a conspicuous fashion.

With sexual isolation we are made most aware of the evolution of differences in behaviour for their own sake as an aid to discrimination. It is worth noting, however, that sometimes selection favours the opposite tendency and will cause animals to evolve similarities, again for their own sake, with other species. Where colour patterns are concerned the phenomenon of Müllerian mimicry is a familiar example. Many distasteful insects (e.g. caterpillars of the cinnibar moth, ladybird beetles) or those with stings (e.g. wasps, bumble-bees) are coloured with conspicuous stripes or blotches of black and yellow or black and orange. The same colour patterns are found in some fish, amphibians, lizards and snakes—in every case associated with powerful defence mechanisms of some kind. Selection will favour a distasteful or poisonous animal having a conspicuous colour and also showing conspicuous behaviour, or at least not trying to hide itself. Predators learn rapidly as a result of one unpleasant experience and avoid contact with others of the same type. Müller suggested that it was no coincidence that so many diverse animals used the same type of warning colouration. The obvious conclusion is that they have evolved to resemble one another in order to benefit from each other's warning effect. A predator avoiding one type of distasteful prey will also avoid others without having to sample and learn each one separately—clearly an advantageous situation for members of any of the prey species.

Convergence of colour markings is also found among bird species which form mixed flocks for feeding (we shall discuss the significance of flocks more fully in Chapter 8). Moynihan[359] describes a number of

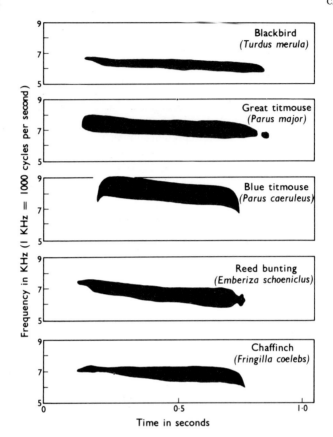

Fig. 6.11 Sound spectographs of the alarm calls of five species of passerine bird from three different families. They are remarkably similar in form and pitch and they share the property of being difficult to locate. (From Marler and Hamilton,[333] 1966 *Mechanisms of Animal Behavior.* John Wiley, New York and London.)

examples from the American tropics, e.g. tanagers, warblers and finches of western Panama which form mixed flocks and are largely black and/or yellowish sometimes with white markings. Birds from these species tend to assemble and move together and the common markings are probably specially evolved to maintain the cohesiveness of the group. Moynihan refers to this as 'social mimicry'.

In Chapter 3 we mentioned that many small birds respond to the alarm calls of other species as well as their own, and that these calls have converged during evolution to provide a common sign stimulus. Marler[329] discusses the characteristics of an 'ideal' alarm call which must carry as far as possible whilst providing the least clues to a predator on the localization of the calling bird. Amongst other things this means that the call should be of constant pitch and both begin and end gradually. It is the abrupt onset of sounds which allows their localization, since one way vertebrates locate a sound source is by comparing the time at which the sound hits each ear. Fig. 6.11 shows the sound spectrograms of alarm calls in small birds from three different families which are extremely similar and possess the properties outlined above. They provide a striking contrast to Fig. 6.12 in which we see the territorial display songs of three species of European warbler. Here the requirements of advertisement and sexual isolation have led to marked divergence between close relatives living in the same area.

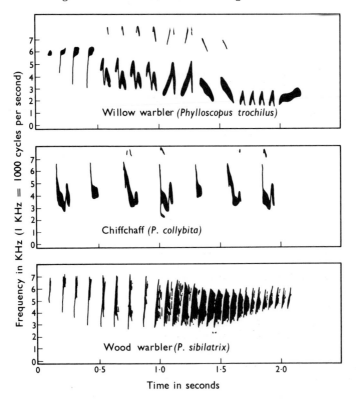

Fig. 6.12 Sound spectographs of the songs of three very closely related species of European warbler. The three are very similar in plumage and share the same habitat, hence the songs probably serve, in part, as a means of sexual isolation. They are strikingly different in form and quality but share the property of being easy to locate (From Marler and Hamilton,[333] 1966 *Mechanisms of Animal Behavior*. John Wiley, New York and London.)

7

Learning

Throughout this book we have had to make reference to learning as it affected the particular aspect of behaviour under discussion; now attention must be focused on learning itself. Learning involves a change in behaviour, often a long-lasting one, and it would have been valid to discuss it earlier as a form of behavioural development. However, the approach to the study of learning and, as we have seen, the people who have been involved in it, have usually been distinct. In this chapter we shall consider the general phenomenon and the different forms it takes, together with some account of how learning ability varies between different animal groups.

Thorpe[457] defines learning as '. . . that process which manifests itself by adaptive changes in individual behaviour as a result of experience'. This definition draws attention to two important features. Firstly, learning normally results in *adaptive* changes, and in Chapter 2 we discussed how learning and instinct are both ways for equipping an animal with a set of adaptive responses to its environment. Normally both methods are found in combination and logically they have much in common. In one case we have the selection of individuals bearing the best genes during the evolution of a population of animals; in the other, the selection of the best responses during the learning process in an individual animal. Russell[410-3] develops the analogy between learning and evolution in a most complete and revealing fashion.

The second important point arising from Thorpe's definition is that, strictly speaking, learning is a *process* which we cannot usually measure directly; we measure what has been remembered as a result of learning. Because we can communicate so easily with human subjects, they are in many respects better material for learning studies than animals. For instance, it is possible to test our memory in two ways; by 'recall', i.e. by reciting or writing down a list of nonsense syllables which we have previously learnt, or by 'recognition', i.e. presented with a set of nonsense syllables which includes those we have learnt, we record which syllables we recognize. Recognition is always an easier task than recall because the situation provides stimuli which, as we say, 'jog our memory' and help the process of recall. If we have trained a rat to run through a maze we cannot ask it to draw a map of its route on a piece of paper. The only way to test what it has learnt and retained is to put it back in the maze and observe its behaviour. If the rat makes mistakes we have no means of knowing whether it failed to learn adequately or learnt but failed to recall.

This difficulty brings us up against the problem of learning mechanisms. What happens in the nervous system when an animal learns something? When a young game-bird crouches on the first exposure to the sound of its parent's alarm call, it must utilize pre-set, inherited pathways in its brain whereby a particular auditory input has easy 'access' to the motor system controlling crouching. When a rat learns to press the bar of a Skinner box new pathways must become established because prior to learning the bar evoked no special response. We know too that mere presentation of the bar to the rat is not enough, there must be reinforcement. Some part of the learning mechanism records the result of the rat's reaction and if this result is 'good' it increases the probability that the reaction will occur again the next time the situation is presented. Further, we know that somewhere in the nervous system there is stored a more or less permanent record of the learning that can be 'consulted' or recalled on future occasions.

The search for this record or memory trace and the physiological mechanisms which underlie its formation has been an active area of research for decades. This is certainly one of the outstanding problems in biology but it has not been easy to make progress. Here we shall confine ourselves to discussing the behavioural aspects of learning. For a good introduction to the problems of learning mechanisms the reader is referred to the extensive series of reviews covering many areas of this research in the volume edited by Rosenzweig and Bennett.[398]

Animal learning studies have, until recently, been dominated by the work of experimental psychologists. Soon after J. B. Watson founded the new 'behaviourism', the learning abilities of the domestic white rat became the subject of a scrutiny which has continued for half a century. Out of this work have grown a number of schools of learning such as those associated with Hull, Tolman and Skinner. Each school has tried to construct a system of behavioural laws which will predict more or less exactly under what conditions learning will occur. We shall not discuss their respective

merits at any length; there are admirable accounts of learning theories
in Munn's[360] classic textbook on the rat, and in Broadbent.[67] Until quite
recently one weakness they have shared has been excessive concentration
upon one or two domesticated species—notably the white rat and the
pigeon.

CLASSIFYING LEARNING

Learning occurs in a great range of different animals under a wide
variety of circumstances. Central to some schools of thought in psychology
has been the idea that there are general 'laws of learning' which apply with
equal force wherever it occurs. This viewpoint has taken some hard
knocks recently. In Seligman and Hager's[431] and Hinde and Stevenson-
Hinde's[224] books, a wide range of authors argue convincingly that an
animal's learning abilities must have evolved to suit its own special requirements just as any other aspect of its behaviour. We cannot expect
learning in honey-bees to resemble that in pigeons except in a very general
way. Perhaps one type of generality which we can extract concerns the
nature of the logical operations involved. When a honey-bee learns that a
black area on a white background denotes a food dish and a pigeon learns
to peck a blue key to release a portion of grain, there is a real sense in
which their learning is of the same type. At any rate it is helpful to try to
distinguish different categories of learning as a convenient way to organize
our thinking. We must certainly recognize that any classification is likely
to be little more than a set of artificial abstractions. They may not correspond to the natural situation particularly if a broad range of animal types
is under consideration.

The classification we shall use here is taken from Thorpe[457] and the
account which follows owes a great deal to his book which includes much
more detail and full references. Although now somewhat out of date, the
particular value of Thorpe's book lies in his complete coverage of the
animal kingdom. His classification of learning, together with the commoner
synonyms found in the literature, is

1. Habituation

'Associative learning'
2. Conditioned Reflex (CR) Type I (= 'classical conditioning' or 'respondent conditioning')
3. 'Trial and error' and CR Type II (much of which is
   called 'instrumental conditioning' or, by Skinner,
   'operant conditioning')

4. Latent learning
5. Insight learning
6. Imprinting

We have already discussed imprinting in Chapter 2 for it now appears more appropriate to treat it as an aspect of development rather than a distinct type of learning. The other learning categories can be taken in turn.

Habituation

It is convenient to deal with habituation first because, in some respects, it is the simplest form of learning. Unlike the other forms, habituation involves not the acquisition of new responses but the loss of old ones. If an animal is repeatedly given a stimulus which is not associated with any reward or punishment, it ceases to respond. Birds soon ignore the scarecrow which put them to flight when it was first placed in a field. A snail crawling across a sheet of glass retracts into its shell when the glass is tapped. After a pause it emerges and continues moving; a second tap causes retraction again but it emerges more quickly. So we proceed with the snail's response becoming increasingly perfunctory until it ceases to respond at all.

A study by Clark[96-7] on the ragworm, *Nereis*, illustrates some of the typical features of habituation. *Nereis* is a marine worm which normally lives in a burrow or tube which it constructs in the mud at the bottom of brackish estuaries. The worm's head and anterior segments protrude from the tube whilst it feeds from the surface of the mud. At such times a variety of sudden stimuli will cause the worm to jerk back rapidly into its tube. In the laboratory Clark could easily get the worms to live in glass tubes in shallow basins of water. He found that jarring the basin (mechanical shock), touching the head of the worm, a sudden shadow passing over and a variety of other stimuli would all cause rapid retraction into the tube, but the majority of worms emerged again within a minute. If these stimuli were repeated at 1 minute intervals the proportion of worms responding fell off until none of them were retracting. Clark found that habituation occurs more rapidly if stimuli were given close together. For example, with a bright flash of light it took less than 40 trials at half-minute intervals, but nearly 80 trials if the interval was 5 minutes. The speed of habituation also depended on the nature of the stimulus; mechanical shock, shadow, touch and light flash all produced their characteristic rates of habituation. Further habituation was to a large extent 'stimulus-specific'; Fig. 7.1 shows how the waning of retraction to repeated mechanical shock is independent from that to a moving shadow. In this respect the behaviour of the worms resembles that of the nestling chaffinches in Prechtl's[383] experiments described on p. 10.

There are a number of other processes which may be confused with habituation because they also lead to a reduction in responsiveness. Results such as those illustrated in Fig. 7.1 eliminate any possibility that the waning of response is due to motivational changes or to muscular

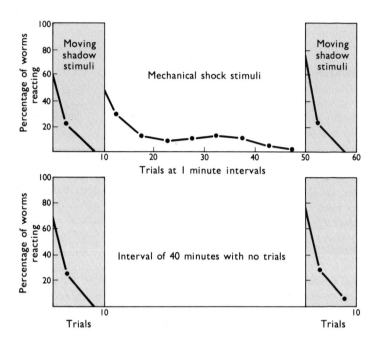

Fig. 7.1 The rates of habituation to two different stimuli in *Nereis*. The response measured is the sudden retraction of the worm into its tube and trials are given at 1-minute intervals. The shaded areas record the responses of a group of 20 worms to a moving shadow. Within 10 trials they have all ceased to respond, but switching to a mechanical shock stimulus (unshaded areas in upper graph) brings back the response in half the worms, and it characteristically habituates more slowly, taking more than 30 trials. The recovery of the response to moving shadow is complete after 40 minutes, whether the habituation trials to mechanical shock intervened (upper graph), or if the worms were simply left alone (lower graph). Clearly habituation is quite independent for these two very distinct types of stimulus. (From Clark,[96] *Anim. Behav.*, **8**, 82.)

fatigue, but it is often more difficult to eliminate sensory adaptation. Many sense organs eventually stop responding to repeated stimulation—on p. 114 we discussed the adaptation of the chemo-sensory hairs in the blowfly. *Phormia*. We cease to be aware of our clothes within a minute or so of putting them on because the tactile receptors in the skin cease to respond. In *Nereis*, Clark could also eliminate sensory adaptation as an explanation for the waning. For example, the worm soon ceased to retract when touched by a probe, but clearly still detected the stimulus because it then attempted to seize the probe with its jaws.

Sensory adaptation is usually a short-lived phenomenon; a few minutes without stimulation is usually sufficient for complete recovery. It would not be appropriate to say that we habituate to the feeling of our clothes, and we retain the term for a more persistent waning of responsiveness which must be a property of the central nervous system and not the sense organs.

Clark could detect some recovery from the waning of retraction within an hour or less in *Nereis*, and the worm was completely recovered within 24 hours. We have already mentioned that this waning was stimulus-specific, and it is this type that meets most of the generally accepted criteria for habituation. However there are other types of waning with different time courses and it is difficult to make any firm distinctions on behavioural grounds. Hinde's experiments on the mobbing behaviour of chaffinches described in Chapter 1 (see Fig. 1.5 for example) suggest that there is a dual recovery process following waning. There is a relatively rapid component, complete within 24 hours and also stimulus specific and thus very comparable to the *Nereis* situation. But there is also a long-term component which showed little sign of recovery over the same period, because the response remained greatly reduced. It is doubtful if we are justified in lumping these two components under the term 'habituation'. In nature sensory adaptation, short and long term waning will probably all occur together in some situations so it is impossible to define habituation with any degree of rigidity.

Habituation-like phenomena are found in every group of animals from the Protozoa upwards. They appear to be almost a general property of living material and certainly all the typical characteristics of habituation and recovery can be exhibited at the level of single neurons and neuro-muscular junctions (see Bruner and Kennedy,[75] and reviews in Horn and Hinde[232]).

Habituation is an important process for adjusting an animal's behaviour to its environment. Small prey animals such as *Nereis* cannot spend too long skulking in their tubes—they must normally be out feeding. Although it is best to retreat rapidly to sudden shadows not every shadow means a predatory fish is overhead, it might just as well be a floating piece of sea weed. In general, frequent shadows are more likely to be sea weed than predators and it is adaptive to cease responding when a repeated stimulus has no attendant consequences.

Habituation also plays an important part in the development of be-haviour in young animals which are often threatened by a wide range of predators and may begin by showing escape responses to anything large which moves. Rapidly they learn to ignore leaves moving in the wind and other neutral stimuli. Schleidt's[420] work on the alarm response of young turkeys described on p. 84, illustrates this point well. The inherited pecking response of newly-hatched chicks is at first directed towards any small object that contrasts with the background. Again responses to inappropri-ate objects quickly habituate and at the same time the chick learns positively which objects are food.

Conditioned reflex Type I *Changing old response*

The term 'conditioned reflex' is inseparable from the name of the great Russian physiologist, I. P. Pavlov,[378] whose school was active around the turn of the century. Pavlov's influence on Russian behaviour studies is still very great but, perhaps because the behaviour theory he developed has attracted little favour here, his influence in the West has been less. Pavlov and Sherrington were working at the same time but from completely different viewpoints. Sherrington studied the organization of reflexes in the isolated spinal cord of dogs and cats, having deliberately cut off influences from the higher centres. Pavlov worked with intact animals and considered that, just as simple reflexes are a property of the spinal cord, so conditioned reflexes are the particular property of the higher centres of the brain especially the cerebral hemispheres. Pavlov's aim was to study 'the physiology of higher nervous activity', but most of his experiments were, in modern terms, pure experimental psychology. Indeed, Pavlov was really one of the founders of experimental psychology; he was applying objective techniques to the study of learning some years before J. B. Watson.

Pavlov's classical experiments with dogs often involved the 'salivary reflex'. Dogs salivate when food is put into their mouths and Pavlov could measure the strength of their response by arranging a fistula through the cheek from the salivary duct, so that drops of saliva fell from a funnel and could be counted. A hungry dog was placed on a stand, restrained by a harness and every precaution was taken to exclude disturbances. In this position it could be given various controlled stimuli such as lights, sounds or touch, and meat powder could be puffed into its mouth through a tube. A standard quantity of meat powder caused the secretion of a certain amount of saliva. Now Pavlov preceded each ration of powder by—say—the sound of a metronome ticking. At first, this stimulus caused no response, save that the dog pricked up its ears momentarily. However, after five or six pairings of metronome followed by food, saliva began to drip from the dog's fistula soon after the metronome started and *before* the meat powder arrived. Eventually the amount of saliva produced to the metronome alone was the same as that which was given to the meat powder.

The dog had learnt to respond to a new stimulus which was previously neutral, and Pavlov called this the 'conditioned stimulus' (CS). The salivation response to the CS is the 'conditioned response' (CR). Prior to learning, only the meat powder or 'unconditioned stimulus' (UCS) produced salivation as an 'unconditioned response' (UCR).

Pavlov found that almost any stimulus could act as a CS provided that it did not produce too strong a response of its own. With very hungry dogs even painful stimuli, which initially caused flinching and distress, quite soon evoked salivation if paired with food.

The CR is formed by the *association* of a new stimulus with a reward or 'positive reinforcement' (a term which we shall examine more closely later). A CR for withdrawal can also be formed by associating the CS with a punishment or 'negative reinforcement'. An electric shock to the foot causes a dog to lift its paw; if a metronome is paired with the shock, the dog soon raises its paw to the sound alone.

Conditioned reflexes of this type have been observed in many different animals from arthropods to chimpanzees. For example, birds learn to avoid the black and orange caterpillars of the cinnibar moth after one or two trials which reveal their evil taste. They associate this with the colour pattern and generalize (see p. 242) from cinnibar caterpillars to wasps and other black and orange patterned insects. Because predators generalize it is advantageous for different distasteful insects to resemble one another—the phenomenon of Müllerian mimicry (see p. 227).

In nature we rarely observe a conditioned reflex in as 'pure' a form as in the laboratory. Bees do not just learn to associate a colour with the nectar reward, they also learn the position of the group of flowers with respect to their hive and learn at what time of day the nectar secretion is highest. Even Pavlov in his scrupulously controlled environment found that his dogs learnt more than one particular response to one particular stimulus. A hungry dog familiar with his laboratory's methods would run ahead of the experimenter into the test room and jump up on to the stand with every sign of expectancy.

Trial and error   *making new response to new situation*

In the CR Type I the animal starts out with a *response*—the UCR linked to its UCS—which subsequently becomes attached to a novel stimulus; hence the term 'respondent conditioning' which is sometimes used for this type of learning. In contrast to this situation we may have one in which the animal is motivated by thirst, hunger or fear as before but has no UCS to evoke an appropriate UCR. Rather, the animal is exploring or showing appetitive behaviour during the course of which it performs spontaneously a variety of motor patterns, sniffing, walking, looking round etc. Suppose now one of these patterns is followed by reinforcement—e.g. a hungry animal receives food—then, if this association is repeated a number of times the animal learns to perform this pattern regularly in this particular situation.

An example will make this clearer and suggest why this type of learning is justifiably called 'trial and error'. In his pioneer experiments on learning

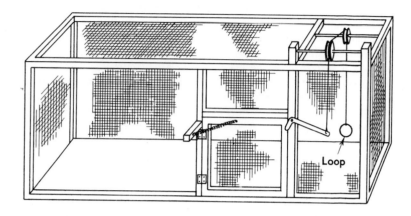

Fig. 7.2 One of Thorndike's problem boxes. A cat is confined inside the cage and must learn to pull the string loop to open the door. (From Maier and Schneirla,[322] 1935, *Principles of Animal Psychology*. © 1935. By permission of McGraw-Hill, New York and Maidenhead.)

Thorndike used a variety of 'problem boxes'; one of these is shown in Fig. 7.2. It is a cage which can be opened from inside only by depressing the lever. A cat is shut in and tries hard to escape, it moves around restlessly and after a time—by chance—it steps on the lever and the door opens. The second trial may be a repetition of the first, and the third, but soon the cat concentrates more attention on the lever and eventually it moves swiftly across the box and presses the lever as soon as it is confined. The name 'trial and error' obviously fits this type of learning very well. The cat learns to eliminate behaviour which led to no reward and increases the frequency of behaviour which is rewarded, but in the early stages there is little system in its activity—the first reward is obtained by pure chance.

We have already described the use of a Skinner box, which is basically a problem box of a convenient form in which an animal learns by trial and error that pressing a bar yields a small reward. Because the animal's own 'spontaneously generated' behaviour has been instrumental in its gaining a reward, such learning is often called 'instrumental conditioning' (Skinner[437] also uses the term 'operant conditioning'), but it is no different in principle from trial and error. In this type of conditioning there is no CS, unless we consider that the proprioceptive feed-back from the muscles during the performance of the CR—bar pressing for instance—is recorded as a pattern of stimulation which signals the reward. In the course of time other features of the trial and error situation may come to act as conditioned stimuli which signal to the animal that the response is now 'appropriate' and will yield reward.

It is not at all clear how fundamental is the distinction between trial and error and the CR Type I. There is some clear evidence (see Moore[354]) that pigeons working in Skinner boxes become classically conditioned to the key and treat it 'as if' it were food or drink. If associated with food, they peck the key sharply as they peck at grain. When water is provided they press the partly open bill against the key and make the sucking movements of drinking. Such behaviour is less clearly observable in rats or monkeys working in operant conditioning situations, but it may be there. Certainly we cannot regard trial and error and CR Type I as completely distinct categories. Konorski[275] has argued that the former is best regarded as a CR Type II and also refers to this class a special type of learning in the Pavlovian situation. A dog on the stand has its left front foot lifted and then meat powder is blown into its mouth. After a few trials the dog *spontaneously* lifts its foot when hungry and placed on the stand. Here we are simply directing the nature of the animal's CR, and the problem box or Skinner box situation in which the animal spontaneously performs the rewarded act is obviously more natural.

Trial and error learning will often be involved when animals modify their appetitive behaviour to obtain food, shelter or a mate. Most commonly it will be mixed with classical conditioning (CR Type I) because both new stimuli and new behaviour patterns must be learnt. Learning to run a maze is a familiar example when a rat may learn to use a combination of clues, some visual depending on stimuli perceived at the choice points, others 'kinaesthetic', i.e. learning whether to turn to right or left.

Trial and error is probably the most appropriate category under which to classify the learning of new motor skills. Young mammals and birds, for example, may perfect the co-ordination of their movements by practice, often during sessions of play with their siblings or parents.

Some characteristics of associative learning

Because both types of associative learning have much in common and are relatively simple, this is a good point at which to summarize what general features of the learning process they exemplify. Most of these features were first clearly described by Pavlov from his work on the CR Type I. As mentioned earlier there are real problems in drawing up 'laws of learning' which have general validity and this account will have to be highly simplified. Where there are marked deviations from the general rule they will be noted.

Contiguity

In the great majority of conventional learning situations it has been

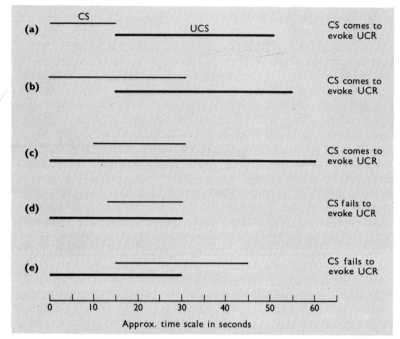

Fig. 7.3 The effect of the sequence of stimuli upon the formation of a conditioned reflex. In each case the upper, thin line denotes the duration of the conditioned stimulus (CS), the lower, thick line that of the unconditioned, reinforcing stimulus (UCS). The results are given on the right ; note that the CS must not end with or persist beyond the UCS if a positive conditioned reflex is to be established. If it does so, then the CS will remain neutral and may even tend to inhibit the response to the UCS. (From Konorski,[275] 1948, *Conditioned Reflexes and Neuron Organisation*. University Press, Cambridge.)

found that if a conditioned stimulus or a new response, such as lever pressing, is to become associated with a reinforcement, the two must occur close together. Figure 7.3 shows some of Pavlov's findings. Conditioning occurs most readily when the CS starts before the UCS and goes on to overlap it. It is very difficult to get a CR if the CS ends more than about a second before the UCS begins, or if the CS goes on after the UCS has stopped.

Now in trial and error or instrumental conditioning the CR always *precedes* the reinforcement but, again, the latter must follow rapidly if learning is to occur. Skinner[437] found that delays of only a few seconds between a rat pressing the bar and the delivery of food greatly slowed the rate of learning the response. The deleterious effects of delayed reward can often be overcome by introducing a 'secondary reinforcement'. Suppose the rat learns that a reward is delivered when a light comes on in the box (light and reward must overlap as discussed above), then it will learn to press the bar to switch the light on. Light becomes a secondary reinforcement or a 'bridging stimulus' between the response and the primary reinforcement—food. Bridging stimuli are useful for training animals—as in a circus—when it is often difficult to present a reward immediately after the response is made.

However there are certain types of associative learning which consistently occur when reinforcement is delayed for a matter of hours. Barnett[28] describes how wild rats only nibble at small amounts of any novel foods that appears in their territory. If it proves edible, they will gradually take more on successive nights until they are eating normally. If it is poisonous, and they survive, they avoid it completely on subsequent occasions. This type of behaviour is highly adaptive and makes poisoning rats no straightforward task. The interesting feature for our discussion is the delay that must ensue between a rat tasting poison bait (always made superficially palatable with sweet substances) and any subsequent ill effects. Few rat poisons take less than an hour to produce effects. Laboratory findings have confirmed this ability. Not only will rats learn to avoid tastes associated with sickness that sets in at least an hour later, if deprived of the vitamin thiamine they will learn to choose a diet containing it, although many hours must elapse before they can feel its benefits. Rozin and Kalat[407] provide an interesting review in which they link the specializations of learning which are involved in such 'specific hungers' with the rat's ability to avoid poison baits.

It is only when a new *taste* acts as the CS that reinforcement can be so delayed. In one ingenious experiment, Garcia and Koelling[161] supplied rats with a drinking tube containing saccharin-flavoured water so arranged that when they licked the tube, bright lights flashed on. During these sessions the animals were irradiated with X-rays which made them sick about an hour later. Subsequently the rats avoided saccharin taste but did *not* avoid flashing lights. Conversely if they were given flashing light plus immediate electric shock to the feet whenever they licked the tube, they subsequently avoided the light but still licked at saccharin.

Rats are in some way 'prepared' to associate taste with sickness after a single trial and a long delay,* but visual stimuli and sickness are not con-

* Garcia and his co-workers[189] have put this effect to practical use in the field of conservation. Sheep farmers in the western U.S.A. are troubled by coyotes killing their stock and correspondingly they have tried to eliminate the coyotes. When a sheep's carcase is injected with lithium chloride the coyotes which come to feed on it become violently sick later. They subsequently avoid the smell or taste of sheep and switch their attention to other prey. Farmers report a marked reduction in sheep killing following the treatment of their ranches with a few doped carcasses. Sheep and coyotes may be able to co-exist!

nected. Conversely light and electric shock are easily associated if they occur close together, but taste and shock are not.

In the past it has commonly been assumed that contiguity plus reinforcement is all that is required to associate any stimulus with any response. However experiments such as these (and many more, ably reviewed by Seligman[430]) suggest that animals approach learning situations with a good deal of built-in bias. This bias usually relates to the natural requirements which their behaviour has evolved to meet. Pigeons easily learn to peck at a key for food reward but cannot learn to do so in order to switch off an electric shock to the feet. However they will readily lift their wing (part of the natural defence pattern) to switch it off. We discussed in Chapter 2 (p. 57) evidence suggesting that animals may inherit a tendency to learn particular things. Breland and Breland[66] in a paper which they mischievously entitled '*The Misbehavior of Organisms*' (note Skinner's title) give a number of examples from their own efforts to train animals to perform tricks for commercial purposes. They found that the laws of contiguity and reinforcement are not sufficient to overcome powerful inborn tendencies to behave in particular ways. Thus chickens persisted in scratching at the ground even though it hampered their getting a food reward. One is reminded of Dilger's hybrid love-birds (see p. 53) which could not prevent themselves from making abortive tucking movements and as a result could scarcely ever manage to carry material to the nest.

Repetition

Although, as we have seen, one-trial learning can occur especially with strong punishment as a reinforcement, most associative learning improves with repetition. Pavlov found the amount of saliva produced to the CS steadily increased with each trial which was reinforced until it was at the same level as that given to the UCS. The rat learning a maze steadily reduces its errors with successive trials until it runs unhesitatingly through to the goal box where food is found. The number of errors or the time taken to reach the goal box on each trial can be used to plot learning curves such as those in Fig. 7.4, which shows typical records of maze learning.

Repetitive reinforcement eventually produces a maximum response beyond which we cannot show that learning has improved in terms of performance on any one trial. However, the longer we go on reinforcing a response beyond this maximum ('overtraining'), the more resistant it becomes to 'extinction', i.e. the animal will go on responding longer if reinforcement is removed.

Generalization and discrimination

If Pavlov conditioned a dog to salivate when a pure tone of—say—1,000

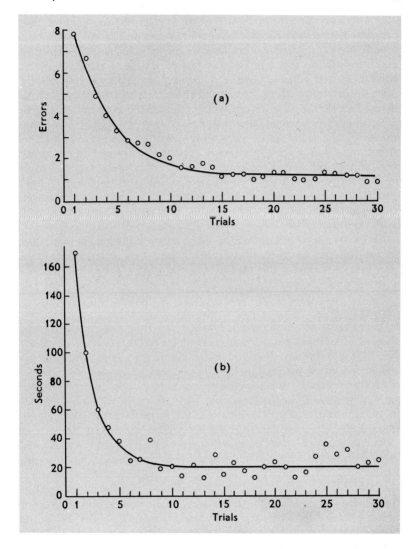

Fig. 7.4 The learning curves of rats learning a multiple T maze plotted (a) by errors, and (b) by time to reach the goal box. Each circle represents the mean value for 47 animals. (From Woodrow,[513] 1942, *Psychol. Bull.*, **39**, 1.)

cycles was sounded, it would also salivate but to a lesser extent when other tones were given. It *generalized* its responses to include stimuli similar to the conditioned one and the more similar they were the more the dog salivated. The opposite process to generalization is *discrimination*. Dogs naturally discriminate to some extent or they would salivate equally to all sounds, but their discrimination becomes refined after repeated trials when only one particular tone is followed by reward. We can accelerate discrimination if, as well as rewarding the right tone, we slightly punish the dog when it salivates to others. This 'conditioned discrimination' method has been of enormous value for measuring the sensory capacities of animals. After training to one particular stimulus—it may be a colour, brightness, shape, texture, sound, smell, weight, etc.—we then test to see how far the animal can discriminate this stimulus from others. We present it together with another stimulus of the same type and reward only responses to the former, perhaps giving slight punishment for incorrect responses. The two stimuli are made increasingly similar until there comes a point beyond which the animal can no longer learn to discriminate between them. This marks the limit of its sensory capacities as measured by its behaviour.

To give but three examples from many hundreds, this method was used by von Frisch[156] in his classical studies of the colour vision of bees, it was also used to examine the touch sensitivity of the octopus,[494] and the chemical senses of fish.[78]

Reinforcement

The concept of reinforcement has been a central one for the learning theories of Hull and Skinner and some objections to its universal application have already been discussed under 'contiguity'. Hull considered that reinforcement was essential for learning and that it consisted of a reduction in 'drive' or 'need'. A hungry animal reacts to a stimulus or performs an activity because in the past they have been associated with a reduction in its hunger. If this reward is removed, the learnt response gradually extinguishes.

This argument is sometimes used in reverse; the occurrence of learning in a particular situation is taken as evidence of drive reduction. Miller and Kessen[350] find that hungry rats learn to visit the alleyway of a maze where they can drink sweetened milk, in preference to a similar alleyway where milk is directly injected into their stomachs. However, they will learn to visit the latter if the alternative alleyways have no food. Miller and Kessen assume that hunger drive is reduced more by the act of drinking milk than by having it placed in the stomach, but that this also reduces drive. Similarly Myer and White[361] conclude that some rats have a spontaneous aggressive drive, because they will learn a simple maze with the opportunity to kill a mouse for a reward. We discussed other examples relating learning to aggressive motivation in Chapter 4, p. 137.

In its simplest form the drive-reduction hypothesis runs into trouble with 'avoidance conditioning'. Animals will learn to avoid punishment if it is signalled by some stimulus. A rat runs off an electrified grid when a light flashes once it has learnt that this signals that a shock will follow in 5 seconds. It will continue to respond in this way indefinitely and never gets another shock, hence the name 'avoidance conditioning'.

Here, on the face of it, is the persistence of a learnt response in the absence of any reinforcement, i.e. shock. Hullians can get round this difficulty by proposing that the initial shocks induced a state of 'anxiety' which continues to be aroused each time the rat sees the light. It is the reduction of this anxiety or 'secondary drive' which is rewarding. But if learning always entails drive reduction, we might repeat a question asked in Chapter 4, how many drives are there? Monkeys will learn to press a bar which moves aside a shutter for a few seconds and allows them to watch an electric train working! Do we ascribe this learning to a reduction in the 'curiosity drive'? As we shall see, the drive-reduction hypothesis has further trouble in accounting for learning during the exploration of a new environment (so-called 'latent learning') and for imprinting. There is a good discussion of this problem as a whole in Watson.[487]

Although we may reject the idea that a 'conventional' reward is necessary for all learning, we must remember that its effectiveness for simple associative learning is not in doubt. Loucks[311] stimulated a part of the motor cortex in the brain of dogs and thereby induced them to move a leg. He paired this stimulus with the sound of a buzzer, for more than 600 trials in some cases, without getting any trace of a leg movement to the sound given alone. Clearly, in this situation, mere association of the two events was inadequate to produce learning. Loucks then gave a small food reward following the sound and the induced leg movement, and within 6 trials the dog moved its leg spontaneously to the buzzer alone.

Olds[373] has found that electrical stimulation in some areas of the brain is 'rewarding' in itself. Rats learn to press a bar which delivers a brief electrical pulse to one of a number of sites in the hypothalamus, for example. Sometimes this self-stimulation is so intensely rewarding that a rat will press over 7,000 times per hour until it falls asleep exhausted! Presumably brain sites which produce this effect when stimulated are some part of the mechanisms whereby normal rewards are registered in the brain.

Many studies have been made to determine how the strength of motivation affects learning. Up to a certain level, learning for a food reward may be improved and accelerated by increasing hunger, but there usually comes a point at which very high motivation begins to interfere with learning. In anthropomorphic terms the animal is so desperate to get at the reward that he cannot 'give his mind' to the problem at hand.

Extinction

If we cease to reward a conditioned response it becomes reduced and eventually disappears. It is much easier to extinguish classical CRs than instrumental CRs although we do not know why this should be so. After a rat has learnt to press a bar for food, the proportion of rewards to presses can be reduced to as low as 1 in 100 in some cases, and the rat goes on pressing. If rewards are stopped altogether it is a very long time before the response finally extinguishes.

Pavlov realized that an extinguished CR did not just disappear and leave the animal as it was before conditioning started. In the first place, if we simply leave the animal alone for a few hours and then give it the CS again, the CR returns, i.e. it shows *spontaneous recovery*. This recovery is not back to the original level and the response extinguishes more rapidly, but this process of a pause followed by spontaneous recovery can be repeated several times.

A second way of reviving an extinguished response is to give a novel stimulus along with the CS. A dog which has had its conditioned salivary response to a bell extinguished, salivates again if a light flashes as the bell is sounded. Similar results have been obtained with rats in Skinner boxes. Pavlov called this process 'disinhibition' because he regarded extinction as another new learning process which inhibited the original CR. Neutral stimuli presented with the CS early in the original acquisition of the CR often 'inhibit' it temporarily and reduce its strength. Similarly, perhaps the neutral stimulus disinhibits an extinguished CR by inhibiting the new learning that takes place during extinction.

Having discussed some of the more important properties of associative learning, we can now complete our examination of Thorpe's learning classification. The remaining categories are certainly more complex and less clearly defined than the earlier ones.

Latent learning

Thorpe defines latent learning as '. . . the association of indifferent stimuli or situations without patent reward'. By this he implies that the actual process may involve associative learning, but of course the key distinction is the lack of any obvious reward or drive reduction and the fact that what is learnt may not be obvious at the time—it remains 'latent' or hidden. Latent learning in its natural setting will often result from an animal exploring its new surroundings. Suppose we put a maze up against a rat's home cage so that he can wander into it at will. The rat is not hungry or thirsty, nor does he find any food or water in the maze, but he explores through it, sniffing into corners, running down blind alleys, retracing his steps and so on. Does he learn anything about the maze as he explores? According to strict Hullian theory he cannot, because he has received no reinforcement for doing so. Tolman's theory on the other hand, lays less emphasis on reinforcement and predicts that learning would occur. The test comes when we train our rat to run through the maze when hungry to get a food reward at the other end. Does he perform better than a rat put hungry into the maze for the first time? If so then his previous learning was 'latent' in that it was not revealed until we gave him a chance to put it to good use in finding food. Munn[360] provides an excellent discussion of the dispute between the Hullians and the Tolmanians which continued for some years in the experimental psychology journals.

To a biologist the whole controversy seems quite unreal because it is so obvious that animals do learn about their environment when they explore. A detailed knowledge of the geography of their home area will often mean the difference between life and death to a small mammal or bird as a predator swoops down. Barnett[27-8] discusses the nature of exploratory behaviour and concludes that it can be distinguished from the apperitive behaviour of a specific motivational state. There are times when the animal deliberately seeks out new stimuli and 'explores' in the true sense. Information it gains about its environment in this way can later be employed when it is searching for food or a mate, for example. Some workers have proposed that there is an 'exploratory drive' (which might be reduced during learning) but there is no good evidence that exploratory behaviour is organized in the same way as hunger and thirst. For example, Halliday[197] found that a rat's tendency to explore a new situation is not reduced if it has just been exploring something else—if anything its exploration increases.

We know that some birds and mammals 'explore' in the sense outlined above. In the other groups all we know for certain is that animals learn, often in fantastic detail, the geography of their home area. Many insects make special 'orientation flights' during which they make a 'fix' of the home area's position relative to the sun and landmarks near by. If a honey-bee colony is shut up in the hive and moved to a new site, a large proportion of the workers make orientation flights when they first leave the hive in its new position. They hover outside the entrance hole and then circle, gradually increasing their distance from it before flying off. During this orientation flight, lasting only 1 or 2 minutes, they learn the new position in sufficient detail to be able to return from long foraging flights. Many of the Hymenoptera have this amazing orientation ability because they build nests which require repeated attention and the insect needs to return many times to precisely the same spot. It might be possible to fit learning of this type into a drive-reduction framework but there seems little profit in the exercise. This ability has been built into the insect's nervous system by natural selection and we shall need to study it as a problem in ontogeny and evolution as well as learning. Thorpe gives numerous examples of similar orientating ability in animals as diverse as limpets, fish and newts.

Insight learning

Insight appears to us as the highest form of learning. Everyone can recall occasions when the solution to a problem has 'come in a flash', perhaps as the climax to several minutes of concentrated thinking. It is obviously difficult to demonstrate conclusively that there are similar processes going on in animals. Most workers have used the term 'insight' when, for instance, they observe animals solving problems very rapidly, too rapidly for normal trial and error. At least, too rapidly for the animal to carry out actual trials, but there is the possibility that it is 'thinking' about them and trying them out in its brain. This would imply that the animal can form ideas and 'reason', and studies on animal reasoning seem doubtfully distinct from those on insight.

Maier (see Maier and Schneirla[322]) defines reasoning as '... the ability to combine spontaneously two or more separate or isolated experiences to form a new experience, which is effective for obtaining a desired end'. Maier and others have tested for the existence of reasoning in the rat using a number of methods, mostly involving the animal making detours or taking short cuts through mazes. Figures 7.5 and 7.6 with their captions provide two typical examples, and there are two general features of such

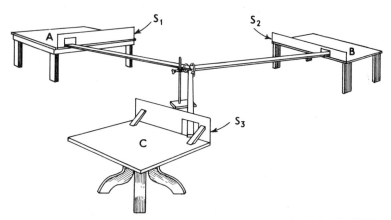

Fig. 7.5 One type of apparatus used by Maier to test 'reasoning' in the rat. The pathways are 8 feet long and the small tables vary in size, shape, and character. S_1, S_2 and S_3 are wooden screens placed on the tables to obstruct vision from one to the other. After exploring the three tables and runways, the rat is fed, let us say, on table A. It is then, let us also assume, placed on table C. After reaching the joint origin of the three paths, the animal now has a choice between A and B. If it chooses A, it is credited with a correct response. Exploration precedes each test and the rat is started from different tables from test to test. In a group of such tests, a score of 50% would occur by chance; some rats score much better than this. (From Maier,[321] 1932, *J. Comp. Neurol.*, **56**, 179.)

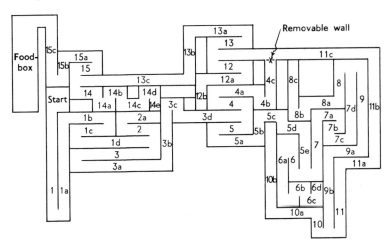

Fig. 7.6 A complex maze used by Shepard to test 'reasoning'. After rats have learned the maze, the section indicated by X is removed, thereby causing a previous blind alley to become a short cut. Having discovered the change whilst running along 11c, and exploring from there a little into 4c, some rats entered 4 (and thence 4a, 4b, 4c) instead of 5 on the next trial. (From Maier and Schneirla,[322] 1935, *Principles of Animal Psychology*. © 1935. By permission of McGraw-Hill, New York and Maidenhead.)

experiments. Firstly, they all include a period of exploration prior to testing which will mean that final performance depends to a great extent on latent learning. Secondly, we judge performance largely by speed of solution: if the animal makes mistakes and has to explore further we assume it is using a simpler, trial and error form of learning.

It is perhaps unwise to read too much into ordinary detour experiments because some insects show extraordinary ability to handle this type of problem. Thus Hebb and Williams[207] suggest that a good measure of 'intelligence' is obtained by using a series of movable barriers on an open space with a fixed starting-point and goal. Once it has become thoroughly accustomed to the space and goal the barriers are moved between each trial to present the animal with a series of simple detour problems (see Fig. 7.7). Scores can be given for directness of runs between start and goal box. Modifications of this technique have been extensively used by experimental psychologists. Although of course it has not been tried, there seems little doubt on the preliminary evidence given by Thorpe,[455] that the wasp *Ammophila* would score as highly as a dog on this test! In the natural situation, *Ammophila*'s starting-point is the place where it has stung a caterpillar, the goal is its nest to which it drags the prey. It has never made the journey on land before, although it has made orientation flights around

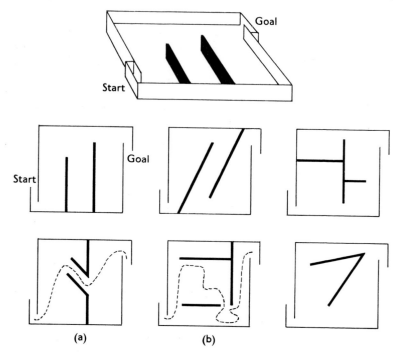

Fig. 7.7 One type of situation suggested for measuring animal 'intelligence'. A simple enclosed arena, illustrated at the top, has moveable barriers. The diagrams below show a series of different detour problems with which the animal can be presented in turn. The route shown on (a) would rate a high mark, that on (b) a poor one. The marks from a number of problems are averaged to give an 'intelligence score'. (From Hebb and Williams,[207] 1946, *J. gen. Psychol.*, **34**, 59.)

the nest. If whilst it is moving, barriers are placed in its path, it usually changes direction *instantly*, so as to carry itself smoothly around the barrier with the least effort. *Ammophila*'s remarkable ability to detour is probably based on orientation with respect to the sun, and one would hesitate to call it 'insight'. Even so it must make us reconsider our judgment of how 'insightful' are the detour abilities of higher animals.

The classic example of insight in animals came initially from the work of Köhler[271] on chimpanzees. Presented with a bunch of bananas too high to reach, they would pile up boxes to make a stand for themselves or fit two sticks together in order to pull down the bananas. Often they arrived at this solution quite suddenly, although they benefited by previous experience of playing with boxes and sticks (latent learning) and showed considerable trial and error when actually building a stable pile of boxes.

Köhler's chimpanzees were using knowledge obtained in one context (something of the properties of sticks and boxes) and applying it in another. There is no question but that apes and other primates can show true reasoning on occasions. Many dog owners will cite examples of their pets doing likewise; this is possible, but we must take care to exclude other explanations. Thorpe[457] discusses some other types of behaviour in birds and mammals which are fair evidence of reasoning, and the reader is referred to Chapters 6 and 7 of his book.

THE COMPARATIVE STUDY OF LEARNING

Comparative psychology has a long history, although it has often been criticized with some justification—notably by Beach[42] in a famous paper entitled 'The Snark was a Boojum'—for abandoning its comparative aims and concentrating almost entirely on two animals, the white rat and the pigeon. A far wider range of animals is now under scrutiny and we might hope that a comparative survey of learning capacities would yield information both on the evolution of learning and on the generality of any 'laws of learning' which have been put forward.

One approach has been to examine the correlation between brain development and learning ability. Dethier and Stellar[133] include an excellent introductory survey of nervous system structure through the various animal groups, and they discuss how far behavioural complexity can be linked with brain development. The prevalence of learning increases through the vertebrate series and we can roughly equate this increase with the evolution of a large brain. It is not, of course, just brain size that is important; whales and elephants have larger brains than men but smaller, though considerable, learning powers. It is less easy to construct a series among the invertebrates, which are much more heterogeneous, but higher insects and cephalopods have the largest brains of their respective phyla, Arthropoda and Mollusca, and also the greatest capacity for learning.

Again, within the vertebrates the most dramatic aspect of brain evolution is the growth of the cerebral hemispheres, especially their cortex, which reaches a climax in the primates. However, we must avoid any simple equation of cerebral cortex with learning ability. The birds have in the past often been underestimated because, although their brains are relatively large, those parts homologous to the mammalian cerebral cortex are small. But birds have evolved along a line separate from the mammals for over 200 million years. They have evolved another type of brain structure, and their learning ability is in some respects second only to the primates. A converse example has already been mentioned on p. 23 for amongst the insects the Diptera and Hymenoptera have similar brain development but very different degrees of learning ability.

It is clear that brain structure alone is inadequate as a guide to learning abilities and to study the evolution of learning we need to compare how different animals perform on particular behavioural tests. We are immediately faced with a number of problems. Some concern the selection of representative animal types for our evolutionary analysis. It is all too easy, to refer to 'higher' and 'lower' animals, and within the vertebrates, for example, we often find the sequence fish, amphibian, reptile, bird, mammal quoted as an evolutionary scale of increasing complexity of behaviour and increasing learning capacity.

The construction of such a scale from living representatives of each class ignores the actual course of evolution. We have just mentioned the completely separate histories of the birds and the mammals. All the living vertebrates are equally distant in time from their common ancestors and all are specialized for their particular mode of life. We should be naïve to expect the learning capacities of a modern teleost fish (e.g. a goldfish, commonly used in learning studies) to reflect accurately those of the ancestral fish from which the teleosts and other vertebrates diverged some 400 million years ago. Hence the construction of a valid phylogeny of behaviour is fraught with difficulties because behaviour does not fossilize. Hodos and Campbell[227] provide a vigorous critique of the phylogenies that have sometimes been constructed by comparative psychologists.

A further problem for comparative learning studies is that of devising truly comparable situations for testing different animals which vary so widely in their sensory capacities and manipulative ability. The procedures needed to measure discriminative conditioning in an octopus, a honey-bee and a rat have to be very different and we can no longer be sure that problems are of equal difficulty or that the animals 'see' them in the same way. Motivation and reinforcement present further problems; we have seen (p. 245) that the level of motivation affects rate of learning and may, indeed, determine whether the animal learns at all. How can we equate level of hunger motivation in a rat and a fish? The latter may live for weeks without food, the former only days. It is just as difficult to equate reinforcement between different animals. A small piece of food may be an excellent reinforcement for a hungry mammal, but mean much less to a fish and less still to an annelid. It is perhaps easier to equate punishment since all animals 'dislike' electric shocks. Even here there are difficulties because shock or fear affects the behaviour of animals in such diverse ways. As we discussed earlier (see p. 242) animals come to learning situations with a good deal of built in bias. To equate the effects of punishment we require some knowledge of the animal's natural responses in fearful situations. For example, avoidance conditioning apparatus (see p. 244) usually requires an animal to run off an electrified grid when shocked, and subsequently to learn that a buzzer or a light flash signals that the shock is coming. However, some rodents do not readily run when alarmed, they freeze and may stay motionless for long periods even when receiving electric shocks. Such an animal will appear very slow to learn when compared to one that runs easily. From the experimenter's point of view active animals are usually much more amenable in almost any type of learning situation (see Denny and Ratner[129] for a good review of these problems).

Nevertheless with some ingenuity and much caution it is possible to devise test situations which probably give diverse types of animals a fair chance to demonstrate their abilities. For example, Bitterman[51] describes 'Skinner boxes' for fish and turtles which provide food rewards; Schneirla (see Maier and Schneirla[322]) and Vowles[479] both found that the best reinforcement for maze learning in ants was to get back to their nest. For newts[476] and some annelids,[98] the best reward was a brief period in which the unwilling subjects were simply left in peace!

Despite the problems involved in its study, the evolution of learning remains a fascinating topic. It is reasonable to examine our classification of learning to see whether the different categories have any phylogenetic validity, i.e. can they be ranked in order of appearance on the evolutionary scale, at least as far as this scale can be represented by living examples?

Undoubtedly the oldest type of learning is habituation which, as we discussed earlier, is found in all animals. It is easy to understand the selective advantage it confers because habituation saves animals wasting time on stimuli which are unimportant, and this will be advantageous no matter how limited is their repertoire of responses.

Some form of associative learning must have been the next step in evolution, but it appears to be a large one. Associative learning does share certain features with habituation—for any type of learning to occur the animal must be able to distinguish familiar from unfamiliar situations—but apart from this it is hard to construct evolutionary intermediates. However the phenomenon usually called 'sensitization' may represent a stage in the evolution of associative learning and is, in some ways, the opposite of habituation. We speak of sensitization when an animal's responsiveness to a variety of stimuli is increased following a reward or punishment. For example, Evans[148] found that the worm *Nereis* (see p. 233) will occasionally emerge from its tube when a light is flashed, and in one experiment 21% of worms did so. However following a single food reinforcement—given in the absence of light—more than 60% of the worms emerged to the next flash given 30 minutes later. Similarly an octopus which has just been fed is far more likely to attack cardboard shapes suspended in its tank than is an unfed animal. Sensitization works even more effectively with punishment. Following electric shock to their feet, both cockroaches and rats will flee from a variety of novel stimuli which have never been associated with shock and which do not elicit escape from unsensitized animals.

To evolve true associative learning the animal must begin to discriminate between stimuli that occur close to the reinforcement and those that do not. Clearly a heightened responsiveness at the time of reinforcement will help it to do this. Unequivocal associative learning has been shown for the arthropods and the cephalopod molluscs (the octopus and its relatives) among the invertebrates. Results that seem to indicate such learning in annelids and flatworms are—so far—almost all susceptible to other explanations and often sensitization is one such (see Evans,[148] and Thorpe and Davenport[458] for discussion of the problems of interpretation here).

Amongst the vertebrates, we might hope by a good series of learning tests to be able to record stages in the evolution of 'intelligence', and speed of learning—a quality much admired in human school children—might seem to be one useful measure. However a brief survey of the literature shows that speed alone does not tell us much. Ants and rats for example, show very comparable speed when first learning a fairly complex maze. Gellerman[167] describes in detail experiments in which two chimpanzees and two 2-year-old children were learning that a food reward was associated with a white triangle on a black square and not with a plain black square. One child learnt in a single trial, but the other took 200 trials and both the chimpanzees took over 800 trials to reach the criterion of 19 correct trials out of 20. On a comparable test most rats would learn in 20 to 60 trials, though admittedly they would usually be mildly punished for wrong responses as well as rewarded for correct ones. Discrepancies of this kind abound both within and between species and we have no reliable evidence that speed of initial learning for simple associative problems varies within the vertebrates, or even between them and the higher invertebrates.

However, speed is only one aspect of learning, we might also ask *what* is learnt. For instance, Fig. 7.8 shows that, although the chimpanzee may take longer to learn the triangle discrimination outlined above, he learns more about 'triangularity' than does the rat. One aspect of 'intelligence' is the ability to strike a reasonable balance between generalization and discrimination in tests of this type. Similarly, if we consider more complex forms of learning then we can at least note some gradation of ability within the vertebrates.

Testing for **learning sets** has been useful in this respect. If an animal can form a learning set it means that it can learn not just a problem, but the principle behind it and can steadily increase its learning speed when given a series of similar problems. Harlow[201] has described the basic technique with primates.

A monkey is presented with a pair of dissimilar objects—a matchbox and

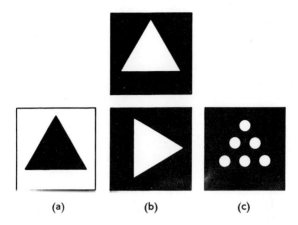

(a)   (b)   (c)

Fig. 7.8 The concept of 'triangularity'. Trained to respond to the top figure, a rat makes random responses to any of the lower figures. A chimpanzee responds to (a) and (b), but makes random responses to (c). A 2-year-old human child recognizes 'a triangle' in (a), (b) and (c). (From Hebb,[206] 1958, in *Biological and Biochemical Bases of Behaviour*, H. F. Harlow and C. N. Woolsey, eds. University of Wisconsin Press, Madison. By permission of the Regents of the University of Wisconsin.)

an egg-cup, for example. The matchbox, no matter where it is placed, always covers a small food reward, the egg-cup never has a reward. After a number of trials the monkey picks up the matchbox straight away. Now the objects are changed, a child's building block is rewarded, a half-tennis-ball unrewarded. The monkey takes about the same time to learn this, again the objects are changed, and so on. After some dozens of such discrimination tests the monkey learns each discrimination much more rapidly, though viewed as an individual problem it is just as difficult as the first one. Eventually after 100 or so tests the monkey presented with a pair of objects lifts one, if it yields a reward he chooses it for all subsequent trials, if it is unrewarded he chooses the other (rewarded) object on the next and all succeeding trials. He has learnt the principle of the problem or, in Harlow's terminology, he has formed a 'learning set'.

This is one type of learning set based on successive trials of discrimination. Perhaps a simpler version of the same type is the 'repeated reversal' problem. Here we train the animal to select object A in preference to object B. Once learnt, object B is now rewarded and A unrewarded; when this first reversal is learnt, the reward is again given with A, and so on. If the animal gets progressively quicker at learning each reversal this again implies it has learnt a principle.

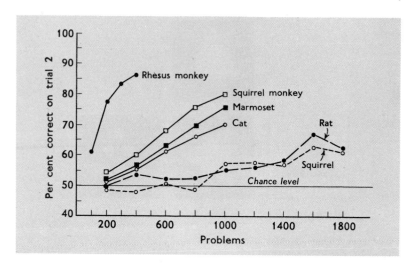

Fig. 7.9 The rate at which various mammals can form discrimination learning sets. With each new problem the animal's choice on the first trial has to be random, but if it has learnt the principle behind the problems, trial 2 should be correct. Note how long it is before the scores of rats or squirrels on trial 2 become better than chance or 50%. Many monkeys reach almost 100% within 400 problems. (From Warren,[484] 1965, *The Behavior of Non-human Primates*, vol. 1. Academic Press, New York and London.)

The ability to form learning sets was once regarded as a property of the higher mammals only, but we now know this is not the case. Warren[484] summarizes comparative data from fish, turtles, birds, rats, cats and primates. Although further research may prove otherwise, it looks as if this criterion does separate the fish, which cannot form learning sets, from all the other vertebrates. Mackintosh[319] has provided good evidence that the octopus can form repeated reversal sets, so in this respect the ability of a higher invertebrate exceeds that of a lower vertebrate. No insect has yet been tested in this way.

Although speed of learning a simple discrimination does not vary much between groups, the speed with which learning sets are acquired does change dramatically. Figure 7.9 shows how much faster a monkey acquires a set than a rat, which shows no improvement at all for some 800 successive problems but thereafter slowly improves.

With learning sets and with more complex problems, the superiority of primates is undoubted but it seems to be a quantitative superiority. We have no evidence that they possess any abilities which are not foreshadowed earlier in the mammalian series.

One cannot help feeling vaguely dissatisfied by this conclusion. The rather circumscribed artificiality of many learning experiments does not seem to do justice to the incredible flexibility and ingenuity of the higher primates, such as chimpanzees. Here we are undoubtedly influenced by their similarity to ourselves and particularly because they can manipulate objects as we do. We may underestimate the intelligence of other animals because they lack good hands and good eyesight. Only recently have we become aware of the remarkable intelligence of dolphins because their structure and environment is so different from our own.

As mentioned in the previous chapter (p. 200) studies on the behaviour of chimpanzees in the field have revealed that they can select and fashion simple tools. Goodall[169] has described how they will chew up leaves to use as a sponge for getting water out of crevices. They also select twigs, and peel off their side branches until they make a probe of a size suitable for extracting termites from their nests. Such observations should not, perhaps, surprise us because it now seems certain that some of man's own Australopithecine ancestors were using bones as weapons against other animals and fashioning crude flint tools. Their brain size was about 650 to 680 ml which is little more than that of the modern gorilla (le Gros Clark[99]).

From this stage to that of the human with a brain of some 1,500 ml is an enormous jump and some people argue that men possess abilities which are not even dimly foreshadowed in the behaviour of apes. Nevertheless man's ancestors were at the ape stage relatively recently, and there were successful intermediates like *Homo erectus* whose brain size was 1,000 to 1,200 ml. Harlow[202] points out that man's use of symbolic language gives him a great advantage over the non-lingual apes, but this gap does not imply that there was anything other than a step-wise progression to the human condition. Perhaps the modern apes are only just below the level for language. Kellogg[264] describes several unsuccessful attempts that have been made to teach chimpanzees to speak. Nobody has succeeded in getting more than one or two recognizable words out of them. But this is probably due to their inability to articulate properly (the human larynx and buccal cavity are considerably specialized for speech, see Lennenberg[290]) and because communication by gesture is much more natural for them. Recently, in fact, two remarkable long term educational exercises with individual chimpanzees have succeeded in proving that they can use a symbolic language with remarkable facility. The Gardners[164-5] have used the language of gesture developed for the use of deaf people; Premack[384] has used coloured shapes to represent objects and words. In both cases the experimenters have concentrated on training one female chimp, not just to recognize objects and associate them with gestures or coloured-shape symbols, but also to use these same gestures or symbols to ask for things and communicate with the human experimenter. Both chimps have acquired a useful 'vocabulary' of many dozens of 'words'. There can be no doubt of their remarkable capacity to attach names to objects and to some actions, such as 'come gimme' and 'play peekaboo', and modifying phrases such as 'more'. Debate continues on the exact nature of the chimpanzees achievement, do they have some sense of sentence construction and of simple grammar? The Gardners suggest that they can match the abilities of young children over the first years of their learning to speak.[163, 165]

These experiments are important and offer fascinating possibilities although they require a formidable investment of time and patience. In the Gardners' laboratory the chimpanzees have human companions through all their waking hours, and no form of communication other than sign language is ever used. Now sign language is being taught to small groups of young chimpanzees. What will they 'say' to each other when provided with such an extended means of communication? Even the results already obtained have forced us to revise completely our estimates of what a 600 ml brain can achieve.

8

Social organization

So far we have surveyed some topics which illuminate the organization of behaviour within the individual animal. However we have frequently had to discuss aspects of behaviour which involve the interaction between individuals, as in aggressive behaviour and courtship. In fact virtually all animals exist in pairs or in larger groups for at least part of the time. For some species, sociality is the dominating feature of their whole lives and we must now consider the social life of animals in its own right.

The cohesiveness and co-ordination of animal societies are often their most striking feature. It is easy to lose sight of the individual in such groups, thus failing to recognize that the forces which shape its behaviour are the same as those for more solitary species. Natural selection operates upon individuals and the responses of a social animal to the other members of the group will evolve to its own best advantage.

The term 'social organization' refers to populations and not to individuals and defines the nature of the interactions between members of a species. In some instances—the various social insects for example—social organization is fairly rigid and species-specific. In vertebrates, as we shall discuss later, it is a much more dynamic phenomenon, and may vary with changing conditions. Certainly use of the term is not restricted to highly social animals. Tigers which usually live and hunt alone in large territories, avoiding contact with others save for breeding, and honey bees which spend their entire life in a dense colony, both provide examples of social organization, albeit of widely different types.

It is perfectly valid to refer to any interaction between one individual of a species and another as 'social behaviour'. This is the criterion adopted by Tinbergen[463] and Dimond[137] in their books but here, whilst reviewing social organizations of diverse types, we shall concentrate most attention on what might be called 'animal societies'. A true society will involve more than a mated pair or a mother and her offspring; it will mean a stable group whose members inter-communicate extensively and bear some relatively permanent social relationship to one another.

Within the animal kingdom there are an enormous variety of groupings, not all of which meet the criteria outlined above. Two contrasted examples of undoubted societies are provided by ants and elephants. The ants that make up a colony live in a common nest which they have constructed together. There is a stable relationship between the queen and her sterile daughters which may be differentiated into two types—often called workers and soldiers—distinguished both by their structure and their role in the life of the colony.

African elephant society is much less closely organized. A herd of 40–50 will generally be led by an old female—the matriarch. Her offspring of 2 or even 3 generations will move closely with her, this smaller group comprising her family. Other families will associate with that of the matriarch to form what Douglas-Hamilton[138] calls a 'family unit'. The young bulls stay with their families at first but move off as they become mature and live rather solitarily, though usually quite close to other bulls, sometimes associating with them. They form temporary contacts with family units and then mate with the cows in oestrus. Within the family unit elephants may live in one another's company for 40 or 50 years.

Compared with these true societies the organization within a flock of birds or a school of fish is much less complex, although individuals may stay together for months. When we come down to a swarm of water fleas gathered in some area rich in food, or a mass of *Drosophila* collected on some rotting fruit, then quite clearly we are not justified in using the term 'society'. *Drosophila* and water fleas form **aggregations** because they are attracted to a common food source; they certainly react to one another's presence, for example by spacing themselves out so that they do not touch, but they do not constitute a society.

THE ADVANTAGES OF GROUPING

We have already suggested that the forces of natural selection will shape an individual's responses within a group just as they do for any other aspect of its adaptation to its environment. Consequently we may expect that all animal groups, whether aggregations or true societies, confer an advantage on the members that constitute them.

The work of Allee[6] and his school was of great importance in revealing numerous examples of the way in which an aggregation benefits the individuals which comprise it. Sometimes the advantages relate almost entirely to the physical size of a group. Water fleas cannot survive in alkaline water, but the respiratory products of a large group of them are sometimes sufficiently acid to bring down the alkalinity to acceptable levels. Thus a group can survive where a few individuals could not. *Drosophila* cultures do not do well if they begin with too large a number of eggs—the resulting larvae are undernourished—but they fare equally badly with too few eggs, because there are not enough larvae to break up the medium and make it soft enough for them to feed. It is advantageous for a female to lay her eggs close to those of others because her own offspring will benefit.

Bird flocks and schools of fish exemplify groups which, though lacking some of the attributes of a society, often go far beyond simple aggregations. Physical factors may still count as with emperor penguins which huddle closely together as they stand incubating their eggs during the antarctic winter. Heat is conserved and birds on the outside move more than those in the centre thereby leading to mixing and a reasonable distribution of shelter. However there is often a degree of interaction between the individuals which goes far beyond the physical.

One of the most obvious advantages of a cohesive group whose members respond to each other's behaviour is protection against predators. With a number of animals on the alert, the approach of a predator is less likely to go undetected and one alarm signal will suffice for all. Table 8.1 gives some actual measurements relating to vigilance in starling groups of various sizes. Powell[382] had starlings foraging in an area where he could arrange the sudden display of a stuffed hawk. He measured the frequency with which birds kept in groups or singly interrupted feeding to look up, scanning the sky around, and also their latency to take flight after the hawk appeared. The flocks effectively take off together, although the bird which

Table 8.1 Feeding, surveillance and reaction times of single and grouped starlings. Further details in text, (From data in Powell[382]).

	Single birds	Groups of 5 birds	Groups of 10 birds
Percentage of time available spent feeding	53	70	88
Percentage of time available spent in surveillance	47	30	12
Reaction time to flying hawk model	4·1 secs	Not measured	3·2 secs

Fig. 8.1 The response of a flock of starlings to the approach of a bird of prey. (From Tinbergen,[460] 1951. *The Study of Instinct*. Oxford University Press, London.)

first spots the hawk presumably initiated the move and commonly gives an alarm call as it takes off. Notice that birds in flocks look up less frequently and take off quicker and even half a second's lead may mean the difference between life and death as a hawk swoops down. Decreased necessity for vigilance must leave more time available for feeding and other activities. Even when a predator puts the group to flight being grouped offers further advantages for it can take concerted evasive action. Fig. 8.1 illustrates this for a flock of starlings and exactly similar behaviour is shown by some fish which bunch together at the least alarm. Predators rarely attack an individual in a close group and their commonest stratagem is to make swoops towards the group which may cause it to scatter, when the predator singles out an isolated animal.

Colonial nesting birds—gulls and terns for example—may provide formidable opposition to an invading predator such as a fox by their mobbing attacks, even though each is responding more or less individually to defend its own territory. More highly organized societies may have even more positive defences against predation; we have already mentioned the protection against leopards which baboons secure by living in groups—several adult male baboons united against it are more than any leopard will face. Fig. 8.2 shows the protective formation of a herd of musk oxen at the approach of wolves, with the young animals sheltered by the adults.

Protection from predation is only one factor conferring advantage upon groups. A colony of prairie dogs can by their continuous grazing keep a large area free of tall grasses thus allowing the smaller herbaceous plants on which they feed to grow more freely. Foraging in flocks may enable birds to find food more easily both because the group will disturb the habitat more effectively and stir up prey animals, and also because the sight of one bird feeding successfully attracts others to the spot. Krebs, MacRobert and Cullen[277] showed that when one member of a great tit flock finds a food item, the others rapidly alter their searching strategies and concentrate their attention both on the general area and the type of niche in the trees where the food was found. Colonial nesting also conveys a foraging advantage in some cases, especially if food supplies in the local environment are patchy in their distribution. Krebs[276] studied great blue heron colonies and found that birds which had been unsuccessful on a foraging trip would hang around in the colony until they could follow other birds who left quickly after returning from the feeding grounds. These were the birds who had been successful in locating a local shoal of fish. The colonial nesting of the red-winged blackbird probably conveys a similar advantage. It may also be one of the functions of the huge communal roosts which crows, starlings and other birds use during winter months. Groups which know of a good feeding area will leave the roost early to return to it and can

Fig. 8.2 The defensive formation of a group of musk oxen on the Canadian tundra. When a predator approaches, they bunch with the older animals at the front, facing the threat. (Photo by D. Wilkinson from Information Canada Photothèque.)

be followed by others. Over the whole winter every individual is likely to have benefited because the large numbers will mean that a far higher proportion of the feeding areas around the roost will have been exploited (Zahavi[517]).

Predators may also benefit from group life in some instances. Lions, hyaenas and cape hunting dogs rely on co-operation between members of a group to hunt their prey. Their stratagems may involve some individuals driving prey towards others hidden in cover or as with the dogs, taking it in turns to run down an antelope to the point of exhaustion. Kruuk[280] gives an excellent review of the different types of social hunting strategies adopted by carnivores.

Another factor which may make group life advantageous concerns the stimulation and synchronization of breeding. In Chapter 4 we discussed experiments which showed that the hormonal cycles of some female birds and mammals are affected by the presence of males. Large numbers of sea-birds nest in dense colonies, even though they may vigorously defend their own small territory within the group. Fraser Darling[119] was among the first to point out that in such colonies a great deal of stimulation must result from the displays, both visual and auditory, which are a constant feature of a large group early in the breeding season. The effects of such stimulation, coming from neighbours as well as from a bird's own mate, will tend to accelerate and synchronize the reproductive cycle within the colony. This, Darling suggested, was advantageous because it reduced the vulnerable breeding season to a minimum. If it leads to synchrony in egg laying and hatching then predators, who can only eat a limited number of prey each day, will be 'swamped' by the sudden brief flush of eggs and chicks so that the chances of an individual nest being attacked are reduced.

Experimental verification of the so-called 'Fraser Darling effect' is not easy to come by, but Coulson's[105] remarkable long-term studies of kitti-wakes go some of the way. He has shown that birds breeding in the most densely-packed central areas of a nesting colony are more successful than those in occupying more peripheral nest sites. They lay more eggs and lay them earlier and there seems little doubt that heightened social stimulation in the central area is one factor contributing to these effects. The centre exerts great attraction and early in the season kittiwakes compete fiercely for sites in this area.

Finally in our discussion of the advantages of group life, we must briefly consider the views of Wynne-Edwards set out in his important book *Animal Dispersion in Relation to Social Behaviour*.[514] Wynne-Edwards takes a radically different view from that expounded earlier in this chapter. Here we have discussed the benefits which social life confers upon the indivi-duals which make up the group. Wynne-Edwards suggests that this is not the whole story and that commonly during the evolution of social life, the 'selfish' needs of the individual have been sacrificed for the benefit of the group as a whole. Only in this way, he suggests, can animals live within the resources that their environment provides. If individuals always reproduced at the maximum rate they would rapidly eat out their food supply and Wynne-Edwards asserts that this almost never happens in natural circumstances. Groups have evolved in which breeding rate is linked to the available resources. Animals maintain their breeding rate below the maximum and the most important regulatory mechanisms are to be found in their social systems. Because groups are stable and to a large extent reproductively isolated from others, selection has operated on the groups themselves, weeding out those which did not adapt to match their resources.

There are certainly great difficulties with the concept of group selection. Suppose, as Wynne-Edwards suggests, a colony of birds lives within its food supply by laying fewer eggs than the maximum. In fact many sea birds do lay only a single egg and do not replace it if they lose it. Clutch size must have an hereditary basis and if there is enough food in the short-term, how could selection prevent the spread of genes which caused birds to lay 2 eggs? Even if we assume the 1-egg-only convention could evolve—and it is hard to see how this happens unless it benefited not only the colony but the individuals—it would seem to be an inherently unstable situation because new mutations would constantly threaten it in the face of short-term individual gains. (Wiens[498] provides a good discussion of this issue and of Wynne-Edwards' ideas generally.)

In his later writings Wynne-Edwards[515] himself recognizes the implausibility of group selection mechanisms. Straightforward individual advantage augmented by kin selection together with their combined effects on inclusive fitness (see p. 210) provide a more plausible set of concepts with which to explain the evolution of much social behaviour and apparent altruism. Nevertheless by emphasizing the relationship between behaviour and the structure of animal populations Wynne-Edwards was something of a pioneer and his ideas, even when later discounted, have been influential in much of the more recent work on social behaviour.

One clear example of the kind of effect he was proposing comes from the work on red grouse populations on the heather moorlands of Scotland. Here Watson[486] and his co-workers have shown that population size is effectively determined by territorial behaviour. In autumn the cock birds fight over the breeding areas (and we now know that the richer the habitat the harder they fight) and a proportion of them succeed in winning territories. The unsuccessful birds are driven away into the valley bottoms. Only the territory holders can breed the next spring, the surplus birds are rendered effectively sterile and, in any case, have a far higher mortality in the unfavourable habitat into which they are forced. Territorial behaviour limits the effective population size and links it to the current resources of the moorland.

It is never easy to obtain good evidence for the interrelations between behaviour and population control. A few season's work on a species is rarely sufficient; it may well take 25 years to get a reliable answer. The value of Wynne-Edwards hypothesis is that it forces us to think of sociality as adaptation which transcends individual responses—although dependent upon them—thus enabling a higher order of adaptiveness to emerge. In other words adaptation can involve not just animals becoming social, but also the nature of their social organization.

SOCIOBIOLOGY

This is certainly a concept central to several of the new approaches to social behaviour which are often lumped together under the name 'sociobiology'. Wilson's remarkable book with this title[507] surveys the whole range of social life across the animal kingdom and he defines sociobiology as '. . . the systematic study of the biological basis of all social behaviour'. This is a very broad definition and certainly most of the past work on animal societies could be included. What gives a fresh and exciting impetus to the new approach are its links with quantitative ecology, genetics and evolution. Sociobiology often looks at the functions of social behaviour and tries to measure, or at least estimate, selective values and suggest how the particular type of society observed adapts its members to their environment. In one sense there is nothing new in this and ethologists have always concentrated on the function and evolution of behaviour. Nevertheless, as Crook[109] pointed out in an important review, ethologists also tended to concentrate on causal mechanisms within individuals or social interactions within pairs or single families. Their original concepts had to be widened to study the more complex types of social interaction within organized groups. The concept of kin selection has been of particular importance in sociobiological thinking because it links directly the selective forces acting on an individual to the social interactions it has with other members of its group. We have already discussed kin selection in Chapter 6 and further examples will emerge in our account of social systems.

The linking of social behaviour and ecology began with attempts to correlate the size of groups with their habitat (e.g. Crook[110] and Ito[253]). There are some quite striking regularities for often the group size of birds, ungulates and primates living in open grassland or plains resemble one another more than each does its own relatives living in forest habitats. In open habitats communication is easy, groups wander quite widely and tend to be large. In dense forest a large group would have formidable problems in maintaining its cohesion; accordingly groups tend to be small and many forest animals are solitary. The large groups of hippopotamus in open lakes and rivers and of red deer in the Scottish Highlands can be compared with the solitary pigmy hippopotamus and the moose of dense forests. Even within a species we can sometimes see the same trend. Prides of lions rarely amount to more than a family party of 6 to 8 when they are living in thick bush but in open savannah groups of 30 are not uncommon.

This relationship clearly holds to a certain extent but we now know of too many exceptions to be satisfied with a classification of habitats into such broad categories. In many cases there is evidence that the social system is adapted to habitat and feeding ecology in much finer detail. Clutton-Brock[101] studied two related monkey species, the red and the black and white colobus, which both live in forest, often in the same forest. They even feed off the same trees, two species of which made up the main bulk of the food for both types of colobus. The red colobus lives in large bands of 40 or more which range widely over an area of about one square kilometre. The black and white has small troops of 5 to 10 individuals and they actively defend a small territory of less than one fifth the size of the red. Clutton-Brock accounts for these contrasts using his detailed observations of the feeding of both species. Although they feed on the same tree species, they do not take the same parts. Red colobus eat a high proportion of shoots, flowers and fruit. Black and white colobus do take these but also eat mature leaves which are avoided by the red. Clutton-Brock points out that the food supplies of the red will be scattered and 'patchy' for they will have to move around as trees come into leaf and flower. For black and white however, food is more constant for they can stick to one group of trees and utilize them all the year round. It appears that both species require about the same area of forest per animal (50 red moving over their range of 1 km² is roughly equivalent to 10 black and white on their 0·2 km² territory). Clutton-Brock argues that the red lives in large bands rather than territorial groups because it will be easier for a large group to exploit their patchy food supply—we have discussed a similar situation for titmouse flocks earlier in this chapter.

Much of Clutton-Brock's argument is speculative; it has to be in all cases where we observe an end-result and then, without the benefit of any fossil evidence, try to say how it evolved. Nevertheless he can cite other closely related monkey species whose ecology and behaviour seem to vary in the same way. This is obviously a profitable way for sociobiological research to proceed for theories can be tested for their ability to predict relationships in other animal species not yet studied.

Another very extensive survey of social organization and ecology has been made for the antelopes. Jarman[255] has tried to include information from more than 70 species, not all of them studied in much detail, but again it is possible to detect a pattern and make predictions. He divides the antelopes into 5 classes on the basis of their 'feeding styles'—the way they crop their food and the variety they eat. Some browse on the tender shoots of herbaceous plants, others graze; they also vary in the degree to which they specialize in one type of grass or many.

Jarman finds that there are very clear relationships between feeding style, group size and social structure. To take two extremes, at the one end of the scale there are the strictly territorial duikers which live solitarily or in pairs. They browse on shoots from a wide variety of plants with scattered distribution. They may be contrasted with wildebeeste which live in large groups which sometimes merge into enormous herds during migration. They feed on grasses of diverse types but only when they are at a particular growth stage, so they have to move with the cycle of the pastures. This migratory life has consequences for the reproduction of wildebeeste and, as we shall discuss later (p. 283), forces the males to set up temporary territories where the current grazing is good.

Jarman's conclusions are very convincing and provide a clear picture of the way every aspect of behaviour—feeding, mating, predator defence and so on—all fit together and relate to the individual's struggle to harness the resources of its habitat to the best advantage.

We can now turn from this general discussion of the evolution and adaptiveness of social organization to consider some of the behavioural principles involved. The insects and the vertebrates are well contrasted groups from which to illustrate them.

SOCIAL INSECTS

The form of social life exhibited by some of the insects is in many respects so distinct from that of other animals that we can justifiably treat them separately. Nevertheless for all their distinctiveness in two key aspects of social life—division of labour among the members of a society, and communication between them—the social insects exemplify principles which can be applied to all societies.

Amongst the insects we find some of the most completely solitary animals. This is because they may lack even the most basic of social contacts, that between parent and offspring. The short life span of many insects coupled with seasonal reproduction, often means that parents have died long before their offspring emerge from larval life. The digger wasp whose life history was given in outline in Chapter 2, p. 22, has only one brief with another member of her species when she mates; apart from is totally solitary.

It is obvious that one essential for the development of social life is to lengthen the life span so that there is contact between old and young individuals. The aggregations formed by insects such as cockroaches and earwigs contain such overlapping generations. The life span of a cockroach may be a year or more, three quarters of which is development when it passes through a series of nymphal stages, becoming gradually more like the adult insect. During this period cockroaches of all ages live together in a loose aggregation near sources of shelter and food. Some cockroaches incubate their eggs inside the female's body and bear live young which remain in contact with the mother for some hours after birth. This contact may be important for the survival of the young because, not only are they extremely vulnerable to cannibalism when first born, but they may pick up nutrients from the mother's body surface.

We know that this nutritional factor is of major importance in the termites (which are related to the cockroaches) because, feeding largely on wood, they rely on symbiotic protozoa living in their gut to digest cellulose. These protozoa are acquired by the young termites when they feed on fresh faecal matter from the adults, and they can be transmitted only in this way. Perhaps this essential overlap between the generations was one basis for the evolution of the termites' elaborate social life.

True societies with organized structure are found in two orders of the insects, the Isoptera or termites as just mentioned and the Hymenoptera, the ants, bees and wasps. There is a huge literature on social insects which have attracted man's attention for centuries. Here we can only give a very general account of some of their characteristics which are most relevant to a discussion of social organization. For more details and for an introduction to the whole literature Wilson's masterly survey[506] is unmatched. In his other book Wilson[507] also provides a shorter account in which the social insects are described in the context of sociobiological ideas. Butler,[82] von Frisch[156] and Lindauer[297] are all excellent accounts of different aspects of that most-studied of all insects, the honey-bee.

The unique feature of an insect society is that although it commonly contains thousands of individuals, they are all closely related to each other and constitute, with certain minor exceptions, a single family. Reproduction is generally confined to one female (the queen) and the other members of the society are her offspring which remain sterile. Living in a common nest which they construct, they assist the queen to produce more workers like themselves and eventually more reproductive castes.

This term, caste, is well suited to describe the division of labour within insect societies. It implies a rigid, limited role in society largely determined by one's upbringing. Certainly the most important factor determining caste is diet. In bees, wasps and termites all eggs laid by the queen are potentially equal, but most larvae are fed a restricted diet and develop into workers. There is evidence that when queen ants are laying rapidly their eggs are 'worker biased' and develop accordingly no matter how the larvae are fed. But for most of the time their eggs are also equipotential and only the richly fed individuals develop into the reproductive castes. Termites have both males and females represented in all castes (i.e. there are 'king' termites, which remain with the queen and mate with her several times during her egg-laying period), but in the Hymenoptera all the workers are female. The males (or drones as they are called in bees) constitute a separate caste. They are produced at the same time as new queens from unfertilized eggs (see p. 211).

In most social insects, with the notable exception of the honey-bee which reproduces colonies by swarming, new colonies are founded by a single queen (or pair in termites). She begins the construction of the nest and rears the first batch of workers herself. These then take over the tasks of extending the nest and bringing food, and the queen usually stays in the nest laying eggs from this point on. In termites the gradual growth of the young through the various nymphal stages, as in the cockroach, means that it can play its part as a working member of the colony right from the start. In the Hymenoptera the helpless larvae require protection and feeding by adult workers until following metamorphosis they themselves emerge as fully-grown adults.

The tasks performed by the worker castes vary greatly in detail, but in most colonies they cover the main categories of foraging, rearing the young, nest construction, attending the queen as she moves about laying eggs and guarding the colony. In termites and ants this last task is sometimes the sole responsibility of a special soldier caste which has enlarged jaws or other weapons. Less specialized workers carry out a range of the other duties with varying degrees of attachment to any one task. In bumble-bees very small workers are sometimes produced because of poor nutrition as larvae, and these tend to stay in the nest all their lives, never going out to forage. Food supplies are much more regulated in honey-bee colonies, workers are all of one type and perform all the tasks listed above. Division of labour in the hive is based upon the age of the workers, but it operates quite flexibly.

A honey-bee worker lives for about 6 weeks as an adult and her activities are to some extent synchronized with her physiology. Thus she spends the first 3 days cleaning out cells, and then begins feeding the older larvae on a mixture of pollen and honey which she picks up from the storage cells in the hive. During this period the pharyngeal or 'nurse' glands in her head have been developing. They secrete the so-called 'royal jelly', and from about the 6th to 14th day of her life the worker feeds this secretion to the younger larvae and any queen larvae in the hive. (Royal jelly is fed to all larvae for a brief period early in their development, but those larvae intended to become queens develop in a larger cell and are fed royal jelly throughout.) The worker's wax-secreting glands on the abdomen become active from the 10th day and at the same time the pharyngeal glands begin to regress. She gradually changes her behaviour from feeding larvae to cell construction. From about the 18th day she may leave the hive occasionally for a few brief orientation flights (see p. 247). At this age she may be found guarding the hive entrance and inspecting incoming bees. From 21 days of age onwards the worker is primarily a forager, bringing back nectar pollen and water, and usually remains so for the rest of her life—2 to 3 weeks.

This is the general sequence, but it can be modified to suit the needs of the colony which are certainly not fixed but vary with the flower crop, the temperature, the age of the colony and many other factors. The sequence of activities can be modified because of the remarkable communication system which exists between the members of a honey-bee society.

Fig. 8.3 shows a record kept by Lindauer[297] of the activities of one individual worker throughout her life. The sequence just described can be discerned, but the most conspicuous activities are two not previously mentioned, 'resting' and 'patrolling'. Clearly one way in which the individual is made aware of the needs of the colony is by personal inspection. The worker moves about the hive exploring empty cells, the food storage areas, the edges of comb where new construction is proceeding and the brood area where the larvae are developing. In this way she is stimulated to initiate activity of her own. If there are a large number of underfed larvae, the pharyngeal glands of an older forager regenerate and she moves back to feeding. If food is short, young workers prematurely become foragers.

The most obvious way in which information on the state of the colony is transmitted involves contact between the workers. A resting bee is easily aroused by the activities of other workers nearby. There is incessant contact between bees on the combs and incoming foragers are approached and solicited for food. It is not just from foragers that food is taken, any pair of workers that meet may proffer or solicit food. One bee extends its tongue and rubs its antennae along the antennae of the facing worker, who then regurgitates a drop of nectar between her mandibles. In a second or two this is lapped up and the two bees move on. Within a few moments the receiver may become the donor as she, in turn, is solicited or proffers food to another.

Nixon and Ribbands[366] fed a small quantity of sugar solution containing radioactive phosphorus to 6 foraging bees from a colony of 24,600 bees. Within 5 hours radioactivity could be detected in 62% of foragers and 18%

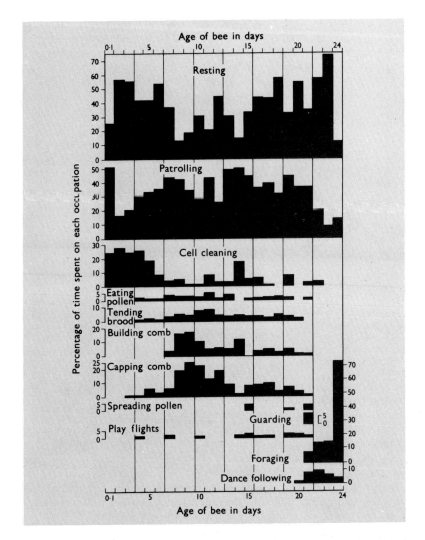

Fig. 8.3 Lindauer's complete record of the tasks performed by one individual worker honey bee throughout her life. The records are classified according to the type of task. One can recognize the age-determined succession of cell cleaning, brood care, building, guarding and foraging. Note, however, the large amount of time spent in patrolling the interior of the hive and in seeming inactivity. (From Lindauer,[297] 1961, 1971. *Communication among Social Bees.* By permission Harvard University Press, Cambridge, Mass. © President and Fellows of Harvard College.)

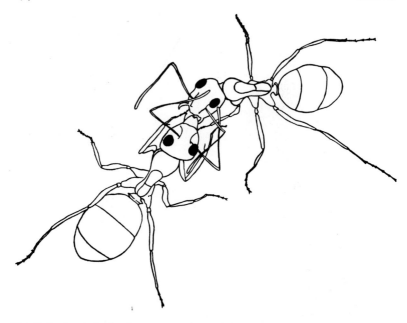

Fig. 8.4 Food sharing between workers of the ant, *Formica fusca*. The worker on the left is receiving food from that on the right which has regurgitated a drop of liquid from its crop and offers it between the outstretched mandibles. (From Wallis, D.I. 1962. *Animal Behaviour*, **10**, 105–11.)

of bees tending brood. After 29 hours 76% of foragers and 43% of those in the brood chamber were radioactive. Thus within a day the food collected by 6 bees had been shared between many thousands.

Such incessant food transmission is basic to communication in all insect societies (Fig. 8.4 shows the analogous behaviour of ant workers). Not only is every individual kept directly informed of the state of food supplies in the colony, but pheromones (see p. 90) are also circulated. The development of worker termites is controlled by pheromones produced by the king and queen. Similarly the queen honey-bee secretes a pheromone ('queen substance') which both suppresses the ovaries of the workers and prevents them from rearing new queens. The queen is always surrounded by attendant workers who lick her body, subsequently proferring food to other workers. The level of the pheromone must be kept up and once the source is cut off its concentration rapidly drops below the critical level. The effectiveness of the incessant food sharing in circulating queen substance is shown by the fact that some workers in the brood area of the hive exhibit changed behaviour within an hour or two of a colony losing its queen. They begin construction of an 'emergency' queen cell in which one of the youngest larvae, destined in the normal course of events to become a worker, is fed royal jelly throughout its larval life and becomes a queen to replace her mother.

In nearly all the social insects there is further communication between the foragers about sources of food. In ants and the group of 'stingless' bees (Meliponinae) communication is by purely chemical means.[505] When leaving a food source to return to the colony, ants lay scent trails and the Meliponines dip down from their foraging flights at intervals of a few yards to mark leaves or the ground with scent from special mandibular glands. Subsequently other foragers are attracted by such trails and can be shown to find the food more easily.

We have already discussed the evolution of the 'dance language' of honey-bees. This is certainly the most remarkable example of communication between social insects because of the detailed information which it conveys about the distance and direction of a food source. One vital factor in the operation of the dance is the close attention paid to incoming foragers by other workers. As she moves up on to the comb the forager is contacted many times and she can gauge the needs of the colony by the eagerness with which her preferred nectar load is accepted. The persistence of a forager's dance on the comb is related to the richness of the food source which she has discovered. Accordingly other foragers who may be resting in the hive are more likely to be attracted towards rich sources. On the other hand if her food source was not a rich one, the incoming forager may not dance. She is quite likely to stay in the hive and subsequently be attracted by the dances of more successful foragers.

Although other social insects lack the detailed communication system of the honey-bee, their foraging is not at random. As we mentioned, incoming foragers may lay scent trails and other workers become aroused to go out and forage by the highly excited behaviour shown by a forager who has discovered a rich food source. Thus social insects tend to forage as a team which directs its effort where it will be most profitable.

Throughout this brief survey of social insect behaviour we have used descriptive terms which may give the impression that their life is organized in an intelligent way. We must not confuse adaptiveness with intelligence here. The beautiful adaptiveness of the social insect colony and the control it gives them over their environment are based upon a relatively simple series of responses to other workers in the colony and to the nest itself. Within the limits of their remarkable capacities the social organization of the insects is flexible, as when honey-bee workers change their normal sequence of tasks in response to a sudden requirement in the colony. However the type of response the social insects can make is itself determined within inherited limits which are characteristic of the species. All honey-bees respond the same way, so do all ants of a particular species and so on.

This consistency of social organization within a species is highly characteristic of the insects and we shall not find it so well marked in other animals.

SOCIAL ORGANIZATION IN VERTEBRATES

Very few vertebrates are as solitary as the digger wasp; they live far
longer and often have well developed parental care, both factors which
ensure that there is overlap of the generations. Nevertheless they exhibit
a very wide diversity of social organization and furthermore, in striking
contrast to the social insects, such organization is not rigidly species specific. Thus the answer to the question, what is the social organization of the
house mouse? is not fixed. It depends upon the density of the population,
its age and sex structure and a number of other factors.

Recently there has been a remarkable increase in studies of vertebrate
social organization, and the majority have been made on wild populations
in the field. In our general discussion we shall try to bring together
information from a wide range of work. More detailed discussion and
fuller references to the literature can be found in Wilson[507] (a complete
general survey), Crook[107] and Wood-Gush[512] (for birds) and Chapter 4 in
Ewer's book[149] (for mammals).

Territory and dominance hierarchy

The role played by individuals within most vertebrate societies depends
on their age and sex, but it is far less rigidly determined than in the social
insects. McBride,[312] in an important discussion of social organization based
on studies of the fowl, uses the term 'caste' to represent the different stages
specialized in 'function, behaviour and way of life' through which the
bird passes. But the very fact that a bird's caste changes as it grows up,
and between seasons when it is adult, shows that caste cannot retain its
complete insect meaning. McBride's aim is to direct attention to the
relatively proscribed nature of a bird's role at any one stage. In the majority
of vertebrates, as in the fowl, the individual's role and the interactions it
makes with other members of the social group quite commonly involve
assertions of rank or dominance. Territory and hierarchy represent in some
ways two extremes of such organization.

We have discussed various aspects of territory before, especially in
Chapter 5, and at first sight territoriality may seem to be the antithesis of
social behaviour. Territory takes many forms, and that illustrated for
willow warblers in Fig. 5.1 represents a fairly typical case. Each male
defends a substantial area which will include an adequacy of food for himself and eventually a mate and young, so that he will rarely leave his territory. It impinges closely on the territories of others and, where the

habitat permits, nearly all the ground is occupied. Territories of this type are also found in some mammals, notably carnivores, where territory is defined by scent posts marked with urine or special glandular secretions.

Whilst the territory holders are largely solitary, the whole group of territories certainly constitutes a social organization if not a society and the fact that territories are grouped will, for example, influence the behaviour of females seeking mates. In sea birds, such as gannets, gulls and terns, this grouping is much more obvious. The territory is no longer for feeding but simply a small area around the nest. The territories are tightly packed and the birds form a coherent colony.

The common feature of all territories is that they confer complete dominance on the owner within the boundaries. There he is secure and the boundary of his territory marks the point at which his dominance begins to give way to that of his neighbour. The hierarchical organization of dominance does not relate to a fixed area, but to rank order between a group of individuals living in a common area. The group will move around together and may indeed defend a communal territory, but within the group some animals are dominant over others. This means that they can displace more subordinate individuals for food or a mate, for example.

Schjelderup-Ebbe developed the concept of hierarchy from his work on flocks of birds (see his summary[419]). He observed that a definite 'peck-order' developed amongst small groups of chickens confined together in a pen. As the birds disputed amongst one another one gradually emerged as the dominant who could displace all the others. Below her was a second-ranking bird who could dominate all except the top bird and so on down the group until at the bottom was a bird displaced by every other in the flock. Fig. 8.5 shows one example of a peck-order of this very precise type —a linear hierarchy as it is sometimes called. The hierarchy develops as the birds dispute and involves a good deal of fighting in the early stages, but once established it is as much an hierarchy of submission as one of dominance. Subordinates usually defer without question at the approach of a more dominant bird. Note that individual recognition is a prerequisite of stable hierarchies and Schjelderup-Ebbe found that they failed to form in large flocks where disputes continued as recognition failed.

Hierarchical structure of some type is very widespread amongst the vertebrates and in some invertebrates too, but it rarely assumes such a perfect regularity as that shown in Fig. 8.5. Divergences from the 'pure' type can take several forms. The hierarchy may not be linear, and triangular relationships may develop where A displaces B who displaces C, but C can displace A. In primate groups it is not uncommon for high-ranking animals to co-operate to assert their dominance over others. This may make

	Y	B	V	R	G	YY	BB	VV	RR	GG	YB	BR
Y												
B	22											
V	8	29										
R	18	11	6									
G	11	21	11	12								
YY	30	7	6	21	8							
BB	10	12	3	8	15	30						
VV	12	17	27	6	3	19	8					
RR	17	26	12	11	10	17	3	13				
GG	6	16	7	26	8	6	12	26	6			
YB	11	7	2	17	12	13	11	18	8	21		
BR	21	6	16	3	15	8	12	20	12	6	27	

Fig. 8.5 A perfect linear hierarchy established within a group of twelve hens. Each bird is marked by colour rings on its legs whose initials identify it. The number of times each bird pecked another flock member is given in the vertical columns (e.g. Y pecked B 22 times and V 8 times) whilst the number of pecks received from another is given in the horizontal rows, (e.g. VV received 19 pecks from YY and 8 pecks from BB). Note that no bird was *ever* seen to peck an individual above it in rank : hierarchies as perfect as this are probably rare in nature (From Guhl, A. M. 'The Social Order of Chickens.' *Scientific American.* © 1956 by Scientific American. All rights reserved).

it difficult to separate them in the hierarchy. Sometimes it may not be possible to distinguish any ranking below the top animal, i.e. one animal dominates all the rest who can only be lumped together as 'subordinate'. This is often the result if male mice are kept in a fairly crowded group. It also occurs if male sticklebacks are placed in a tank with inadequate space. The first one to develop into reproductive condition usually takes the whole space as his territory and all the others are driven into submission.

Territory and hierarchy are obviously not completely distinct types of social organization—as we shall see some systems which might be called territorial-hierarchical do occur—but they are useful descriptive terms around which to organize our discussion.

Perhaps the simplest social structure is that of solitary animals in which both sexes hold a separate territory, whose boundaries break down only to allow reproduction. Hamsters are strongly territorial and the female will only allow a male to approach her for an hour or two at the very peak of oestrus. The European robin also has territory in both sexes and female robins, like males, sing throughout the winter. During spring female song subsides and pairing takes place, usually on the male's territory which both birds then defend.

Leyhausen[294] has described an interesting variation of territory amongst solitary cats. The domestic cat has a fairly well defined territory with certain areas which are marked with urine, linked by a network of favourite paths. However the territories are not held exclusively; neighbouring cats avoid one meeting another by using common areas at different times. They learn one another's habits and this apportioning of access to an area by time breaks down only when males compete for an oestrus female. The territory of a cat would normally provide an adequate food supply in the wild. It is of great interest that Eaton[140] has recently found a similar time-sharing system in the African cheetah. Small groups of cheetahs, probably derived from a family and sometimes containing more than one male, move about together. The males scent mark frequently and examine scent posts for traces of other cheetahs. If the scent is fresh the group alters its direction of march, but it continues if the scent is more than about 24 hours old. In this way meetings between groups are reduced to a minimum and available space is shared efficiently.

Amongst some birds and mammals not all individuals hold territories of the same type. There is a combination of hierarchy and territory such that dominant animals hold larger territories or hold them in the most preferred areas. Familiar examples are provided by the 'lek' system of birds such as the black grouse and the ruff. The lek is an area of ground divided into a cluster of territories held purely for mating. Males compete for desired positions near the centre of the lek and here territories are small. Less successful males may hold large territories at the periphery of the lek, but females are attracted by the displays and high activity of males in the central areas and it is these males who have most success in mating.

Fig. 8.6 shows a map of territories on a black grouse lek with 5 small territories at the centre and a number of large territories at the periphery whose boundaries are sometimes indistinct. Kruijt and Hogan[279] found that central males arrived first at the lek soon after dawn each morning and stayed for several hours. These males probably kept the same territories from season to season, but eventually were replaced by younger birds from the periphery. Each year some young birds became established at the periphery but a male was usually several years old before he could hold a central territory. Visiting females tend to alight in the territories of

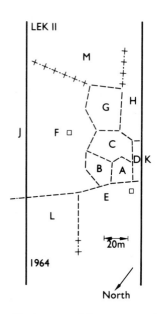

Fig. 8.6 Territories on a black grouse lek on pasture land in the Netherlands observed during 1964. The two vertical black lines represent the position of fences which are sometimes adopted by the birds also as territorial boundaries. Crosses mark uncertain boundary lines. There is a central area occupied by cocks A, B, C, D and G on small territories. The peripheral birds can occupy much larger areas. (From Kruijt, and Hogan,[279] 1967. *Ardea*, **55**, 203.)

central males, who display vigorously to the sight or sound of grouse flying over the lek. Thus the younger peripheral males are largely excluded from mating, although if they are persistent and return to the lek each season, their turn will come. Once she has mated, the female grouse leaves the lek area to nest and rear the young alone.

There is certainly a regular relationship between the type of social organization and the strategies of the two sexes in reproduction—the type of mating system. Territorial animals are usually monogamous and the pair stay together for the season and co-operate to rear their young. The great majority of birds live this way and the young are helpless at hatching and often require large quantitites of high protein food—insects, fish, part-digested seeds, crop milk etc.—which is more than a single parent can provide. A few mammals are monogamous, notably some carnivores where again the mother alone cannot catch enough food to rear the young, (see Kleiman's[268] review of monogamy in mammals).

A female's best strategy in this situation is clear, she must choose a male who is holding a good territory for he thereby effectively 'proves' that he is fit and can help her rear young. The male must also try to choose a good female, although it is much less obvious how he can do so. Perhaps his attacks on the female when she first enters his territory (see p. 187) are an aid to his choice. It has always been somewhat puzzling that even in widely sexually dimorphic birds the male still attacks a potential mate. By staying put or by persistence in returning she could reveal that she is fit and in good breeding condition.

The stakes are balanced quite differently in polygamous, or more correctly polygynous species, where one male mates with several females. Here, as on the grouse lek we have just described, there is commonly a hierarchy among the males and subordinate animals are excluded from mating. A number of birds are polygynous and so are the majority of mammals. In all such animals males are emancipated from care of the young. In birds this is usually because the young of polygynous species are active and can feed themselves from hatching, protected by the female. They are herbivorous species where food is relatively abundant and easy to find. In mammals the female has a far greater commitment to the young because she alone can feed them. As on the lek, the male's best strategy is to mate with as many females as possible. They may not all be equally fit, but he can mate many times and loses far less from multiple matings than he stands to gain. The males compete fiercely for females and a female must choose a victorious male who has thus proven his fitness relative to other males. Unlike the monogamous male on his territory, the polygynous male offers his temporary mate nothing but his genes. It is of course amongst such species that sexual selection has operated most consistently. The intense competition for the attention of females has led to the evolution of considerable size differences between the sexes, especially in mammals, and to the elaboration of male plumage or other adornments such as antlers (see Chapter 6, p. 198).

Here we have described only two of the many types of mating system. Emlen and Oring[144] discuss the whole range in more detail and relate them to the ecology of the species concerned.

VARIATIONS IN SOCIAL STRUCTURE WITHIN SPECIES

We mentioned earlier that vertebrate societies can vary within a species according to environmental conditions. It is interesting that, although the lek displays of birds have been known for centuries, recent work has shown that some mammals have a strikingly similar type of social organization. Further, the form in which it occurs appears to vary with population density. Fig. 8.7 shows a map of territories held by males of the kob

0   100   200 metres (approximate scale)

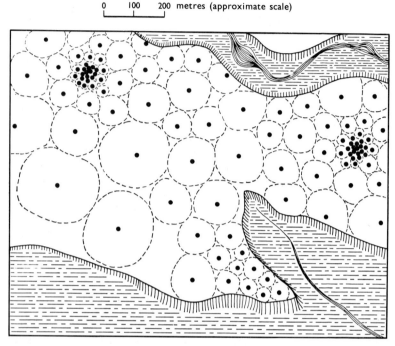

Fig. 8.7 Territorial map of males of the Uganda kob on an area of grassland raised above two swampy river bed areas, indicated by shading. The boundaries are only very approximate, but each black dot represents the centre of a territory, close to where a male usually stands. There are two closely packed 'territorial grounds' and a third may be forming on the 'peninsula' at the lower part of the map. (After Leuthold,[292] 1966. *Behaviour,* **27**, 215.)

antelope of East Africa which were studied by Leuthold.[292] In each local population of kob, which may number around 15,000, a favoured area of land is occupied by adult, territorial males. Younger males and females move about in small bachelor herds as do females with their young. As males become sexually mature they tend to leave the herds and become territorial, especially during the dry season when grass is short and does not restrict movement or visibility. As can be seen in Fig. 8.7, territories vary greatly in size. Some—which Leuthold calls 'single territories'—are 1–200 metres in diameter, others are much smaller. The smallest territories are clustered tightly together to form a 'territorial ground'. Here each male is defending an area less than 30 metres across and there are frequent disputes and various advertising displays. The ground is trampled flat and bare.

Males may occupy single territories for weeks and, since they can get a good proportion of their food on the spot, spend most of their time there. There is not usually much competition for these large peripheral territories but competition becomes much stronger as one approaches a territorial ground. There is quite a rapid turnover of males in these small territories in part, at least, because lack of food and rest forces males to leave. Males leaving the territorial areas for any length of time rejoin all-male herds in the vicinity.

Female herds move through the territorial areas, and those in oestrus detach themselves and approach males. Although they are courted by males on single territories, almost all copulations occur with males in the territorial grounds. The situation is therefore strikingly analogous to that on the black grouse lek. If the analogy is to be exact, then single territories should be held predominantly by young males who would gradually win their way into the central group, but this does not necessarily seem to be the case. Leuthold found that the age structure of groups of males holding single territories was the same as that for males from the territorial ground. He suggests that the territorial group phenomenon is related in some way to a high population density of kob. In some other areas where kob are less abundant, single territories seem to be the rule and oestrus females mate with males holding them.

Certainly there are numerous clear examples of changes in social organization with density. Often crowding leads to a more hierarchical type of structure. Experimental studies on animals as diverse as sunfish (Erickson[145]) and mice (Davis[120]) have shown that males form typical hierarchies under high density but lapse into territoriality when space and cover permit. Owen-Smith[376] suggests that it is high population density thanks to protection which—for once—has worked, that accounts for the rather anomalous territorial structure of white rhinoceros in a South African game reserve. Adult bulls are strictly territorial but many territories now include a 'subsidiary' bull, who always defers to the territory holder and whose presence is largely ignored. This may be one stage in the breakdown of the territorial system to form an hierarchy. Since territories are normally about 2 km^2 and there are now about 5 rhinos (of all ages and both sexes) per km^2 there is obviously pressure on the normal social structure.

Territoriality and density are not always related in this way—everything depends on the general ecology of the species concerned. Estes[147] finds that wildebeeste tend to be more territorial in areas of rich food supply. Here densities are high, but the paramount factor is adequate food which allows the animals to be sedentary. Male wildebeeste always defend territories for breeding but in poorer areas these are only transitory. For much of the time the animals have to be on the move following grass where it is to be found.

They form large herds, usually consisting of a mixed group with associated all-male herds, from which latter males detach themselves to take up temporary territories as conditions permit.

For many vertebrates the environmental changes which they meet with changed social organization are the regular, seasonal ones. A wide range of birds which are territorial in the breeding season form flocks in the winter, sometimes combining with other species. The selective advantages of moving and feeding as a group were discussed earlier. During the more rigorous winter conditions these outweigh the advantages of a territory for each individual, whose selective value is largely associated with reproduction and rearing the young. With most passerine birds there is little social structure within the winter flocks. They move together, respond to one another's flight and alarm calls and roost together, but this is all. Chaffinches do have a tendency to form one-sex flocks and we know from the experimental work of Marler[328] that males tend to dominate females in winter and displace them at feeding sites. There is a reversal of this dominance in spring and females tend to displace males when the flocks break up and males disperse to set up territories.

Some geese and ducks retain family groups at least during migration and perhaps in the winter flocks. The fowl, studied in a truly feral state on an island off Queensland, Australia, alternates between a territorial system in the breeding season and a more hierarchical flock structure during the winter. McBride et al.[312] found that only the dominant or alpha male of each flock sets up a territory in spring. He mates with several of the hens from his flock during the season. During winter after the young birds of the year have returned to the flocks, the alpha male and his harem move about over a home range with numbers of subordinate males staying at the periphery of the group, often moving between the home ranges of different alpha males. The alpha male leads his flock in every sense. He it is who initiates all movements of the group, particularly across open ground, and his posture is normally more alert than those of his females. He is the first to give alarm calls and may even approach predators—feral cats—whilst the others take cover.

Seasonal changes in behaviour are just as marked in some mammals. The classic study *A Herd of Red Deer* by Fraser Darling[118] remains one of the most complete of any ungulate species. The sexes normally remain separate when adult and the female herds, which include young animals of both sexes, move within a preferred home range which in summer extends on to the higher feeding grounds. There will usually be one dominant old female who acts as leader. The herds of stags are smaller and may be loosely associated with the female herds, although they often move over a different home range. Lincoln, Youngson and Short[295] in their studies of deer on the Hebridean island of Rhum have shown that there is a dominance hierarchy in the male herds. Males challenge each other with lowered antlers and the largest males become dominant. Nevertheless the herd stays together as a unit through the winter. In early April the antlers are shed and immediately the aggression takes on a new form and males rise on their hind legs and 'box' with their hooves. New antlers begin growing but remain soft and tender in 'velvet', which is not shed until August. Now the stags become increasingly aggressive and fighting with the new antlers increases until about mid-September when the male herds break up. Males go off singly and begin roaring and gathering hinds who are coming into oestrus and the rut begins. Oestrus females are attracted to displaying stags who gather a harem, and they defend their group of females rather than any particular area. The rutting season is relatively brief and soon the females return to their herd for the winter move on to lower ground. The stags shed their antlers and once more join together.

The striking changes in social behaviour which accompany the onset of the breeding season in the vertebrates we have hitherto been describing, are less marked in those species where the group co-operate to hunt and to defend themselves. Wolves, for example, hunt in packs and maintain a very stable social structure based on an extended family unit. There is usually one dominant male leader, but several other adult males may be included. Primates also tend to retain a uniform social structure throughout the year and in conclusion we may turn to consider their social behaviour in more detail.

PRIMATE SOCIAL ORGANIZATION

Studies on the primates have accelerated more rapidly than those on the social behaviour of any other type of animal. Workers from a number of disciplines, zoologists, psychologists, anthropologists and sociologists have converged on this group, both for their intrinsic interest and in the hope that they will provide information which is relevant to speculations on the origins of human societies. The great emphasis in the recent primate work has been on studies of natural communities in the field. Certain primates, notably the rhesus monkey, have been familiar laboratory animals for some time, but it is generally accepted that studies on captive communities are inadequate by themselves.

Here we can attempt only a brief survey of the primate work. There are now two journals devoted mainly to behavioural studies, *Primates* and *Folia Primatologia* and a rapidly growing range of books and symposium volumes. (For an introduction to this range the reader is referred to Altmann,[7] De Vore,[135] Jay,[257] Jolly,[260] Napier and Napier,[362] and Rowell.[405])

Primates live in a wide variety of habitats. We tend perhaps to think of them as tropical animals, but they were more widely distributed in the

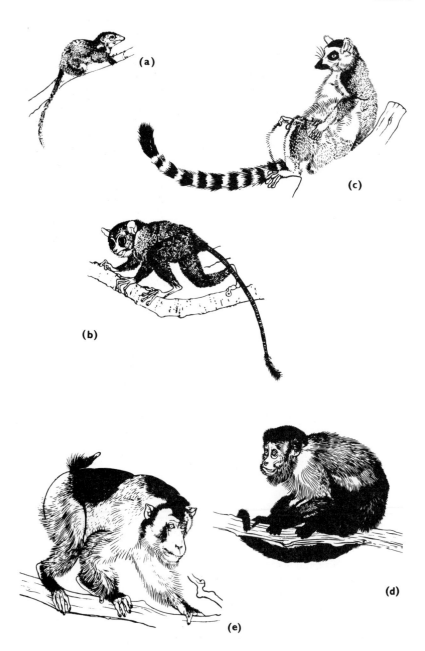

(f)   **(g)**

Fig. 8.8 Some diverse living representatives of the primates. Although they do not in any way represent a phylogenetic series, the higher primates and man did have ancestors which were rather similar to their more primitive relatives illustrated here. (a) A tree shrew (*Tupaia*)—probably best regarded as a specialized insectivore rather than a primate, but certainly resembling the stock from which the early primates arose; (b) the tarsius (*Tarsius*) of the Philippines; (c) the ring-tailed lemur (*Lemur*) of Madagascar; (d) a South American capuchin monkey (*Cebus*); (e) the pig-tailed macaque of S.E. Asia (*Macaca*) and two of the apes—man's closest relatives—(f) the gibbon (*Hylobates*) and (g) the chimpanzee (*Pan*). (After Le Gros Clark,[99] 1965. *History of the Primates*, 9th edn. Brit. Mus. (Nat. Hist.), London, not drawn to scale.)

recent past and two of the macaques, the Barbary 'ape' of the Atlas mountains and the Japanese macaque live in areas where snow and frost are regular every winter. The majority of primates are arboreal, some of them like the spider monkeys of South America and the colobus monkeys of Africa, exclusively so. However a number have returned to ground living, such as the baboons and the patas monkey of Africa, whilst the chimpanzee and the gorilla also spend a lot of time on the ground. In general the more arboreal primates are fruit and leaf eaters, the ground dwellers tend to be more omnivorous. Although the gorilla is exclusively vegetarian, chimpanzees and baboons will eat meat if they get the opportunity; the former have been seen to hunt and kill monkeys for food.

The living primates exhibit a tremendous range of morphological types, from the primitive lemurs which retain a long muzzle, a moist nose and claws on one of their hind toes—little modified from the ancestral primate types—up through the monkeys to the great apes and man (see Le Gros Clark[99] and Fig. 8.8). Throughout this series we can observe certain trends; the enlargement of the brain, the development of the grasping hand and in contrast to many other mammals, the great reliance on colour vision as a dominant sense for exploration and communication.

It seems certain that from very early in their history the primates were social animals moving around in groups whose organization was stable. Jolly[260] makes this point from her study of lemurs; here in these primitive primates we already find small mixed troops (12-20 individuals) which include several adult males—a very typical primate grouping. There is a dominance hierarchy within the troop (and in some lemurs females rank more highly than males) but it remains as a permanent, cohesive unit. Lemurs have group territories within their mixed woodland habitat whose boundaries are often remarkably stable. They are marked by scent in some species—and are defended by calling, which is usually sufficient to cause the neighbouring troop to retreat without further threat or fighting.

Within the troop there are frequent minor disputes but serious fights are rare outside the breeding season. This is very brief in most lemurs—at most two weeks—and it is at this time that subordinate males seriously challenge the older dominant ones. Apart from this short period of strife, much of lemur social life is characterized by non-aggressive interactions between individuals—indeed it is pedantic to avoid use of the term 'friendly'. There is always close contact between a mother and her infant who clings continuously to her at first and is carried around everywhere. As it grows older other adults approach and play with the infant, as they also play with each other. Lemurs have thick, dense fur and groom frequently. Mothers groom their infants and adults frequently groom each other—this being one of the commonest types of friendly contacts between individuals.

In the behaviour of lemurs we can detect most of the elements which characterize all primate societies, although there are many variations on the theme. These variations involve group size, territoriality and intra-group relationships to mention some of the more obvious ones.

Some of the smallest groups are found in two of the apes—the gibbon and the orangutan. The orang of Borneo and Sumatra appears to be an exceptionally solitary animal; it is rare to see more than 2 or 3 together moving through the high canopy of tropical rain forest and lone males are common. Little is known about its social interactions but they are certainly sparse—it remains something of an enigma for an animal of such high intelligence. Gibbons live as small family parties, a monogamous pair and their young offspring. They are territorial and hoot loudly when they detect other gibbons in the vicinity: usually a contest of calls is sufficient to maintain the boundaries.

Other dense forest primates such as the howler and several other South American monkeys live in moderate sized groups, 12–30 or so. Often too they tend to avoid other groups and have their own preferred home range but unlike the gibbons, they cannot really be said to defend a territory. In a large home range in forest it is impossible for a group of monkeys to survey and patrol boundaries in the way—say—a bird can on its territory. As Carpenter[85] says of howler monkeys, '(they) do not defend boundaries or whole territories; they defend the place where they are, and since they are most frequently in the familiar parts of their total ranges, these areas are most frequently defended—typically by interchanges of roaring at approaching or approached animals.' Other forest monkeys have analogous calls which are used rather as in gibbons for defence of a territory. Black and white colobus, whose territory and feeding behaviour was described on p. 268, have roaring calls and there is a good deal of hostile interaction with neighbouring troops.[331] Their territory is small enough to be defendable, but the larger home range of the red colobus is not so strictly delimited nor will interactions with neighbours be so frequent.

Inter-group avoidance seems to be less marked in the more terrestrial primates, although it is difficult to make generalizations. Certainly there can be a considerable amount of overlap between the home ranges of Indian langurs, baboons and chimpanzees for example. Hall and De Vore[195] report several instances of two groups of baboons intermingling during food foraging with no sign of hostility. Some of the largest groups are to be found in the terrestrial primates—baboon troops range up to 90 strong and the gelada baboon of the high Ethiopian plateau sometimes has herds of 400, although there are many social sub-units within a party of this size.[108]

One cannot fail to be impressed and fascinated by the high level of intercommunication that goes on between the members of a primate group. Each individual is constantly responsive to the movements, gestures and calls of others. Naturally, we must be cautious about assumptions that the primate societies are more organized than those of—say—ungulates or carnivores. Because we ourselves are primates we find it much easier to identify elements in their communication system. In particular the mobility of their faces and the way they watch one another's faces for information on mood and intentions. Nevertheless there are, on the most objective criteria, good grounds for the assumption. Two factors that contribute to this situation are the high learning ability of primates and their extended period of infancy, to which is coupled considerable longevity. The larger

Fig. 8.9 Friendly grooming in the Barbary macaque. The juvenile male on the right is about 3 years old ; he is grooming a 'subadult' male aged about 4½, who ranks above him in the social hierarchy. A 10 week old baby is being held by the subadult male. (Photo by John Deag.)

primates commonly live for 20 or 30 years and this means that a young primate grows up to take its place in a group where—literally—everybody knows everybody else from long experience in their company. (There are several good reviews of both visual and vocal communication in Altmann.[7])

We shall be discussing below aggression, and dominance hierarchies based upon it, but before doing so it is most important to remember that primate societies are as much characterized by positive interactions as negative ones. Overt aggression is not common under natural circumstances and there are many friendly contacts between animals as when they invite grooming or offer to groom another. Mutual grooming, which we first mentioned in the lemurs, is very important as a placatory gesture in primates (see Fig. 8.9). Often a dominant animal will 'allow' itself to be groomed by a subordinate following a brief threat to which the subordinate has deferred. Sexual presentation as an appeasement gesture has already been discussed on pp. 98 and 179, it is very common in baboons and chimpanzees and is made by males or females towards a dominant animal who threatens, or even if the subordinate wants to pass close to the dominant one, see Fig. 8.10.

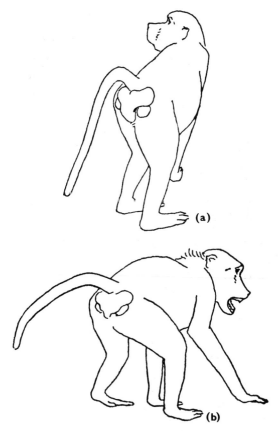

Fig. 8.10 Sexual presentation in the chacma baboon, (a) 'genuine' presentation by an oestrus female, she characteristically looks back over her shoulder towards the male, (b) appeasement presentation towards a threatening dominant animal, the general stance is the same but the facial expression is one of fear. (After Bolwig, N., *Behaviour*, **14**, 136–65, 1959.)

The pattern of grooming relationships in a group is often a good measure of its detailed structure. It will show us which animals associate together and is a good index of the cohesive forces which maintain the group as a real biological unit. Fig. 8.11 taken from Sade's work[414] on rhesus macaques and illustrates just one small sub-unit of a group—that which hinges around a single old female. The other individuals recorded are her children and grand-children. A very high proportion of all their grooming is to their mother or siblings—this genealogical unit really acts as a social unit

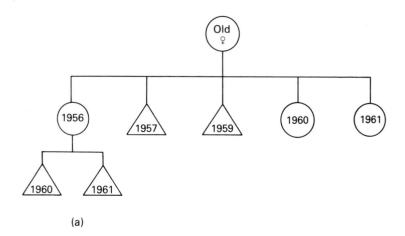

(a)

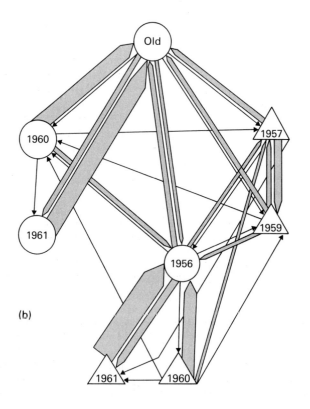

(b)

too and associates together. Only the old female herself spends much of her time grooming other individuals outside this group. Sade found that as animals matured or changed their status the grooming patterns also changed. Yet it remains true that in many primate societies the matriarchal group is much the most constant.

Fig. 8.11 represents one kind of 'sociogram'—a diagram which illustrates relationships. Many different types of sociogram can be constructed (they are well reviewed in Hinde,[219] Chapter 23) based upon various behavioural measures, depending on what type of relationship one wishes to study. A dominance hierarchy, such as that shown in Fig. 8.12, is one familiar example based upon agonistic interactions.

Certainly the role of aggression in primate societies has attracted a great deal of attention. As may be imagined, the non-human primates have been used as evidence by both sides in the dispute about the nature of human aggression which was discussed in Chapter 4. Primates vary greatly in the degree of fighting which is observed both within and between groups. Even within a single species there may be considerable variations; langur populations in northern India are far more peaceful than those in the south. The population densities are much greater in the areas where the aggressive groups live and this is probably one factor involved. Fighting is very common in crowded zoo colonies which rarely give sufficient space for subordinate animals to keep out of the way of more dominant ones. Russell and Russell[409] lay great emphasis on the effects of density and consequent stress upon aggression in both human and non-human societies. It is clearly a crucial factor under certain circumstances but, as we shall see when discussing different types of primate social organization below, some depend more than others on overt aggression for their maintenance and stability quite irrespective of density.

Much of the work on primate behaviour has emphasized the importance of dominance hierarchies in their social structure. Dominance is always based on the threat of physical violence although it is not always employed. A dominant or high ranking animal is often defined as one whose be-

Fig. 8.11 (a) The genealogy of one matrilineal group of rhesus macques from a free-ranging troop studied by Sade. Circles represent females, triangles males and dates of birth are inscribed. The age of the old female was not known. (b) A sociogram illustrating the grooming relationships within this group during 1961. Arrows point from groomer to groomed and their thickness represents the proportion of total grooming time spent grooming that animal. The thinnest arrows represent 1–10% of grooming time, and that from the youngest male (born 1961) to his mother (1956) represents 100%. Most animals within this group spent at least half their grooming time within it. The old female herself has, of course, more grooming relationships outside it. Mothers tend to groom their youngest offspring the most. Siblings, particularly the two brothers 1957 and 1959 also groom each other a great deal. (Modified from Sade,[414] 1965, *Amer. J. Phys. Anthropol.*, **23**, 1).

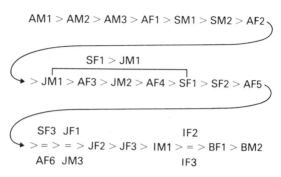

Fig. 8.12 The agonistic hierarchy recorded by Deag in a troop of wild Barbary macaques in the Middle Atlas Mountains of Morocco. It is based upon the distribution of threat and avoidance between pairs of individuals. The hierarchy is reasonably linear, especially at the upper end where adult males are dominant over all other animals. Amongst the younger animals there is less strict linearity and this reflects changing status as animals grow up. Baby animals are not much involved in the hierarchical system and rarely give or receive threats. There is one clear exception to linearity higher up. Sub-adult female 1 threatened Juvenile male 1 and he avoided her. A=adult, S=sub adult, J=juvenile, I=infant, B=baby, M = male, F = female (Modified from Deag,[123] 1977, *Animal Behaviour*, **25**, 465.)

haviour is not limited by others whilst a subordinate animal is so limited. The limitations imposed by rank are diverse. For instance, having high rank in a group might determine access to food, preferred resting places and to females. When they come into oestrus, females may move close to the dominant male and form a temporary 'consort relationship' with him.

The hierarchies which have been described are often reasonably linear and in some macaques and baboons, for example, all adult males rank higher than females. Fig. 8.12 shows the situation recorded in one wild troop of Barbary macaques: they showed an almost perfect linear hierarchy.

A number of primate workers have criticized the way in which the concept of dominance has been used. Rowell[406] quotes examples of the imprecise definition of dominance and points out that the different types of supposed advantages conferred by high rank—access to food, to oestrus females etc.—may not always be correlated. She suggests that hierarchies, where they exist, are as much hierarchies of submission as of aggression. Certainly it has often been observed that while rank is initially determined by threat and fighting, once established it is maintained as much by the deference of subordinate animals as by any display of threat by the dominants. Rowell suggests that dominance hierarchies are not common in wild primate groups. The fact that they are almost invariably described from captive groups she claims is a reflection of the crowding and unnatural stress to which the latter are always subjected.

There can be little doubt that many of Rowell's criticisms are justified. As mentioned above, primate colonies in zoos *are* overcrowded by natural standards. When obvious dominance hierarchies have been observed in the wild they could sometimes result from human interference. For example the rigid hierarchies described for Japanese monkey troops are probably an artifact of their particular situation. The detailed work on these monkeys has been greatly helped by the success of the Japanese workers in getting them to visit artificial stations where food is regularly provided (see Frisch[155]). Some troops stay close to these stations and providing food in one spot is likely to reinforce dominance structure. Goodall[169] found just the same thing when she provided caches of bananas for her chimpanzees. The food acts as a focus which all animals are trying to approach, the dominant males sit at the best places and the rest of the group tends to space out according to rank. A less rigid system probably prevails when food has to be searched for individually.

Despite these valid criticisms, it remains the case that dominance hierarchies *can* be observed in undisturbed wild primates. Deag's study[123] of a troop of Barbary macaques revealed the hierarchy already illustrated in Fig. 8.12. He discusses how such a regular system of rank might arise from all individuals behaving to their own best advantage. The cohesion of the troop is vital for everyone's survival and some predictability about social interactions will be advantageous not only to the top animals but to low ranking ones too. Fighting is reduced, and probably stress also, since subordinate animals can keep clear of others to whom they will predictably lose in competitive encounters. Ranks change with time and subordinates will come to take over high ranking positions as previously dominant animals become older.

Certainly the nature of the ranking system is such that age and sex determine to a great extent the role a primate plays in its group. One recalls McBride's use of the term 'caste' to apply to an analogous situation in the fowl. However there are, as might be expected, further complications in the primate societies. For example, the Japanese scientists working on *Macaca fuscata* and those working with free-ranging groups of the rhesus monkey (*M. mulatta*) have found that young monkeys tend to 'inherit' the rank of their mother. Thus sons of mothers who rank high have a better chance than most of gaining rank amongst the males (see Sade[415]). High-ranking animals tend to support one another in asserting their dominance. As males become mature they begin to challenge the leaders and individual groups have now been kept under observation for long enough to record new males taking up dominant positions and gaining access to females. We now know that many of the newly successful males have moved across from neighbouring groups.

Primate groups are never very large and since they remain such a tight social unit there is potentially a serious inbreeding problem. This is avoided by the transfer of newly mature males between groups. Packer[377] and Hausfater[205] have studied this process in baboons and Lindburg[299] in rhesus macaques. They find that males rarely if ever breed in their natal troop. There is a good deal of disputing as young males become sexually mature and this is the commonest time for them to leave. Rowell[404] has also seen interchange between groups of baboons which she could not relate to dominance disputes. Packer records males moving more than once; one male had been a member of four different troops. Such exchanges are of extreme importance biologically because otherwise the very stability of primate social structures might threaten their fitness.

Sometimes the rank of an animal seems to affect not just its role in the group, but the actual position it takes up on the ground. Fig. 8.13 is taken from Hall and De Vore's work[195] with baboons and shows a group on the march. The most dominant males are in the centre, close to the females with small infants. Two males are in close consort relationships with females in oestrus, whilst less dominant adult males and young animals move peripherally. For baboons living in open country with the constant threat of predation, this kind of spacing could ensure the best protection for the females and young. Hall and De Vore describe how at any threat to the troop the largest males rush to investigate.

Not all observers have found such regular spacing patterns and differences may relate to habitat. The baboon troops studied by Rowell[404] were living in more forested country and she found that they invariably reacted to danger by avoidance rather than defence. She did not observe such regular positions as Hall and De Vore, although larger males did tend to lead and bring up the rear as the troops moved.

The primates we have used for examples so far have all exhibited a social organization commonly called 'the multi-male group' in which there are usually several adult males. This is perhaps the commonest type of primate group but we have already mentioned some others such as the family groups of gibbons. Some primate societies are based on 'one-male groups' which consist of a single adult male, several females and their offspring. This is bound to leave a surplus of adult males and these usually form all-male bands whose members occasionally challenge those with harems. This type of organization is found in the plains-living patas monkey, the gelada and the hamadryas baboon (*Papio hamadryas*). In the latter two, numbers of one-male groups are associated together, particularly in the gelada where hundreds of animals move around together in a loose herd structure.[108] The hamadryas baboon lives in semi-desert regions of north east Africa, and the groups are fairly independent during the day, spreading out to comb the area for food. They come together as a troop at night to sleep on cliffs which provide protection from predators.

Fig. 8.13 A troop of baboons on the march through open country. The adult males are distinguished by their large size and well developed manes. Females with babies more in the centre. The two females in oestrus (shown by dark hind parts) move in consortship with the most dominant males. Further explanation in the text. (From Hall and De Vore,[195] 1965, in *Primate Behaviour: Field Studies of Monkeys and Apes*. Holt, Rinehart and Winston, New York and London.)

The hamadryas baboon has been the subject of some remarkable field experiments by Kummer and his collaborators.[281–3] Their results provide an excellent demonstration of the subtlety and flexibility of primate behaviour. In the one-male groups or 'units' as Kummer calls them, the male closely herds his 3 or 4 females. They move with him at all times and are not allowed to stray. Unattached females are rapidly acquired by males, so it is more striking that there is scarcely any poaching of females between units. This remains the case even if one male is completely dominant over another in terms of access to food or resting places—the subordinate's females remain his own.

Kummer, Götz and Angst[283] set up some simple cage experiments using wild caught baboons. If two males were kept together and an unattached female introduced, they would threaten and sometimes fight each other but eventually the female became attached to one male—usually, of course, the dominant one. However if, in full sight of the dominant male, the subordinate was briefly caged alone with a female, he would pair with her. Now when introduced to the cage the dominant would make no attempt to capture the female. Further, such a male introduced to an established pair appears ill at ease and tends to sit with his head turned away from them. There is complete social inhibition of poaching between males of a troop.

The herding of females into a close unit is accomplished by threat. If as she forages a female strays too far from the male he first threatens her and, if she does not return to his side, chases and bites her. Customarily females return to the male at once when he 'stares' at them—a low intensity threat gesture. The yellow baboon, *Papio cynocephalus* is a close relative, and in fact hybridizes with *P. hamadryas*. This is the baboon studied by Hall and De Vore, whose troop structure was described above. Females in *cynocephalus* troops do not associate with particular males save at their brief oestrus period when they form a temporary consort relationship with a dominant male. For most of the time they move freely and associate with many different members of the troop.

Kummer[282] transplanted a few *cynocephalus* females into a troop of *hamadryas*. Very quickly males began to herd them into their own units. At first the females wandered away and, if they responded at all to a male's threat, it was to flee. This of course produced the opposite effect to that which they intended. The male instantly pursued them and drove them back to his harem. Kummer found that within a few hours the *cynocephalus* females had learnt their harsh lesson and stayed close to their male.

This result forces us to recognize as a real possibility that some of the variation in social organization which we observe between primate species is of cultural and not genetic origin. The young *hamadryas* baboon grows up in a one male unit, the young *cynocephalus* in the less restricted social climate of the larger group. They may have similar potential and yet develop quite differently. In Chapter 6 we discussed the cultural transmission of minor items of behaviour such as food washing habits, but clearly with the long infancy of primates cultural effects can extend much beyond this.

In fact we have already mentioned considerable variations in behaviour even between different groups of the same species, as with the baboons of open habitats as compared with those from more forested areas (see p. 296). In the langur (*Presbytis entellus*) of India there are conspicuous differences between the organization of some groups in the south and more northerly populations. In the former one-male groups are common with attendant all-male groups, whereas to the north multi-male groups are the rule (see Yoshiba[516]).

How far could such differences be adaptive? The southern populations are much more densely packed into the available habitat, and natural predators have largely been removed by human influence. In addition the climate is more severe than in the north, particularly in summer when there are long periods of drought. Is population density the basic factor behind all the behavioural differences, or does social organization provide more specific adaptations to environmental circumstances? It will require long term studies to provide answers to many of the obvious questions.

In the southern langur populations, which show a one-male group type of social structure, males from outside may try to usurp the male leader and take over his females. If one succeeds in doing so he usually kills any infants in the group (see Sugiyama[448] and Hrdy[236]). This infanticide at first seems wildly maladaptive but it makes sense in terms of the male's own fitness. By removing infants which are being suckled, the females will be brought back into oestrus and the new male can father his own children instead of helping to support the offspring of the previous male.

If primate selfishness is easy to explain in terms of individual selection, kin selection has often been suggested as the basis of much co-operative or apparently altruistic behaviour. For example, Hrdy and Hrdy[237] explain the active defence of their troop by elderly female langurs in such terms. They argue that as fertility declines with age and since the old females will have daughters and grandchildren in the troop, self-sacrificing defence may be a good strategy for perpetuating one's genes.

Certainly if kin selection is to work effectively animals should be able to identify their close relatives and discriminate in favour of them. Primate groups which are long-lived and stable are ideally suited to make this possible. All young primates stay with their mother for several years and the matrilineal group is particularly stable. Hence, even in promiscuous species where it can rarely be possible to identify one's father, it is always clear who are one's half-siblings. In some species with well developed·one-male groups such as the hamadryas baboon, full siblings can be identified with fair certainty. A number of detailed studies are now in progress which examine degrees of co-operation and altruism between individuals of known genetic relationship. Some baboon groups have now been under regular observation for a decade or more, so that we have information on some individuals from birth until well on into their maturity. Since relatives usually spend a lot of time in each other's company anyway, it may be difficult to separate out the effects of genetic affinity from those of close association as such. Nevertheless the question is an important one and we may certainly hope that the primates will help us to understand the evolution of altruism. (See Clutton-Brock and Harvey[102] for a further discussion of this question.)

Finally in this brief survey of primate social behaviour, we may describe one other type of interrelationship which is as good an example as any of the range of variability which they provide. It has frequently been recognized that the infants in a group represent a powerful focus of attention. Many individuals other than its mother approach an infant, attempting to groom and play with it. The degree to which the mother allows such contacts is surprisingly variable. Langur mothers are very permissive and allow other females to take their babies from time to time; sometimes even females from other troops when they intermingle. Rhesus macaque mothers scarcely allow the baby to leave them for several weeks and thereafter restrict its contact with other individuals.

A close relative of the rhesus, the Barbary macaque, behaves quite differently. Mothers allow other individuals take their babies, even within a few days of birth. Young babies can be seen moving about in the group not obviously under the care of any particular adult. Males as well as females seek contact with babies and Deag and Crook[124] describe a most remarkable relationship between males and infants. The start of a sequence is illustrated in Fig. 8.14; a young male encourages an infant to approach and ride on his back. The male will then take the baby up to another male, often a high-ranking adult and 'present' it. The exact function of this behaviour is obscure at present, but its effect is to allow a subordinate animal to stay close to a dominant where, without the baby as a 'buffer', it would not be permitted.

It is interesting to speculate about the effect upon the infants development of the permissive type of regime in the Barbary macaque as compared

Fig. 8.14 A subadult male Barbary macaque and a baby (the same animals were seen in Fig. 8.9) in a typical encounter. The two are sitting close to each other (top), the male turns towards the baby and makes the 'teeth-chatter' face—a friendly expression—and the baby approaches him. The baby leaps onto his back, to be carried off and perhaps 'presented' to a higher ranking male. Further details of such behaviour are given above. (Photos by John Deag.)

to the more restricted maternal care of the rhesus. Analogies with diverse human child-rearing practices spring to mind—perhaps too readily.

Certainly, quite apart from any light they may throw on the origins of human social life (see reviews by Jay,[256] and Washburn and De Vore[485]) the primates offer marvellous material for the study of behavioural development. The experimental approach, which must include cross-fostering infants and cultural transplantation, has scarcely begun. If we allow the primates room to live in the natural habitats to which they have adapted, a combination of laboratory studies and field work is certain to yield fascinating and important results.

References

1. ADLER, N. T. (1974). The behavioral control of reproductive physiology. In *Reproductive Behaviour*, ed. MONTAGNA, W., pp. 259–86. Plenum Publishing Corporation, New York.
2. ALEXANDER, R. D. (1960). Sound communication in Orthoptera and Cicadidae. In *Animal Sounds and Communication*, ed. LANYON, W. E. and TAVOLGA, W. N., pp. 38–92. Publ. No. 7, Amer. Inst. Biol. Sci., Washington D.C.
3. ALEXANDER, R. D. (1962). Evolutionary change in cricket acoustical communication. *Evolution*, **16**, 443–67.
4. ALEXANDER, R. D. (1962). The role of behavioral study in cricket classification. *Syst. Zool.*, **11**, 53–72.
5. ALEXANDER, R. D. and BIGELOW, R. S. (1960). Allochronic speciation in field crickets and a new species, *Acheta veletis*. *Evolution*, **14**, 334–46.
6. ALLEE, W. C. (1938). *The Social Life of Animals*. Norton, New York.
7. ALTMANN, S. A., ed. (1967). *Social Communication among Primates*. University of Chicago Press, Chicago and London.
8. ALTMANN, S. A. (1962). A field study of the sociobiology of rhesus monkeys, *Macaca mulatta*. *Ann. N.Y. Acad. Aci.*, **102**, 338–435.
9. ANDERSSON, B. (1953). The effect of injections of hypertonic NaCl solutions into different parts of the hypothalamus of goats. *Acta physiol. scand.*, **28**, 188–201.
10. ANDERSSON, B. and MCCANN, S. M. (1955). A further study of polydipsia evoked by hypothalamic stimulation in the goat. *Acta physiol. scand.*, **33**, 333–46.
11. ANDERSSON, B. and MCCANN, S. M. (1955). Drinking, antidiuresis and milk ejection from electrical stimulation within the hypothalamus of the goat. *Acta physiol. scand.*, **35**, 191–201.
12. ANDREW, R. J. (1956). Some remarks on behaviour in conflict situations, with special reference to *Emberiza* spp. *Br. J. Anim. Behav.*, **4**, 41–5.
13. ANDREW, R. J. (1956). Intention movements of flight in certain passerines and their use in systematics. *Behaviour*, **10**, 179–204.
14. ANDREW, R. J. (1969). The effects of testosterone on avian vocalizations. In *Bird Vocalizations*, pp. 97–130 (see Ref. 216).
15. ANDREW, R. J. (1972). The information potentially available in mammal displays. In *Non Verbal Communication*, ed. HINDE, R. A., pp. 179–206. Cambridge University Press, London.
16. ARCHER, J. (1973). Tests for emotionality in rats and mice: a review. *Anim. Behav.*, **21**, 205–35.
17. ARDREY, R. (1966). *The Territorial Imperative*. Atheneum, New York.
18. AUSTIN, C. R. and SHORT, R. V. (1972). *Reproduction in Mammals*. Book 3. *Hormones in Reproduction*. Cambridge University Press, London.
19. AUSTIN, C. R. and SHORT, R. V. (1972). *Reproduction in Mammals*. Book 4. *Reproductive Patterns*. Cambridge University Press, London.

20. BAERENDS, G. P. (1975). An evaluation of the conflict hypothesis as an explanatory principle for the evolution of displays. In *Function and Evolution in Behaviour*, pp. 187–227 (see Ref. 23).
21. BAERENDS, G. P. (1976). The functional organization of behaviour. *Anim. Behav.*, **24**, 726–38.
22. BAERENDS, G. P. and BAERENDS-VAN ROON, J. M. (1950). An introduction to the study of the ethology of cichlid fishes. *Behaviour Suppl.*, **1**, 1–243.
23. BAERENDS, G., BEER, C. and MANNING, A. eds. (1976). *Function and Evolution in Behaviour*. Clarendon Press, Oxford.
24. BAERENDS, G. P. and KRUIJT, J. P. (1973). Stimulus selection. In HINDE, R. A. and STEVENSON-HINDE, J., eds. *Constraints on Learning*, pp. 23–50. Academic Press, London and New York.
25. BARLOW, G. W. (1977). Modal action patterns. In *How Animals Communicate*, ed. SEBEOK, T. N., pp. 98–134. University of Indiana Press, Bloomington.
26. BARLOW, H. B. (1961). The coding of sensory messages. In *Current Problems in Animal Behaviour*, pp. 331–60 (see Ref. 459).
27. BARNETT, S. A. (1958). Exploratory behaviour. *Br. J. Psychol.*, **49**, 289–310.
28. BARNETT, S. A. (1963). *A Study in Behaviour*. Methuen, London.
29. BARNETT, S. A. (1964). The concept of stress. *Viewpoints in Biology*, **3**, 170–218. Butterworth, London.
30. BARNETT, S. A. and BURN, J. (1967). Early stimulation and maternal behaviour. *Nature, Lond.*, **213**, 150–52.
31. BASTOCK, M. (1956). A gene mutation which changes a behavior pattern *Evolution*, **10**, 421–39.
32. BASTOCK, M. (1967). *Courtship: a Zoological Study*. Heinemann Educational, London.
33. BASTOCK, M. and MANNING, A. (1955). The courtship of *Drosophila melanogaster*. *Behaviour*, **8**, 85–111.
34. BATESON, P. P. G. (1964). Effect of similarity between rearing and testing conditions on chicks' following and avoidance responses. *J. comp. physiol. Psychol.*, **57**, 100–3.
35. BATESON, P. P. G. (1966). The characteristics and context of imprinting. *Biol. Rev.*, **41**, 177–220.
36. BATESON, P. P. G. (1971). Imprinting. In *Ontogeny of Vertebrate Behavior*, ed. MOLTZ, H., pp. 369–78. Academic Press, New York and London.
37. BATESON, P. P. G. (1976). Specificity and the origins of behavior. *Advances in the Study of Behavior*, **6**, 1–20.
38. BATESON, P. P. G. (1978). Early experience and sexual preferences. In *Biological Determinants of Sexual Behaviour*, ed. HUTCHISON, J. B., pp. 29–53. Wiley, London.
39. BATESON, P. P. G. and REESE, E. P. (1968). Reinforcing properties of conspicuous objects before imprinting has occurred. *Psychon. Sci.*, **10**, 379–80.
40. BATESON, P. P. G. and REESE, E. P. (1969). The reinforcing properties of conspicuous stimuli in the imprinting situation. *Anim. Behav.*, **17**, 692–99.
41. BEACH, F. A. (1942). Analysis of the stimuli adequate to elicit mating behavior in the sexually inexperienced male rat. *J. comp. Psychol.*, **33**, 163–207.

42. BEACH, F. A. (1950). The Snark was a Boojum. *Am. Psychol.*, **5**, 115–24.
43. BEACH, F. A. and LEVINSON, G. (1950). Effects of androgen on the glans penis and mating behavior of castrated male rats. *J. exp. Zool.*, **114**, 159–71.
44. BEER, C. G. (1975). Was Professor Lehrman an ethologist? *Anim. Behav.*, **23**, 957–64.
45. BEKOFF, M. (1977). Social communication in canids: evidence for the evolution of a stereotyped mammalian display. *Science, N.Y.*, **197**, 1097–99.
46. BELLOWS, R. T. (1939). Time factors in water-drinking in dogs. *Am. J. Physiol.*, **125**, 87–97.
47. BENNET-CLARK, H. C. (1970). The mechanism and efficiency of sound production in mole crickets. *J. exp. Biol.*, **52**, 619–52.
48. BENNET-CLARK, H. C. (1970). A new French mole cricket, differing in song and morphology from *Gryllotalpa gryllotalpa* L. (Orthoptera: Gryllotalpidae). *Proc. R. ent. Soc. Lond.* (B), 39, 125–32.
49. BENTLEY, D. R. and HOY, R. R. (1970). Postembryonic development of adult motor patterns in crickets: a neural analysis. *Science, N.Y.*, **170**, 1409–11.
50. BENZER, S. (1973). Genetic dissection of behavior. *Scient. Amer.*, **229** (6), 24–37.
51. BITTERMAN, M. E. (1965). The evolution of intelligence. *Scient. Am.*, **212** (1), 92–100.
52. BLAIR, W. F. (1955). Mating call and stage of speciation in the *Microhyla olivacea—M. carolinensis* complex. *Evolution*, **9**, 469–80.
53. BLAIR, W. F. (1958). Mating call in the speciation of anuran amphibians. *Am. Nat.*, **92**, 27–51.
54. BLAKEMORE, C. and COOPER, G. F. (1970). Development of the brain depends on the visual environment. *Nature, Lond.*, **228**, 477–78.
55. BLEST, A. D. (1957). The function of eyespot patterns in the Lepidoptera. *Behaviour*, **11**, 209–56.
56. BLEST, A. D. (1957). The evolution of protective displays in the Saturnioidea and Sphingidae (Lepidoptera). *Behaviour*, **11**, 257–309.
57. BLEST, A. D. (1961). The concept of ritualization. In *Current Problems in Animal Behaviour*, pp. 102–24 (see Ref. 459).
58. BLÜM, V. and FIEDLER, K. '(1965). Hormonal control of reproductive behavior in some cichlid fish. *Gen. Comp. Endocrinol.*, **5**, 186–96.
59. BLURTON-JONES, N. G. (1959). Experiments on the causation of the threat postures of Canada geese. *Rep. Severn Wildfowl Trust*, 1960, 46–52.
60. BLURTON-JONES, N. G. (1968). Observations and experiments on causation of threat displays of the great tit (*Parus major*). *Anim. Behav. Monogr.*, **1**, 75–158.
61. BOWLBY, J. (1969). *Attachment and Loss, I. Attachment*. Hogarth, London.
62. BOWLBY, J. (1973). *Attachment and Loss, II. Separation*. Hogarth, London.
63. BOYD, H. and FABRICIUS, E. (1965). Observations on the incidence of following of visual and auditory stimuli in naïve mallard ducklings (*Anas platyrhynchos*). *Behaviour*, **25**, 1–15.

64. BRADDOCK, J. C. and BRADDOCK, Z. I. (1959). Development of nesting behaviour in the Siamese fighting fish *Betta splendens*. *Anim. Behav.*, **7**, 222–32.

66. BRELAND, K. and BRELAND, M. (1961). The misbehavior of organisms. *Am. Psychol.*, **16**, 681–84.

67. BROADBENT, D. E. (1961). *Behaviour*. Eyre and Spottiswoode, London.

68. BROADHURST, P. L. (1969). Psychogenetics of emotionality in the rat. *Ann. N.Y. Acad. Sci.*, **159**, 806–24.

69. BROADHURST, P. L. (1975). The Maudsley reactive and nonreactive strains of rats: a survey. *Behav. Genet.*, **5**, 299–319.

70. BROADHURST, P. L. and LEVINE, S. (1963). Behavioural consistency in strains of rats selectively bred for emotional elimination. *Br. J. Psychol.*, **54**, 121–25.

71. BRONSON, F. H. and DESJARDINS, C. (1968). Aggression in adult mice: modification by neonatal injections of gonadal hormones. *Science*, *N.Y.*, **161**, 705–6.

72. BROWN, J. L. (1964). The integration of agonistic behavior in the Steller's jay *Cyanocitta stelleri* (Gmelin). *Univ. Calif. Publ. Zool.*, **60**, 223–328.

73. BROWN, J. L. (1975). *The Evolution of Behavior*. Norton, New York.

74. BROWN, J. L. and HUNSPERGER, R. W. (1963). Neuroethology and the motivation of agonistic behaviour. *Anim. Behav.*, **11**, 439–48.

75. BRUNER, J. and KENNEDY, D. (1970). Habituation: occurrence at a neuromuscular junction. *Science*, *N.Y.*, **169**, 92–94.

76. BUCKLEY, P. A. (1969). Disruption of species-typical behaviour patterns in F_1 hybrid *Agapornis* parrots. *Z. Tierpsychol.*, **26**, 737–43.

77. BUCKLEY, P. A. and BUCKLEY, F. G. (1972). Individual egg and chick recognition by adult royal terns (*Sterna maxima maxima*). *Anim. Behav.*, **20**, 457–62.

78. BULL, H. O. (1957). Conditioned responses. In *The Physiology of Fishes*, 2 vols. ed. BROWN, M. E., pp. 211–28. Academic Press, New York.

79. BURGHARDT, G. M. (1970). Defining 'communication'. In *Advances in Chemoreception*, Vol. 1, eds. JOHNSTON, J. W., MOULTON, D. G. and TURK, A., pp. 5–18. Appleton Century Crofts, New York.

80. BURGHARDT, G. M. (1971). Chemical-cue preferences of newborn snakes: influence of prenatal maternal experience. *Science, N.Y.*, **171**, 921–23.

81. BURGHARDT, G. M. and PRUITT, C. H. (1975). Role of the tongue and senses in feeding of naïve and experienced garter snakes. *Physiol. Behav.*, **14**, 185–94.

82. BUTLER, C. G. (1974). *The World of the Honeybee*, 3rd edn. Collins, London.

83. CARMICHAEL, L. (1926). The development of behavior in vertebrates experimentally removed from the influence of external stimulation. *Psychol. Rev.*, **33**, 51–58.

84. CARMICHAEL, L. (1927). A further study of the development of behavior in vertebrates experimentally removed from the influence of external stimulation. *Psychol. Rev.*, **34**, 34–47.

85. CARPENTER, C. R. (1965). The howlers of Barro Colorado island. In *Primate Behavior*, pp. 250–91 (see Ref. 135).

86. CARTER, C. S. and MARR, J. N. (1970). Olfactory imprinting and age variables in the guinea-pig, *Cavia porcellus*. *Anim. Behav.*, **18**, 238–44.

87. CARTHY, J. D. and EBLING, F. J., eds. (1964). *The Natural History of Aggression*. Academic Press, London and New York.
88. CHALMERS, N. R. (1968). The visual and vocal communication of free living mangabeys in Uganda. *Folia primat.*, **9**, 258–80.
89. CHANCE, E. P. (1940). *The Truth about the Cuckoo*. Country Life, London.
90. CHANCE, M. R. A. (1962). An interpretation of some agonistic postures: the role of 'cut-off' acts and postures. *Symp. zool. Soc. Lond.*, **8**, 71–89.
91. CHENG, M-F. (1973). Effect of ovariectomy on the reproductive behavior of female ring doves (*Streptopelia risoria*). *J. Comp. Physiol. Psychol.*, **83**, 221–33.
92. CHENG, M-F. (1973). Effect of estrogen on the behavior of ovariectomized ring doves (*Streptopelia risoria*). *J. Comp. Physiol. Psychol.*, **83**, 234–39.
93. CHITTY, D. (1960). Population processes in the vole and their relevance to general theory, *Can. J. Zool.*, **38**, 99–113.
94. CHRISTIAN, J. J. (1971). Population density and reproductive efficiency. *Biol. Reprod.*, **4**, 248–94.
95. CLARK, E., ARONSON, L. and GORDON, M. (1954). Mating behavior patterns in two sympatric species of xiphophorin fishes: their inheritance and significance in sexual isolation. *Bull. Am. Mus. nat. Hist.*, **103**, 141–225.
96. CLARK, R. B. (1960). Habituation of the polychaete *Nereis* to sudden stimuli. 1. General properties of the habituation process. *Anim. Behav.*, **8**, 82–91.
97. CLARK, R. B. (1960). Habituation of the polychaete *Nereis* to sudden stimuli. 2. Biological significance of habituation. *Anim. Behav.*, **8**, 92–103.
98. CLARK, R. B. (1965). The learning abilities of Nereid polychaetes and the role of the supra-oesophageal ganglion. *Anim. Behav. Suppl.*, **1**, 89–100.
99. CLARK, W. E. LE GROS (1965). *History of the Primates*, 9th edn. Brit. Mus. (Nat. Hist.), London.
100. CLAYTON, F. L. and HINDE, R. A. (1968). The habituation and recovery of aggressive display in *Betta splendens*. *Behaviour*, **30**, 96–106.
101. CLUTTON-BROCK, T. H. (1974). Primate social organisation and ecology. *Nature, Lond.*, **250**, 539–42.
102. CLUTTON-BROCK, T. H. and HARVEY, P. H. (1976). Evolutionary rules and primate societies. In *Growing Points in Ethology*, eds. BATESON, P. P. G. and HINDE, R. A., pp. 195–237. Cambridge University Press, London.
103. COONS, E. E., LEVAK, M. and MILLER, N. E. (1965). Lateral hypothalamus: learning of food-seeking response motivated by electrical stimulation. *Science, N.Y.*, **150**, 1320–21.
104. COULSON, J. C. (1966). The influence of the pair-bond and age on the breeding biology of the kittiwake gull *Rissa tridactyla*. *J. Anim. Ecol.*, **35**, 269–79.
105. COULSON, J. C. (1968). Differences in the quality of birds nesting in the centre and on the edges of a colony. *Nature, Lond.*, **217**, 478–79.
106. CRANE, J. (1957). Basic patterns of display in fiddler crabs (*Ocypodidae*, genus *Uca*). *Zoologica*, **42**, 69–82.
107. CROOK, J. H. (1965). The adaptive significance of avian social organizations. *Symp. zool. Soc. Lond.*, **14**, 181–218.

108. CROOK, J. H. (1966). Gelada baboon herd structure and movement: a comparative report. *Symp. zool. Soc. Lond.*, **18**, 237–58.

109. CROOK, J. H. (1970). Social organization and the environment: aspects of contemporary social ethology. *Anim. Behav.*, **18**, 197–209.

110. CROOK, J. H. (1970). The socio-ecology of primates. In *Social Behaviour in Birds and Mammals*, ed. CROOK, J. H., pp. 103–66. Academic Press, London and New York.

111. CRUZE, W. W. (1935). Maturation and learning in chicks. *J. comp. Psychol.*, **19**, 371–409.

112. CULLEN, E. (1957). Adaptations in the kittiwake to cliff-nesting. *Ibis*, **99**, 275–302.

113. CULLEN, E. (1960). Experiment on the effect of social isolation on reproductive behaviour in the three-spined stickleback. *Anim. Behav.*, **8**, 235.

114. CULLEN, J. M. (1962). The pecking response of young wideawake terns, *Sterna fuscata. Ibis*, **103**, 162–70.

115. CULLEN, J. M. and ASHMOLE, N. P. (1963). The Black Noddy (*Anous tenuirostris*) on Ascension Island. Part 2. Behaviour. *Ibis*, **103**, 423–46.

116. DAANJE, A. (1950). On locomotory movements in birds and the intention movements derived from them. *Behaviour*, **3**, 48–98.

117. DALY, M. (1973). Early stimulation of rodents: a critical review of present interpretations. *Brit. J. Psychol.*, **64**, 435–60.

118. DARLING, F. F. (1935). *A Herd of Red Deer*. Oxford University Press, London.

119. DARLING, F. F. (1938). *Bird Flocks and the Breeding Cycle*. Cambridge University Press, Cambridge.

120. DAVIS, D. E. (1958). The role of density in aggressive behaviour of house mice. *Anim. Behav.*, **6**, 207–10.

121. DAVIS, W. J. (1976). Organizational concepts in the central motor networks of invertebrates. In *Neural Control of Locomotion*, eds. HERMAN, R. M., GRILLNER, S., STEIN, P. S. G. and STUART, D. G., pp. 265–92. Plenum, New York.

122. DAWKINS, R. (1976). *The Selfish Gene*. Oxford University Press, London.

123. DEAG, J. M. (1977). Aggression and submission in monkey societies. *Anim. Behav.*, **25**, 465–74.

124. DEAG, J. M. and CROOK, J. H. (1971). Social behaviour and 'agonistic buffering' in the wild Barbary macaque *Macaca sylvana* L. *Folia Primat.*, **15**, 183–200.

125. DEAUX, E. and KAKOLEWSKI, J. W. (1970). Emotionally induced increases in effective osmotic pressure and subsequent thirst. *Science, N.Y.*, **169**, 1226–28.

126. DELIUS, J. D. (1973). Agonistic behaviour of juvenile gulls, a neuroethological study. *Anim. Behav.*, **21**, 236–46.

127. DENENBERG, V. H. (1964). Critical periods, stimulus input, and emotional reactivity: a theory of infantile stimulation. *Psychol. Rev.*, **71**, 335–51.

128. DENENBERG, V. H. and ROSENBERG, K. M. (1967). Nongenetic transmission of information. *Nature, Lond.*, **216**, 549–60.

129. DENNY, M. R. and RATNER, S. C. (1970). *Comparative Psychology. Research in Animal Behavior*. Revised Edition. The Dorsey Press, Homewood, Illinois.

130. DETHIER, V. G. (1953). Summation and inhibition following contra-lateral stimulation of the tarsal chemoreceptors of the blowfly. *Biol. Bull.*, **105**, 257–68.

131. DETHIER, V. G. (1957). Communication by insects: physiology of dancing. *Science, N. Y.*, **125**, 331–36.

132. DETHIER, V. G. (1964). Microscopic brains. *Science, N. Y.*, **143**, 1138–45.

133. DETHIER, V. G. and STELLAR, E. (1970). *Animal Behavior*, 3rd edn. Prentice-Hall, Englewood Cliffs, New Jersey.

134. DEUTSCH, J. A. (1960). *The Structural Basis of Behavior*. Cambridge University Press, Cambridge.

135. DE VORE, I., ed. (1965). *Primate Behavior. Field Studies of Monkeys and Apes*. Holt, Rinehart and Winston, New York and London.

136. DILGER, W. C. (1962). The behavior of lovebirds. *Scient. Am.*, **206**, 88–98.

137. DIMOND, S. J. (1970). *The Social Behaviour of Animals*. Batsford, London

138. DOUGLAS-HAMILTON, I. and DOUGLAS-HAMILTON, O. (1975). *Among the Elephants*. Collins & Harvill, London.

139. D'SOUZA, F. and MARTIN, R. D. (1974). Maternal behaviour and the effects of stress in tree-shrews. *Nature, Lond.*, **251**, 309–11.

140. EATON, R. L. (1970). Group interactions, spacing and territoriality in cheetahs. *Z. Tierpsychol.*, **27**, 481–91.

141. EHRHARDT, A. A. and BAKER, S. W. (1974). Fetal androgens, human central nervous system differentiation, and behavior sex differences. In *Sex Differences in Behavior*, eds. FRIEDMAN, R. C., RICHART, R. M. and VAN DE WIELE, R. L., pp. 33–51. Wiley, New York.

142. EHRMAN, L. and PARSONS, P. A. (1976). *The Genetics of Behavior*. Sinauer Associates, Sunderland, Mass., Freeman, Reading.

143. EIBL-EIBESFELDT, I. (1970). *Ethology. The Biology of Behavior*. Holt, Rinehart and Winston, New York.

144. EMLEN, S. T. and ORING, L. W. (1977). Ecology, sexual selection and the evolution of mating systems. *Science, N. Y.*, **197**, 215–24.

145. ERICKSON, J. G. (1967). Social hierarchy, territoriality, and stress reactions in sunfish. *Physiol. Zool.*, **40**, 40–48.

146. ESCH, H., ESCH, I. and KERR, W. E. (1965). Sound: an element common to communication of stingless bees and to dances of honeybees. *Science, N. Y.*, **149**, 320–21.

147. ESTES, R. D. (1969). Territorial behavior of the wildebeeste (*Connochaetes taurinus* Burchell, 1823). *Z. Tierpsychol.*, **26**, 284–370.

148. EVANS, S. M. (1968). *Studies in Invertebrate Behaviour*. Heinemann Educational, London.

149. EWER, R. F. (1968). *Ethology of Mammals*. Logos Press Limited, London.

150. EWING, A. W. (1969). The genetic basis of sound production in *Drosophila pseudoobscura* and *D. persimilis*. *Anim. Behav.*, **17**, 555–60.

151. EWING, A. W. and BENNET-CLARK, H. C. (1968). The courtship songs of *Drosophila*. *Behaviour*, **31**, 288–301.

152. FRAENKEL, G. and GUNN, D. L. (1964). *The Orientation of Animals*. Dover Publications, London.

153. FRANCK, D. (1974). The genetic basis of evolutionary changes in behaviour patterns. In *The Genetics of Behaviour*, ed. VAN ABEELEN, J. H. F., pp. 119–40. North Holland, Amsterdam.

154. FRANZISKET, L. (1953). Untersuchungen zur Spezifität und Kumulierung der Erregungsfähigkeit und zur Wirkung einer Ermüdung in der Afferenz bei Wischbewegung des Rückenmarkfrosches. *Z. vergl. Physiol.*, **34**, 525–38.

155. FRISCH, J. E. (1968). Individual behavior and intertroop variability in Japanese macaques. In *Primates: Studies in Adaptation and Variability*, pp. 243–52 (see Ref. 257).

156. FRISCH, K. von, (1967). *The Dance Language and Orientation of Bees.* The Belknap Press of Harvard University Press, Cambridge, Mass.

157. FULLER, J. L. and THOMPSON, W. R. (1978). *Foundations of Behavior Genetics.* C. V. Mosby, St. Louis, Missouri.

158. GALAMBOS, R. (1961). Changing concepts of the learning mechanism. In *Brain Mechanisms and Learning*, ed. DELAFRESNAYE, J. F., pp. 231–41. Blackwell Scientific, Oxford.

159. GALEF, B. G. (1976). Social transmission of acquired behavior: a discussion of tradition and social learning in vertebrates. *Adv. Study Behav.*, **6**, 77–100.

160. GALUSHA, J. G. and STOUT, J. F. (1977). Aggressive Communication by *Larus glaucescens*. Part IV: Experiments on visual communication. *Behaviour*, **52**, 222–35.

161. GARCIA, J. and KOELLING, R. (1966). Relation of cue to consequence in avoidance learning. *Psychonom. Sci.*, **4**, 123–24.

162. GARDNER, B. T. (1964). Hunger and sequential responses in the hunting behavior of salticid spiders. *J. comp. physiol. Psychol.*, **58**, 167–73.

163. GARDNER, B. T. and GARDNER, R. A. (1975). Evidence for sentence constituents in the early utterances of child and chimpanzee. *J. exp. Psychol.*, **104**, 244–67.

164. GARDNER, R. A. and GARDNER, B. T. (1969). Teaching sign language to a chimpanzee. *Science, N.Y.*, **165**, 664–72.

165. GARDNER, R. A. and GARDNER, B. T. (1975). Early signs of language in child and chimpanzee. *Science, N.Y.*, **187**, 752–53.

166. GEIST, V. (1966). Evolution of horn-like organs. *Behaviour*, **27**, 175–214.

167. GELLERMAN, L. W. (1933). Form discrimination in chimpanzees and two-year-old children: I form (triangularity) *per se. J. genet. Psychol.*, **42**, 3–27.

168. GILBERT, R. M. and SUTHERLAND, N. S., eds. (1969). *Animal Discrimination Learning.* Academic Press, London and New York.

169. GOODALL, J. van LAWICK (1968). The behaviour of free-living chimpanzees in the Gombe Stream reserve. *Anim. Behav. Monogr.*, **1** (3), 161–311.

170. GORBMAN, A. and BERN, H. A. (1962). *A Textbook of Comparative Endocrinology.* Wiley, New York and London.

171. GOTTLIEB, G. (1968). Prenatal behavior of birds. *Q. Rev. Biol.*, **43**, 148–74.

172. GOTTLIEB, G. (1971). *Development of Species Identification in Birds.* University of Chicago Press, Chicago.

173. GOULD, J. L. (1976). The dance-language controversy. *Q. Rev. Biol.*, **51**, 211–44.

174. GOVE, D. and BURGHARDT, G. M. (1975). Responses of ecologically dissimilar populations of the water snake *Natrix s. sipedon* to chemical cues from prey. *J. Chem. Ecol.*, **1**, 25–40.

175. GOY, R. W. and JAKWAY, J. S. (1962). Role of inheritance in determination of sexual behavior patterns. In *Roots of Behavior*, ed. BLISS, E. L., pp. 96–112. Harper, New York.

176. GREEN, M., GREEN, R. and CARR, W. J. (1966). The hawk-goose phenomenon: A replication and an extension. *Psychon. Sci.*, **4**, 185–86.

177. GREEN, R., CARR, W. J. and GREEN, M. (1968). The hawk-goose phenomenon: Further confirmation and a search for the releaser. *J. Psychol.*, **69**, 271–76.

178. GREGORY, R. L. (1961). The brain as an engineering problem. In *Current Problems in Animal Behaviour*, pp. 307–30 (see Ref. 459).

179. GRIFFIN, D. R. (1958). *Listening in the Dark*. Yale University Press, New Haven.

180. GRIFFIN, D. R. (1976). *The Question of Animal Awareness*. Rockefeller University Press, New York.

181. GROSSMAN, S. P. (1960). Eating or drinking elicited by direct adrenergic or cholinergic stimulation of the hypothalamus, *Science*, *N.Y.*, **132**, 301–2.

182. GROSSMAN, S. P. (1967). *A Textbook of Physiological Psychology*. John Wiley & Sons Inc., New York and London.

183. GROSSMAN, S. P. (1968). The physiological basis of specific and non-specific motivational processes. *Nebr. Symp. Motiv.*, **16**, 1–46.

184. GROSSMAN, S. P. (1972). Aggression, avoidance and reaction to novel environments in female rats with ventromedial hypothalamic lesions. *J. comp. Physiol. Psychol.*, **78**, 274–83.

185. GRUNT, J. A. and YOUNG, W. C. (1952). Differential reactivity of individuals and the response of the male guinea pig to testosterone propionate. *Endocrinology*, **51**, 237–48.

186. GUILMET, G. M. (1977). The evolution of tool-using and tool-making behaviour. *Man*, **12**, 33–47.

187. GUITON, P. (1959). Socialisation and imprinting in brown leghorn chicks. *Anim. Behav.*, **7**, 26–34.

188. GUITON, P. (1962). The development of sexual responses in the domestic fowl, in relation to the concept of imprinting. *Symp. zool. Soc. Lond.*, **8**, 227–34.

189. GUSTAVSON, C. R., KELLY, D. J., SWEENEY, M. and GARCIA, J. (1976). Prey-lithium aversions. I: Coyotes and wolves. *Behav. Biol.*, **17**, 61–72.

190. HAILMAN, J. P. (1962). Pecking of laughing-gull chicks to models of the parental head. *Auk*, **79**, 89–98.

191. HAILMAN, J. P. (1965). Cliff-nesting adaptations of the Galapagos swallow-tailed gull. *Wilson Bull.*, **77**, 346–62.

192. HAILMAN, J. P. (1967). The ontogeny of an instinct. The pecking response in chicks of the laughing gull (*Larus atricilla* L.) and related species. *Behaviour Suppl.*, **15**, 1–159.

193. HALDANE, J. B. S. (1954). A statistical analysis of communication in 'Apis mellifera' and a comparison with communication in other animals. *Insectes Sociaux*, **1**, 247–83.

194. HALE, E. B. (1966). Visual stimuli and reproductive behavior in bulls. *J. Anim. Sci.*, **25** (*Suppl.*), 36–44.

195. HALL, K. R. L. and DE VORE, I. (1965). Baboon social behavior. In *Primate Behavior. Field Studies of Monkeys and Apes*, pp. 53–110 (see Ref. 135).

196. HALL-CRAGGS, J. (1962). The development of song in the blackbird (*Turdus merula*). *Ibis*, **104**, 277–300.
197. HALLIDAY, M. S. (1966). Effect of previous exploratory activity on the exploration of a simple maze. *Nature, Lond.*, **209**, 432–33.
198. HALLIDAY, T. R. and SWEATMAN, H. P. A. (1976). To breathe or not to breathe; the newt's problem. *Anim. Behav.*, **24**, 551–61.
199. HAMILTON, W. D. (1964). The genetical evolution of social behaviour. I and II. *J. Theoret. Biol.*, **7**, 1–52.
200. HARDY, A. (1960). Was man more aquatic in the past? *New Scientist*, **7**, 642–45.
201. HARLOW, H. F. (1949). The formation of learning sets. *Psychol. Rev.*, **56**, 51–65.
202. HARLOW, H. F. (1958). The evolution of learning. In *Behavior and Evolution*, ed. ROE, A. and SIMPSON, G. G., pp. 269–90. Yale University Press, New Haven.
203. HARLOW, H. F. and HARLOW, M. K. (1965). The affectional systems. In *Behavior of Nonhuman Primates*, ed. SCHRIER, A. M., HARLOW, H. F. and STOLLNITZ, F., pp. 287–321. Academic Press, New York and London.
204. HARRIS, G. W. and MICHAEL, R. P. (1964). The activation of sexual behaviour by hypothalamic implants of oestrogen. *J. Physiol. Lond.*, **171**, 275–301.
205. HAUSFATER, G. (1975). Dominance and reproduction in baboons (*Papio cynocephalus*). *Contributions to Primatology*, **7**, 1–150. Karger, Basel.
206. HEBB, D. O. (1958). Alice in Wonderland or psychology among the biological sciences. In *Biological and Biochemical Bases of Behavior*, ed. HARLOW, H. F. and WOOLSEY, C. N., pp. 451–67. University of Wisconsin Press, Madison.
207. HEBB, D. O. and WILLIAMS, K. (1946). A method of rating animal intelligence. *J. gen. Psychol.*, **34**, 59–65.
208. HEILIGENBERG, W. (1963). Ursachen für das Auftreten von Instinktsbewegungen bei einem Fische (*Pelmatochromis subocellatus kribensis*). *Z. vergl. Physiol.*, **47**, 339–80.
209. HESS, E. H. (1964). Imprinting in birds. *Science, N.Y.*, **146**, 1128–39.
210. HINDE, R. (1954). Factors governing the changes in strength of a partially inborn response, as shown by the mobbing behaviour of the chaffinch (*Fringilla coelebs*). I. The nature of the response, and an examination of its course. *Proc. R. Soc. B*, **142**, 306–31.
211. HINDE, R. (1954). Factors governing the changes in strength of a partially inborn response, as shown by the mobbing behaviour of the chaffinch (*Fringilla coelebs*). II. The waning of the response. *Proc. R. Soc. B*, **142**, 331–58.
212. HINDE, R. A. (1958). The nest-building behaviour of domesticated canaries. *Proc. zool. Soc. Lond.*, **131**, 1–48.
213. HINDE, R. A. (1959). Unitary drives. *Anim. Behav.*, **7**, 130–41.
214. HINDE, R. A. (1960). Factors governing the changes in strength of a partially inborn response, as shown by the mobbing behaviour of the chaffinch (*Fringilla coelebs*). III. The interaction of short-term and long-term incremental and decremental effects. *Proc. R. Soc. B*, **153**, 398–420.
215. HINDE, R. A. (1960). Energy models of motivation. *Symp. Soc. exp. Biol.*, **14**, 199–213.

216. HINDE, R. A., ed. (1969). *Bird Vocalizations*. Cambridge University Press, London.

217. HINDE, R. A. (1970). *Animal Behaviour*, 2nd. edn. McGraw-Hill, New York and London.

218. HINDE, R. A., ed. (1972). *Non-verbal Communication*. Cambridge University Press, London.

219. HINDE, R. A. (1974). *Biological Bases of Human Social Behaviour*. McGraw-Hill, New York and London.

220. HINDE, R. A. and FISHER, J. (1952). Further observations on the opening of milk bottles by birds. *Brit. Birds*, **44**, 393–96.

221. HINDE, R. A. and ROWELL, T. E. (1962). Communication by postures and facial expressions in the rhesus monkey (*Macaca mulatta*). *Proc. Zool. Soc. Lond.*, **138**, 1–21.

222. HINDE, R. A. and STEEL, E. (1966). Integration of the reproductive behaviour of female canaries. *Symp. Soc. exp. Biol.*, **20**, 401–26.

223. HINDE, R. A. and STEVENSON, J. C. (1970). Goals and response control. In *Development and Evolution of Behavior*, ed. ARONSON, L. R., pp. 216–37 W. H. Freeman, New York.

224. HINDE, R. A. and STEVENSON-HINDE, J., eds. (1973). *Constraints on Learning*. Academic Press, London and New York.

225. HINDE, R. A. and TINBERGEN, N. (1958). The comparative study of species-specific behavior. In *Behavior and Evolution*, ed. ROE, A. and SIMPSON, G. G., pp. 251–68. Yale University Press, New Haven.

226. HIRSCH, J., LINDLEY, R. H. and TOLMAN, E. C. (1955). An experimental test of an alleged innate sign stimulus. *J. comp. physiol. Psychol.*, **48**, 278–80.

227. HODOS, W. and CAMPBELL, C. B. G. (1969). *Scala Naturae*: Why there is no theory in comparative psychology. *Psychol. Rev.*, **76**, 337–50.

228. HOKANSON, J. E. (1969). *The Physiological Bases of Motivation*. John Wiley & Sons, Inc., New York and London.

229. HÖLLDOBLER, B. (1971). Communication between ants and their guests. *Scient. Amer.*, **224** (3), 86–93.

230. HOLST, E. VON and SAINT-PAUL, U. VON (1963). On the functional organisation of drives. *Anim. Behav.*, **11**, 1–20.

231. HOOKER, T. and HOOKER, B. I. (1969). Duetting. In *Bird Vocalizations*, ed. HINDE, R. A., pp. 185–205. Cambridge University Press, London.

232. HORN, G. and HINDE, R. A. (1970). *Short-term Changes in Neural Activity and Behaviour*. Cambridge University Press, Cambridge.

233. HOTTA, Y. and BENZER, S. (1972). Mapping of behaviour in *Drosophila* mosaics. *Nature, Lond.*, **240**, 527–35.

234. HOTTA, Y. and BENZER, S. (1976). Courtship in *Drosophila* mosaics: sex-specific foci for sequential action patterns. *Proc. Natl. Acad. Sci. U.S.A.*, **73**, 4154–58.

235. HOYLE, G. (1970). Cellular mechanisms underlying behavior-neuro-ethology. *Adv. Ins. Physiol.*, **7**, 349–444.

236. HRDY, S. B. (1974). Male–male competition and infanticide among the langurs (*Presbytis entellus*) of Abu' Rajasthan. *Folia Primat.*, **22**, 19–58.

237. HRDY, S. B. and HRDY, D. B. (1976). Hierarchical relations among female hanuman langurs (Primates: Colobinae, *Presbytis entellus*). *Science, N.Y.*, **193**, 913–15.

238. HUBER, F. (1962). Central nervous control of sound production in crickets and some speculations on its evolution. *Evolution*, **16**, 429–42.

239. HUBER, F. (1967). Central control of movements and behavior of invertebrates. In *Invertebrate Nervous Systems*, ed. WEIRSMA, C. A. G., pp. 333–51. University of Chicago Press, Chicago.

240. HUNSAKER, D. (1962). Ethological isolating mechanisms in the *Sceloporus torquatus* group of lizards. *Evolution*, **16**, 62–74.

241. HUNTINGFORD, F. A. (1976). The relationship between anti-predator behaviour and aggression among conspecifics in the three-spined stickleback, *Gasterosteus aculeatus. Anim. Behav.*, **24**, 245–60.

242. HUNTINGFORD, F. A. (1976). The relationship between inter- and intraspecific aggression. *Anim. Behav.*, **24**, 485–97.

243. HUTCHISON, J. B. (1970). Influence of gonadal hormones on the hypothalamic integration of courtship behaviour in the Barbary dove. *J. Reprod. Fert., Suppl.*, **11**, 15–41.

244. HUTCHISON, J. B. (1976). Hypothalamic mechanisms of sexual behaviour with special reference to birds. *Adv. Study Behaviour*, **6**, 159–200.

245. HUXLEY, J. S. (1914). The courtship habits of the great crested grebe (*Podiceps cristatus*). *Proc. zool. Soc. Lond.*, **1914(2)**, 491–562.

246. IERSEL, J. J. A. VAN (1953). An analysis of the parental behaviour of the male three-spined stickleback. *Behaviour Suppl.*, **3**, 1–159.

247. IERSEL, J. J. A. VAN and BOL, A. C. A. (1958). Preening of two tern species. A study on displacement activities. *Behaviour*, **13**, 1–88.

248. IKEDA, K. and KAPLAN, W. D. (1970). Patterned neural activity of a mutant *Drosophila melanogaster. Proc. natn. Acad. Sci. U.S.A.*, **66**, 765–72.

249. IKEDA, K. and KAPLAN, W. D. (1970). Unilaterally patterned neural activity of gynandromorphs, mosaic for a neurological mutant of *Drosophila melanogaster. Proc. natn. Acad. Sci. U.S.A.*, **67**, 1480–87.

250. IMMELMANN, K. (1969). Über den Einfluss frühkindlicher Erfahrungen auf die geschlechtliche Objektfixierung bei Estrildiden. *Z. Tierpsychol.*, **26**, 677–91.

251. IMMELMANN, K. (1969). Song development in the zebra finch and other estrildid finches. In *Bird Vocalizations*, ed. HINDE, R. A., pp. 61–74. Cambridge University Press, London.

252. IMMELMANN, K. (1976). The evolutionary significance of early experience. In *Function and Evolution in Behaviour*, pp. 243–53 (see Ref. 23).

253. ITÔ, Y. (1970). Groups and family bonds in animals in relation to their habitat. In *Development and Evolution of Behavior*, ed. ARONSON, L. R., TOBACH, E., LEHRMAN, D. S. and ROSENBLATT, J. S., pp. 389–415. W. H. Freeman, San Francisco.

254. JANOWITZ, H. D. and GROSSMAN, M. I. (1949). Some factors affecting the food intake of normal dogs and dogs with esophagotomy and gastric fistulas. *Am. J. Physiol.*, **159**, 143–48.

255. JARMAN, P. J. (1974). The social organization of antelope in relation to their ecology. *Behaviour*, **48**, 215–67.

256. JAY, P. (1968). Primate field studies and human evolution. In *Primates: Studies in Adaptation and Variability*, pp. 487–503 (see Ref. 257).

257. JAY, P. C., ed. (1968). *Primates: Studies in Adaptation and Variability*. Holt, Rinehart and Winston, New York and London.

258. JOFFE, J. M. (1965). Genotype and prenatal and premating stress interact to affect adult behavior in rats. *Science, N.Y.*, **150**, 1844–45.

259. JOHNSON, R. N. (1972). *Aggression in Man and Animals*. W. B. Saunders, Philadelphia and London.

260. JOLLY, A. (1966). *Lemur Behavior. A Madagascar Field Study*. University of Chicago Press, Chicago and London.

261. JOLLY, A. (1972). *The Evolution of Primate Behavior*. Macmillan, New York.

262. KEAR, J. (1964). Colour preference in young Anatidae. *Ibis*, **106**, 361–69.

263. KEAR, J. (1966). The pecking response of young coots *Fulica atra* and moorhens *Gallinula chloropus*. *Ibis*, **108**, 118–22.

264. KELLOGG, W. N. (1968). Communication and language in the home-raised chimpanzee. *Science, N.Y.*, **162**, 423–27.

265. KENNEDY, J. S. (1958). The experimental analysis of aphid behaviour and its bearing on current theories of instinct. *Proc. 10th Int. Congr. Ent.*, *Montreal*, 1956, **2**, 397–404.

266. KENNEDY, J. S. (1965). Co-ordination of successive activities in an aphid. Reciprocal effects of settling on flight. *J. exp. Biol.*, **43**, 489–509.

267. KENNEDY, J. S. (1966). Some outstanding questions in insect behaviour. *Symp. Roy. Ent. Soc. Lond.*, **3**, 97–112.

268. KLEIMAN, D. G. (1977). Monogamy in mammals. *Q. Rev. Biol.*, **52**, 39–69.

269. KLOPFER, P. H. (1959). An analysis of learning in young Anatidae. *Ecology*, **40**, 90–102.

270. KLOPFER, P. H., ADAMS, D. K. and KLOPFER, M. S. (1964). Maternal 'imprinting' in goats. *Proc. natn. Acad. Sci. U.S.A.*, **52**, 911–14.

271. KÖHLER, W. (1927). *The Mentality of Apes*, 2nd edn. Kegan Paul, Trench, Trubner & Co., London.

272. KOMISARUK, B. R. (1967). Effects of local brain implants of progesterone on reproductive behavior in ring doves. *J. comp. physiol. Psychol.*, **64**, 219–24.

273. KOMISARUK, B. R., ADLER, N. T. and HUTCHISON, J. (1972). Genital sensory field: enlargement by estrogen treatment in female rats. *Science, N.Y.*, **178**, 1295–98.

274. KONISHI, M. (1965). The rôle of auditory feedback in the control of vocalization in the white-crowned sparrow. *Z. Tierpsychol.*, **22**, 770–83.

275. KONORSKI, J. (1948). *Conditioned Reflexes and Neuron Organization*. Cambridge University Press, Cambridge.

276. KREBS, J. R. (1974). Colonial nesting and social feeding as strategies for exploiting food resources in the great blue heron (*Ardea herodias*). *Behaviour*, **51**, 99–134.

277. KREBS, J. R., MACROBERTS, M. H. and CULLEN, J. M. (1972). Flocking and feeding in the great tit *Parus major*—an experimental study. *Ibis*, **114**, 507–30.

278. KRUIJT, J. P. (1964). Ontogeny of social behaviour in Burmese red jungle fowl (*Gallus gallus spadiceus*) Bonaterre. *Behaviour Suppl.*, **12**, 1–201.

279. KRUIJT, J. P. and HOGAN, J. A. (1967). Social behaviour on the lek in black grouse, *Lyrurus tetrix tetrix* (L.). *Ardea*, **55**, 203–40.

280. KRUUK, H. (1975). Functional aspects of social hunting by carnivores. In *Function and Evolution in Behaviour*, pp. 119–41 (see Ref. 23).

281. KUMMER, H. (1968). Social organization of hamadryas baboons. A field study. *Bibliotheca Primatologica* No. 6, 1–189.

282. KUMMER, H. (1968). Two variations in the social organization of baboons. In *Primates: Studies in Adaptation and Variability*, pp. 293–312 (see Ref. 257).

283. KUMMER, H., GÖTZ, W. and ANGST, W. (1974). Triadic differentiation: an inhibitory process protecting pair bonds in baboons. *Behaviour*, **49**, 62–87.

284. LACK, D. (1966). *Population Studies of Birds*. Oxford University Press, London.

285. LAGERSPETZ, K. (1964). Studies on the aggressive behaviour of mice. *Ann. Acad. Sci. Fenn.*, *B*, **131**, 1–131.

285A. LEE, A. K., BRADLEY, A. J. and BRAITHWAITE, R. W. (1977). Corticosteroid levels and male mortality in *Antechinus stuartii*. In *The Biology of Marsupials*, eds. STONEHOUSE, B. and GILMORE, D., pp. 209–20. Macmillan Press Ltd.

286. LEHRMAN, D. S. (1953). A critique of Konrad Lorenz's theory of instinctive behavior. *Q. Rev. Biol.*, **28**, 337–63.

287. LEHRMAN, D. S. (1955). The physiological basis of parental feeding behaviour in the ring dove (*Streptopelia risoria*). *Behaviour*, **7**, 241–86.

288. LEHRMAN, D. S. (1961). Hormonal regulation of parental behavior in birds and infrahuman mammals. In *Sex and Internal Secretions*, 3rd edn., ed. YOUNG, W. C., pp. 1268–1382. Bailliere, Tindall and Cox, London.

289. LEHRMAN, D. S. (1964). The reproductive behavior of ring doves. *Scient. Am.*, **211**, 48–54.

290. LENNENBERG, E. H. (1967). *Biological Foundation of Language*. Wiley, New York.

291. LETTVIN, J. Y., MATURANA, H. R., MCCULLOCH, W. S. and PITTS, W. H. (1959). What the frog's eye tells the frog's brain. *Proc. Inst. Radio Engrs*, **47**, 1940–51.

292. LEUTHOLD, W. (1966). Variations in territorial behavior of Uganda kob *Adenota kob thomasi* (Neumann 1896). *Behaviour*, **27**, 215–58.

293. LEYHAUSEN, P. (1956). Verhaltensstudien an Katzen. *Z. Tierpsychol.*, *Beiheft*, **2**, 1–120.

294. LEYHAUSEN, P. (1964). The communal organization of solitary mammals. *Symp. zool. Soc. Lond.*, **14**, 249–63.

295. LINCOLN, G. A., YOUNGSON, R. W. and SHORT, R. V. (1970). The social and sexual behaviour of the red deer stag. *J. Reprod. Fertil. Suppl.*, **11**, 71–103.

296. LIND, H. (1959). The activation of an instinct caused by a 'transitional action'. *Behaviour*, **14**, 123–35.

297. LINDAUER, M. (1961). *Communication among Social Bees*. Harvard University Press, Cambridge, Mass.

298. LINDAUER, M. (1976). Evolutionary aspects of orientation and learning. In *Function and Evolution in Behaviour*, pp. 228–42 (see Ref. 23).

299. LINDBURG, D. G. (1969). Rhesus monkeys: Mating season mobility in adult males. *Science*, *N.Y.*, **166**, 1176–78.

300. LISSMANN, H. W. (1963). Electric location by fishes. *Scient. Am.*, **208**, 50–59.

301. LLOYD, J. E. (1965). Aggressive mimicry in *Photuris* firefly *femmes fatales*. *Science*, *N.Y.*, **149**, 653–54.

302. LLOYD, J. E. (1975). Aggressive mimicry in *Photuris* fireflies: signal repertoires by *femmes fatales*. *Science*, *N.Y.*, **187**, 452–53.

303. LORENZ, K. Z. (1937). The companion in the bird's world. *Auk.*, **54**, 245–73.

304. LORENZ, K. Z. (1941). Vergleichende Bewegungsstudien an Anatinen. *J. Orn., Lpz.*, **89**, 194-293. (An English translation appeared in several parts, in Vols. 57-59 of *Avicultural Magazine.*)

305. LORENZ, K. Z. (1950). The comparative method in studying innate behaviour patterns. *Symp. Soc. exp. Biol.*, **4**, 221-68.

306. LORENZ, K. Z. (1952). *King Solomon's Ring.* Methuen, London. (Reprinted 1959 by Pan, London.)

307. LORENZ, K. Z. (1958). The evolution of behavior. *Scient. Am.*, **199**(6), 67-78.

308. LORENZ, K. Z. (1966). *Evolution and Modification of Behaviour.* Methuen, London.

309. LORENZ, K. Z. (1966). *On Aggression.* Methuen, London.

310. LORENZ, K. Z. and TINBERGEN, N. (1938). Taxis und Instinkthandlung in der Eirollbewegung der Graugans I. *Z. Tierpsychol.*, **2**, 1-29.

311. LOUCKS, R. B. (1935). The experimental delimitation of neural structures essential for learning: the attempt to condition striped muscle responses with faradization of the sigmoid gyri. *J. Psychol.*, **1**, 5-44.

312. MCBRIDE, G., PARER, I. P. and FOENANDER, F. (1959). The social organization and behaviour of the feral domestic fowl. *Anim. Behav. Monogr.*, **2**, 127-81.

313. MCFARLAND, D. J. (1965). Hunger, thirst and displacement pecking in the Barbary dove. *Anim. Behav.*, **13**, 293-300.

314. MCFARLAND, D. J. (1966). On the causal and functional significance of displacement activities. *Z. Tierpsychol.*, **23**, 217-35.

315. MCFARLAND, D. J. (1971). *Feedback Mechanisms in Animal Behaviour.* Academic Press, London and New York. *Man*, **13**, 234-51.

316. MCGILL, T. E. (1962). Sexual behaviour in three inbred strains of mice. *Behaviour*, **19**, 341-50.

317. MCGILL, T. E., ed. (1965). *Readings in Animal Behavior.* Holt, Rinehart and Winston, New York and London.

318. MCGREW, W. C. and TUTIN, C. E. G. (1978). Evidence for a social custom in wild chimpanzees? (In press, *Man*.)

319. MACKINTOSH, N. J. (1965). Discrimination learning in the octopus. *Anim. Behav. Suppl.*, **1**, 129-34.

320. MAGNUS, D. B. E. (1958). Experimental analysis of some 'overoptimal' sign stimuli in the mating behaviour of the fritillary butterfly, *Argynnis paphia* L. (Lepidoptera Nymphalidae.) *Proc. 10th Int. Congr. Ent., Montreal*, 1956, **2**, 405-18.

321. MAIER, N. R. F. (1932). Cortical destruction of the posterior part of the brain and its effect on reasoning in rats. *J. comp. Neurol.*, **56**, 179-214.

322. MAIER, N. R. F. and SCHNEIRLA, T. C. (1935). *Principles of Animal Psychology.* McGraw-Hill, New York and Maidenhead.

323. MANNING, A. (1959). The sexual behaviour of two sibling *Drosophila* species. *Behaviour*, **15**, 123-45.

324. MANNING, A. (1961). The effects of artificial selection for mating speed in *Drosophila melanogaster*. *Anim. Behav.*, **9**, 82-92.

325. MANNING, A. (1965). Drosophila and the evolution of behaviour. *Viewpoints in Biology*, **4**, 125-69. Butterworth, London.

326. MANNING, A. (1966). Sexual behaviour. *Symp. R. ent. Soc. Lond.*, **3**, 59-68.

327. MANNING, A. (1971). *Evolution of behavior*. In *Psychobiology: Biological Bases of Behavior*, ed. MCGAUGH, J., pp. 1–52. Academic Press, New York and London.
328. MARLER, P. (1955). Studies of fighting in chaffinches. (1) Behaviour in relation to the social hierarchy. *Brit. J. Anim. Behav.*, **3**, 111–17.
329. MARLER, P. (1959). Developments in the study of animal communication. In *Darwin's Biological Work*, ed. BELL, P. R., pp. 150–206. Cambridge University Press, Cambridge.
330. MARLER, P. (1968). Aggregation and dispersal: two functions in primate communication. In *Primates, Studies in Adaptation and Variability*, pp. 420–38 (see Ref. 257).
331. MARLER, P. (1969). Colobus guereza: territoriality and group composition. *Science, N.Y.*, **163**, 93–95.
332. MARLER, P. (1976). On strategies of behavioural development. In *Function and Evolution in Behaviour*, pp. 254–75 (see Ref. 23).
333. MARLER, P. and HAMILTON, W. J. (1966). *Mechanisms of Animal Behavior*. John Wiley, New York and London.
334. MARLER, P., KREITH, M. and TAMURA, M. (1964). Song development in hand-raised Oregon juncos. *Auk*, **79**, 12–30.
335. MARLER, P. and TAMURA, M. (1964). Culturally transmitted patterns of vocal behavior in sparrows. *Science, N.Y.*, **146**, 1483–86.
336. MARTIN, R. D. (1968). Reproduction and ontogeny in tree-shrews (*Tupaia belangeri*), with reference to their general behaviour and taxonomic relationships. *Z. Tierpsychol.*, **25**, 409–95.
337. MASSERMAN, J. H. (1950). Experimental neuroses. *Scient. Am.*, **182**, 38–43.
338. MAY, D. J. (1949). Studies on a community of willow-warblers. *Ibis*, **91**, 24–54.
339. MAYR, E. (1963). *Animal Species and Evolution*. Oxford University Press, London.
340. MELZACK, R., PENICK, E. and BECKETT, A. (1959). The problem of 'innate fear' of the hawk shape; an experimental study with mallard ducks. *J. comp. physiol. Psychol.*, **52**, 694–98.
341. MENZEL, E. W. (1971). Communication about the environment in a group of young chimpanzees. *Folia Primatol.*, **15**, 220–32.
342. MENZEL, R. (1968). Das Gedächtnis der Honigbiene fur Spektralfarben. I. Kurzzeitiges und langzeitiges Behalten. *Z. vergl. Physiol.*, **60**, 82–102.
343. MICHAEL, R. P. (1966). Action of hormones on the cat brain. In *Brain and Behavior*, ed. GORSKI, R. A. and WHALEN, R. E.; vol. 3, *The Brain and Gonadal Function*, pp. 81–98. University of California Press, Berkeley.
344. MICHAEL, R. P. and SCOTT, P. P. (1964). The activation of sexual behaviour in cats by the subcutaneous administration of oestrogen. *J. Physiol.*, **171**, 254–74.
345. MILLER, N. E. (1941). The frustration-aggression hypothesis. *Psychol. Rev.*, **48**, 337–42.
346. MILLER, N. E. (1951). Learnable drives and rewards. In *Handbook of Experimental Psychology*, ed. STEVENS, S. S., pp. 435–72. Wiley, New York and London.
347. MILLER, N. E. (1956). Effects of drugs on motivation: the value of using a variety of measures. *Ann. N.Y. Acad. Sci.*, **65**, 318–33.

348. MILLER, N. E. (1957). Experiments on motivation. Studies combining psychological, physiological and pharmacological techniques. *Science N.Y.*, **126**, 1271–78.
349. MILLER, N. E. (1965). Chemical coding of behavior in the brain. *Science, N.Y.*, **148**, 328–38.
350. MILLER, N. E. and KESSEN, M. L. (1952). Reward effects of food via stomach fistula compared with those of food via mouth. *J. comp. physiol. Psychol.*, **45**, 555–64.
351. MOLTZ, H. and STETTNER, L. J. (1961). The influence of patterned-light deprivation on the critical period for imprinting. *J. comp. physiol. Psychol.*, **54**, 279–83.
352. MONEY, J. and EHRHARDT, A. A. (1973). *Man and Woman, Boy and Girl.* Johns Hopkins University Press, Baltimore.
353. MONTAGU, M. F. A. (1968). *Man and Aggression.* Oxford University Press, London and New York.
354. MOORE, B. R. (1973). The role of directed Pavlovian reactions in simple instrumental learning in the pigeon. In *Constraints on Learning*, pp. 159–88 (see Ref. 224).
355. MORRIS, D. (1956). The feather postures of birds and the problem of the origin of social signals. *Behaviour*, **9**, 75–113.
356. MORRIS, D. (1957). 'Typical intensity' and its relation to the problem of ritualization. *Behaviour*, **11**, 1–12.
357. MORRIS, D. (1959). The comparative ethology of grassfinches (Erythrurae) and mannikins (Amadinae). *Proc. zool. Soc. Lond.*, **131**, 389–439.
358. MOYNIHAN, M. (1955). Some aspects of the reproductive behaviour in the black-headed gull (*Larus ridibundus ridibundus* L.) *Behaviour Suppl.*, **4**, 1–201.
359. MOYNIHAN, M. (1968). Social mimicry; character convergence versus character displacement. *Evolution*, **22**, 315–31.
360. MUNN, N. L. (1950). *Handbook of Psychological Research on the Rat.* Houghton Mifflin, Boston.
361. MYER, J. S. and WHITE, R. T. (1965). Aggressive motivation in the rat. *Anim. Behav.*, **13**, 430–33.
362. NAPIER, J. R. and NAPIER, P. H., eds. (1970). *Old World Monkeys. Evolution, Systematics and Behavior.* Academic Press, New York and London.
363. NARINS, P. M. and CAPRANICA, R. R. (1976). Sexual differences in the auditory system of the tree frog *Eleutherodactylus coqui. Science, N.Y.*, **192**, 378–80.
364. NELSON, J. B. (1976). The breeding biology of frigatebirds—a comparative review. *The Living Bird*, **14**, 113–56. Cornell Laboratory of Ornithology.
365. NICE, M. M. (1937). Studies in the life history of the song sparrow. 1. *Trans. Linn. Soc. N.Y.*, **4** (Reprinted Dover Publications, New York. T1219, 1964).
366. NIXON, H. L. and RIBBANDS, C. R. (1952). Food transmission in the honeybee community. *Proc. Roy. Soc. B*, **140**, 43–50.
367. NOBLE, G. K. (1936). Courtship and sexual selection of the flicker (*Colaptes auratus luteus*). *Auk*, **53**, 269–82.

368. NOIROT, E. (1964). Changes in responsiveness to young in the adult mouse. IV. The effect of an initial contact with a strong stimulus. *Anim. Behav.*, **12**, 442–45.

369. NOIROT, E. (1969). Changes in responsiveness to young in the adult mouse. V. Priming. *Anim. Behav.*, **17**, 542–46.

370. NORTON-GRIFFITHS, M. (1967). Some ecological aspects of the feeding behaviour of the oystercatcher *Haematopus ostralegus* on the edible mussel, *Mytilus edulis*. *Ibis*, **109**, 412–24.

371. NORTON-GRIFFITHS, M. (1969). The organisation, control and development of parental feeding in the oystercatcher. (*Haematopus ostralegus*.) *Behaviour*, **34**, 55–114.

372. NOTTEBOHM, F. (1972). The origins of vocal learning. *Amer. Nat.*, **106**, 116–40.

373. OLDS, J. (1962). Hypothalamic substrates of reward. *Physiol. Rev.*, **42**, 554–604.

374. OPPENHEIM, R. W. (1968). Color preference in the pecking response of newly hatched ducks (*Anas platyrhynchos*). *J. comp. physiol. Psychol.*, *Monogr. Suppl.* (2), 1–17.

375. OPPENHEIM, R. W. (1974). The ontogeny of behavior in the chick embryo. *Advances in the Study of Behavior*, **5**, 133–72.

376. OWEN-SMITH, N. (1971). Territoriality in the white rhinoceros (*Ceratotherium simum*) Burchell. *Nature, Lond.*, **231**, 294–96.

377. PACKER, C. (1975). Male transfer in olive baboons. *Nature, Lond.*, **255**, 219–20.

378. PAVLOV, I. P. (1941). *Lectures on Conditioned Reflexes*, 2 vols. International Publishers, New York.

379. PAYNE, R. S. and MCVAY, S. (1971). Songs of humpback whales. *Science, N.Y.*, **173**, 585–97.

380. PERDECK, A. C. (1958). The isolating value of specific song in two sibling species of grasshoppers (*Chorthippus brunneus* Thunb. and *C. biguttulus* L.). *Behaviour*, **12**, 1–75.

381. PETTERSSON, M. (1956). Diffusion of a new habit among green finches. *Nature, Lond.*, **177**, 709–10.

382. POWELL, G. V. N. (1974). Experimental analysis of the social value of flocking by starlings (*Sturnus vulgaris*) in relation to predation and foraging. *Anim. Behav.*, **22**, 501–5.

383. PRECHTL, H. F. R. (1953). Zur Physiologie der angeborenen Auslosenden-mechanismen. I. Quantitative Untersuchungen über die Sperrbe-wegung junger Singvögel. *Behaviour*, **5**, 32–50.

384. PREMACK, D. (1971). Language in chimpanzee? *Science, N.Y.*, **172**, 808–22.

385. PROVINE, R. R. (1976). Eclosion and hatching in cockroach first instar larvae: a stereotyped pattern of behaviour. *J. Insect Physiol.*, **22**, 127–31.

386. QUADAGNO, D. M. and BANKS, E. M. (1970). The effect of reciprocal crossfostering on the behaviour of two species of rodents, *Mus musculus* and *Baiomys taylori ater*. *Anim. Behav.*, **18**, 379–90.

387. RÄBER, H. (1948). Analyse des Balzverhaltens eines domestizierten Truthahns (*Meleagris*). *Behaviour*, **1**, 237–66.

388. RALLS, K. (1971). Mammalian scent marking. *Science, N.Y.*, **171**, 443–49.

389. RAMSAY, A. O. (1961). Behaviour of some hybrids in the mallard group. *Anim. Behav.*, **9**, 104–5.

390. RAMSAY, A. O. and HESS, E. H. (1954). A laboratory approach to the study of imprinting. *Wilson Bull.*, **66**, 196–206.

391. RASA, O. A. E. (1969). The effect of pair isolation on reproductive success in *Etroplus maculatus* (Cichlidae). *Z. Tierpsychol.*, **26**, 846–52.

392. REYNOLDS, R. W. (1965). An irritative hypothesis concerning the hypothalamic regulation of food intake. *Psychol. Rev.*, **72**, 105–16.

393. REYNOLDS, V. (1976). *The Biology of Human Action.* W. H. Freeman, Reading and San Francisco.

394. RICE, J. O. and THOMPSON, W. L. (1968). Song development in the indigo bunting. *Anim. Behav.*, **16**, 462–69.

395. RISS, W. (1955). Sex drive, oxygen consumption and heart rate in genetically different strains of male guinea pigs. *Am. J. Physiol.*, **180**, 530 34.

396. ROEDER, K. D. (1967). *Nerve Cells and Insect Behavior*, Revised edn. Harvard University Press, Cambridge, Mass.

397. ROMER, A. S. (1962). *The Vertebrate Body*, 3rd edn. W. B. Saunders, Philadelphia and London.

398. ROSENZWEIG, M. R. and BENNETT, E. L., eds. (1976). *Neural Mechanisms of Learning and Memory.* M.I.T. Press, Cambridge, Mass. and London.

399. ROTII, L. M. (1948). An experimental laboratory study of the sexual behavior of *Aëdes aegypti* (1). *Am. Midl. Nat.*, **40**, 265–352.

400. ROTHENBUHLER, W. C. (1964). Behaviour genetics of nest cleaning in honey-bees. I. Responses of four in-bred lines to disease-killed brood. *Anim. Behav.*, **12**, 578–83.

401. ROTHENBUHLER, W. C. (1964). Behavior genetics of nest cleaning in honey-bees. IV. Responses of F_1 and backcross generations to disease-killed brood. *Am. Zoologist*, **4**, 111–23.

402. ROWELL, C. H. F. (1961). Displacement grooming in the chaffinch. *Anim. Behav.*, **9**, 38–63.

403. ROWELL, T. E. (1962). Agonistic noises of the rhesus monkey (*Macaca mulatta*). *Symp. Zool. Soc. Lond.*, **8**, 91–96.

404. ROWELL, T. (1969). Long-term changes in a population of Ugandan baboons. *Folia Primat.*, **11**, 241–54.

405. ROWELL, T. (1972). *The Social Behaviour of Monkeys.* Penguin Books Ltd, Harmondsworth.

406. ROWELL, T. (1974). The concept of social dominance. *Behavl. Biol.*, **11**, 131–54.

407. ROZIN, P. and KALAT, J. W. (1970). Specific hungers and poison avoidance as adaptive specialisations in learning. *Psychol. Rev.*, **78**, 459–86.

408. RUNDQUIST, E. A. (1933). The inheritance of spontaneous activity in rats. *J. comp. Psychol.*, **16**, 415–38.

409. RUSSELL, C. and RUSSELL, W. M. S. (1968). *Violence, Monkeys and Man.* Macmillan, London.

410. RUSSELL, W. M. S. (1958). Evolutionary concepts in behavioral science; I. Cybernetics, Darwinian theory and behavioral science. *Gen. Syst.*, **3**, 18–28.

411. RUSSELL, W. M. S. (1959). Evolutionary concepts in behavioral science; II. Organic evolution and the genetical theory of natural selection. *Gen. Syst.*, **4**, 45–73.

412. RUSSELL, W. M. S. (1961). Evolutionary concepts in behavioral science: III. The evolution of behavior in the individual animal, and the principle of combinatorial selection. *Gen. Syst.*, **6**, 51–91.

413. RUSSELL, W. M. S. (1962). Evolutionary concepts in behavioral science: IV. The analogy between organic and individual behavioral evolution, and the evolution of intelligence. *Gen. Syst.*, **7**, 157–93.

414. SADE, D. S. (1965). Some aspects of parent-offspring and sibling relations in a group of rhesus monkeys, with a discussion of grooming. *Amer. J. Phys. Anthropol.*, **23**, 1–18.

415. SADE, D. S. (1967). Determinants of dominance in a group of free-ranging rhesus monkeys. In *Social Communication among Primates*, pp. 99–114 (see Ref. 7).

416. SAUNDERS, D. S. (1977). *An Introduction to Biological Rhythms*. Blackie, Glasgow and London.

417. SCHEIN, M. W. and HALE, E. B. (1959). The effect of early social experience on male sexual behaviour of androgen-injected turkeys. *Anim. Behav.*, **7**, 189–200.

418. SCHENKEL, R. (1967). Submission: its features and function in the wolf and dog. *Amer. Zool.*, **7**, 319–29.

419. SCHJELDERUP-EBBE, T. (1935). Social behavior of birds. In *Handbook of Social Psychology*, ed. MURCHISON, C., pp. 947–72. Clark University Press, Worcester, Mass.

420. SCHLEIDT, W. M. (1961). Reaktionen von Truthühnern auf fliegende Raubvögel und Versuche zur Analyse ihrer AAM's. *Z. Tierpsychol.*, **18**, 534–60.

421. SCHLEIDT, W. M. (1973). Tonic communication: continual effects of discrete signs in animal communication systems. *J. theor. Biol.*, **42**, 359–86.

422. SCHLEIDT, W. and SCHLEIDT, M. (1960). Störung der Mutter-Kind-Beziehung bei Truthühnern durch Gehörverlust. *Behaviour*, **16**, 254–60.

423. SCHNEIDER, D. (1966). Chemical sense communication in insects. *Symp. Soc. exp. Biol.*, **20**, 273–97.

424. SCHNEIRLA, T. C. (1965). Aspects of stimulus and organization in approach/withdrawal processes underlying vertebrate behavioral development. *Adv. Study Behav.*, **1**, 1–71.

425. SCHUTZ, F. (1965). Sexuelle Prägung bei Anatiden. *Z. Tierpsychol.*, **22**, 50–103.

426. SCOTT, J. P. (1958). *Aggression*. University of Chicago Press, Chicago.

427. SCOTT, J. P. (1962). Critical periods in behavioral development. *Science, N.Y.*, **138**, 949–58.

428. SCOTT, J. P. and FULLER, J. L. (1965). *Genetics and the Social Behavior of the Dog*. University of Chicago Press, Chicago.

429. SEBEOK, T. (1977). *How Animals Communicate*. Indiana University Press, Bloomington.

430. SELIGMAN, M. E. P. (1970). On the generality of the laws of learning. *Psychol. Rev.*, **77**, 406–18.

431. SELIGMAN, M. E. P. and HAGER, J. L., eds. (1972). *Biological Boundaries of Learning*. Appleton-Century-Crofts, New York.

432. SEVENSTER, P. (1961). A causal analysis of a displacement activity (fanning in *Gasterosteus aculeatus* L.) *Behaviour Suppl.*, **9**, 1–170.

433. SEVENSTER-BOL, A. C. A. (1962). On the causation of drive reduction after a consummatory act (in *Gasterosteus aculeatus* L.). *Archs néerl. Zool.*, **15**, 175-236.

434. SHEPPARD, P. M. (1961). Some contributions to population genetics resulting from the study of Lepidoptera. *Adv. Genet.*, **10**, 165-216.

435. SHERRINGTON, C. S. (1906). *The Integrative Action of the Nervous System.* Scribner's, New York.

436. SHERRINGTON, C. S. (1917). Reflexes elicitable in the cat from pinna, vibrissae and jaws. *J. Physiol.*, **51**, 404-31.

437. SKINNER, B. F. (1938). *The Behavior of Organisms.* Appleton-Century-Crofts, New York.

438. SLUCKIN, W. (1972). *Imprinting and Early Learning*, 2nd edn. Methuen, London.

439. SMITH, J. M. (1975). *The Theory of Evolution*, 3rd edn. Penguin, Harmondsworth.

440. SMITH, N. G. (1967). Visual isolation in gulls. *Scient. Amer.*, **217(4)**, 94-102.

441. SMITH, W. J. (1969). Messages of vertebrate communication. *Science, N.Y.*, **165**, 145-50.

442. SNOW, D. W. (1958). *A Study of Blackbirds.* Allen and Unwin, London.

443. SONNEMANN, P. and SJÖLANDER, S. (1977). Effects of cross fostering on the sexual imprinting of the female zebra finch *Taeniopygia guttata. Z. Tierpsychol.*, **45**, 337-48.

444. SPALDING, D. (1873). Instinct: with original observations on young animals. *Macmillan's Mag.*, **27**, 282-93. (Reprinted in *Br. J. Anim. Behav.*, **2**, 1-11, 1954.)

445. STOKES, A. W. (1962). Agonistic behaviour among blue tits at a winter feeding station. *Behaviour*, **19**, 118-38.

446. STOUT, J. F. and BRASS, M. E. (1969). Aggressive communication by *Larus glaucescens*. Part II. Visual communication. *Behaviour*, **34**, 42-54.

447. STOUT, J. F., WILCOX, C. R. and CREITZ, L. E. (1969). Aggressive communication by *Larus glaucescens*. Part I. Sound communication. *Behaviour*, **34**, 29-41.

448. SUGIYAMA, Y. (1967). Social organization of hanuman langurs. In *Social Communication among Primates*, pp. 221-36 (see Ref. 7).

449. TAYLOR, A., SLUCKIN, W. and HEWITT, R. (1969). Changing colour preference of chicks. *Anim. Behav.*, **17**, 3-8.

450. TEITELBAUM, P. and EPSTEIN, A. N. (1962). The lateral hypothalamic syndrome: recovery of feeding and drinking after lateral hypothalamic lesions. *Psychol. Rev.*, **69**, 74-90.

451. THIESSEN, D. D. and YAHR, P. (1970). Central control of territorial marking in the Mongolian gerbil. *Physiol. Behav.*, **5**, 275-78.

452. THOMPSON, T. I. (1963). Visual reinforcement in Siamese fighting fish. *Science, N.Y.*, **141**, 55-57.

453. THOMPSON, T. I. (1964). Visual reinforcement in fighting cocks. *J. Exptl. Anal. Behav.*, **7**, 45-49.

454. THOMPSON, W. R., WATSON, J. and CHARLESWORTH, W. R. (1962). The effects of prenatal maternal stress on offspring behavior in rats. *Psychol. Monogr.*, **76**, No. 38.

455. THORPE, W. H. (1950). A note on detour experiments with *Ammophila pubescens* Curt. (Hymenoptera: Sphecidae). *Behaviour*, **2**, 257-63.

456. THORPE, W. H. (1961). *Bird Song*. Cambridge University Press, Cambridge.

457. THORPE, W. H. (1963). *Learning and Instinct in Animals*, 2nd edn. Methuen, London.

458. THORPE, W. H. and DAVENPORT, D., eds. (1965). *Learning and Associated Phenomena in Invertebrates*. *Anim. Behav. Suppl.*, **1**, 1–190.

459. THORPE, W. H. and ZANGWILL, O. L., eds. (1961). *Current Problems in Animal Behaviour*. Cambridge University Press, Cambridge.

460. TINBERGEN, N. (1951). *The Study of Instinct*. Oxford University Press, London.

461. TINBERGEN, N. (1952). 'Derived' activities, their causation, biological significance, origin and emancipation during evolution. *Q. Rev. Biol.*, **27**, 1–32.

462. TINBERGEN, N. (1953). *The Herring Gull's World*. Collins, London.

463. TINBERGEN, N. (1953). *Social Behaviour in Animals*. Methuen, London.

464. TINBERGEN, N. (1957). On anti-predator responses in certain birds—a reply. *J. comp. physiol. Psychol.*, **50**, 412–14.

465. TINBERGEN, N. (1959). Comparative studies of the behaviour of gulls (Laridae): a progress report. *Behaviour*, **15**, 1–70.

466. TINBERGEN, N. (1963). On the aims and methods of ethology. *Z. Tierpsychol.*, **20**, 410–33.

467. TINBERGEN, N. (1968). On war and peace in men and animals. *Science, N.Y.*, **160**, 1411–18.

468. TINBERGEN, N., BROEKHUYSEN, G. J., FEEKES, F., HOUGHTON, J. C. W., KRUUK, H. and SZULC, E. (1962). Egg shell removal by the black-headed gull. *Larus ridibundus*, L.; a behaviour component of camouflage. *Behaviour*, **19**, 74–117.

469. TINBERGEN, N., KRUUK, H., PAILETTE, M. and STAMM, R. (1962). How do black-headed gulls distinguish between eggs and egg-shells? *Br. Birds*, **55**, 120–29.

470. TINBERGEN, N. and PERDECK, A. C. (1950). On the stimulus situation releasing the begging response in the newly-hatched herring gull chick (*Larus a. argentatus* Pont.). *Behaviour*, **3**, 1–38.

471. TSCHANZ, B. (1959). Zur Brutbiologie der Trottelume (*Uria aalge aalge* Pont.). *Behaviour*, **14**, 1–100.

472. TSCHANZ, B. and HIRSBRUNNER-SCHARF, M. (1975). Adaptations to colony life on cliff ledges: a comparative study of guillemot and razorbill chicks. In *Function and Evolution in Behaviour*, pp. 358–80 (see Ref. 23).

473. VALENSTEIN, E. S., COX, V. C. and KAKOLEWSKI, J. W. (1968). Modification of motivated behavior elicited by electrical stimulation of the hypothalamus. *Science, N.Y.*, **159**, 1119–21.

474. VALENSTEIN, E. S., COX, V. C. and KAKOLEWSKI, J. W. (1969). Hypothalamic motivational systems: fixed or plastic neural circuits? *Science, N.Y.*, **163**, 1084.

475. VALENSTEIN, E. S., COX, V. C. and KAKOLWESKI, J. W. (1970). Re-examination of the role of the hypothalamus in motivation. *Psychol. Rev.*, **77**, 16–31.

476. VERNON, J. A. and BUTSCH, J. (1957). Effect of tetraploidy on learning and retention in the salamander. *Science, N.Y.*, **125**, 1033–34.

477. VINCE, M. A. (1969). Embryonic communication, respiration and the synchronization of hatching. In *Bird Vocalizations*, pp. 233–60 (see Ref. 216).

478. VOWLES, D. M. (1954). The orientation of ants: I. The substitution of stimuli. *J. exp. Biol.*, **31**, 341–55.

479. VOWLES, D. M. (1965). Maze learning and visual discrimination in the wood ant (*Formica rufa*). *Br. J. Psychol.*, **56**, 15–31.

480. VOWLES, D. M. and HARWOOD, D. (1966). The effect of exogenous hormones on aggressive and defensive behaviour in the ring dove (*Streptopelia risoria*). *J. Endocr.*, **36**, 35–51.

481. WALL, W. VON DE (1963). Bewegungsstudien an Anatinen. *J. Orn., Lpz.*, **104**, 1–15.

482. WALSH, E. G. (1964). *Physiology of the Nervous System*, 2nd edn. Longmans, Green, London.

483. WARD, I. L. (1974). Sexual behavior differentiation: prenatal hormonal and environmental control. In *Sex Differences in Behavior*, eds. FRIEDMAN, R. C., RICHART, R. M. and VAN DE WIELE, R. L., pp. 3–17 Wiley, New York.

484. WARREN, J. M. (1965). Primate learning in comparative perspective. In *Behavior of Nonhuman Primates*, ed. SCHRIER, A. M., HARLOW, H. F. and STOLLNITZ, F., vol. I, pp. 249–81. Academic Press, New York and London.

485. WASHBURN, S. L. and DE VORE, I. (1961). Social behavior of baboons and early man. In *Social Life of Early Man*, ed. WASHBURN, S. L., pp. 91–105. Aldine Press, Chicago.

486. WATSON, A. (1970). Territorial and reproductive behaviour of red grouse. *J. Reprod. Fertility, Supplement*, **11**, 3–14.

487. WATSON, A. J. (1961). The place of reinforcement in the explanation of behaviour. In *Current Problems in Animal Behaviour*, pp. 273–301 (see Ref. 459).

488. WATSON, J. B. (1924). *Behaviorism*. University of Chicago Press, Chicago.

489. WATTS, C. R. and STOKES, A. W. (1971). The social order of turkeys. *Scient. Am.*, **224**(6), 112–18.

490. WEIDMANN, R. and WEIDMANN, U. (1958). An analysis of the stimulus situation releasing food-begging in the black-headed gull. *Anim. Behav.*, **6**, 114.

491. WEIDMANN, U. (1961). The stimuli eliciting begging in gulls and terns. *Anim. Behav.*, **9**, 115–16.

492. WELLS, M. J. (1958). Factors affecting reactions to *Mysis* by newly hatched *Sepia*. *Behaviour*, **13**, 96–111.

493. WELLS, M. J. (1962). Early learning in *Sepia*. *Symp. zool. Soc. Lond.*, **8**, 149–69.

494. WELLS, M. J. (1962). *Brain and Behaviour in Cephalopods*. Heinemann Educational, London.

495. WELLS, P. H. and WENNER, A. M. (1973). Do honey bees have a language? *Nature, Lond.*, **241**, 171–74.

496. WHITE, S. D., WAYNER, M. J. and COTT, A. (1970). Effects of intensity, water deprivation, prior water ingestion and palatability on drinking evoked by lateral hypothalamic electric stimulation. *Physiol. Behav.*, **5**, 611–19.

497. WICKLER, W. (1968). *Mimicry in Plants and Animals*. World University Library, Weidenfeld and Nicolson, London.

498. WIENS, J. A. (1966). On group selection and Wynne-Edwards' hypothesis. *Am. Scient.*, **54**, 273–87.
499. WIEPKEMA, P. R. (1971). Positive feedbacks at work during feeding. *Behaviour*, **39**, 266–73.
500. WILEY, R. H. (1975). Multidimensional variation in an avian display: implications for social communication. *Science, N.Y.*, **190**, 482–83.
501. WILHELMI, U. (1975). Über den Einfluss sozialer Isolation auf die Rangordnungskämpfe männlicher Schwertträger (*Xiphophorus helleri*). *Z. Tierpsychol.*, **38**, 482–504.
502. WILLOWS, A. O. D. (1967). Behavioral acts elicited by stimulation of single identifiable nerve cells. *Science, N.Y.*, **157**, 570–74.
503. WILLOWS, A. O. D., DORSETT, D. A. and HOYLE, G. (1973). The neuronal basis of behavior in *Tritonia*. III. Neuronal mechanisms of a fixed action pattern. *J. Neurobiol.*, **4**, 255–85.
504. WILSON, D. M. (1964). The origin of the flight-motor command in grasshoppers. In *Neural Theory and Modeling*, ed. REISS, R. F., pp. 331–45. Stanford University Press, Stanford, California.
505. WILSON, E. O. (1965). Chemical communication in the social insects. *Science, N.Y.*, **149**, 1064–71.
506. WILSON, E. O. (1971). *The Insect Societies*. Belknap Press of Harvard University Press, Cambridge, Mass.
507. WILSON, E. O. (1975). *Sociobiology. The New Synthesis*. Harvard, The Belknap Press, Cambridge, Mass.
508. WILZ, K. J. (1970). Causal and functional analysis of dorsal pricking and nest activity in the courtship of the three-spined stickleback *Gasterosteus aculeatus*. *Anim. Behav.*, **18**, 115–24.
509. WISE, R. A. (1968). Hypothalamic motivational systems: fixed or plastic neural circuits? *Science, N.Y.*, **162**, 377–79.
510. WISE, R. A. (1969). Plasticity of hypothalamic motivational systems. *Science, N.Y.*, **165**, 929–30.
511. WOOD-GUSH, D. G. M. (1960). A study of sex drive of two strains of cockerel through three generations. *Anim. Behav.*, **8**, 43–53.
512. WOOD-GUSH, D. G. M. (1971). *The Behaviour of the Domestic Fowl*. Heinemann Educational, London.
513. WOODROW, H. (1942). The problem of general quantitative laws in psychology. *Psychol. Bull.*, **39**, 1–27.
514. WYNNE-EDWARDS, V. C. (1962). *Animal Dispersion in Relation to Social Behaviour*. Oliver and Boyd, Edinburgh.
515. WYNNE-EDWARDS, V. C. (1978). Intrinsic population control: an introduction. In *Population Control by Social Behaviour*, eds. EBLING, F. J. and STODDART, D. N., pp. 1–22. Institute of Biology, London.
516. YOSHIBA, K. (1968). Local and intertroop variability in ecology and social behavior of common Indian langurs. In *Primates: Studies in Adaptation and Variability*, pp. 217–42 (see Ref. 257).
517. ZAHAVI, A. (1971). The function of pre-roost gatherings and communal roosts. *Ibis*, **113**, 106–9.
518. ZAHAVI, A. (1974). Communal nesting in the Arabian babbler. *Ibis*, **116**, 84–87.
519. ZARROW, M. X., DENENBERG, V. H. and ANDERSON, C. O. (1965). Rabbit: frequency of suckling in the pup. *Science, N.Y.*, **150**, 1835–36.
520. ZEIGLER, H. P. (1964). Displacement activity and motivational theory. A case study in the history of ethology. *Psychol. Bull.*, **61**, 362–76.

Index

Bold page numbers refer to major entries